Pearson
BTEC National
Applied Scien

Student Boo

Frances Annets
Joanne Hartley
Sue Hocking
Roy Llewellyn
Chris Meunier

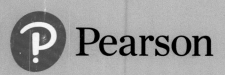 Pearson

Published by Pearson Education Limited, 80 Strand, London, WC2R 0RL.

www.pearsonschoolsandfecolleges.co.uk

Text © Pearson Education Limited 2017
Page design by Andy Magee
Typeset by Aptara
Original illustrations © Pearson Education Limited
Illustrated by Tech-Set Ltd
Cover design by Vince Haig
Picture research by Susie Prescott
Cover photo © kilukilu / Shutterstock.com

First published 2017

19 18 17
10 9 8 7 6 5 4 3 2

British Library Cataloguing in Publication Data
A catalogue record for this book is available from the British Library

ISBN 978 1 292 13413 0

Acknowledgements

We would like to thank Ann Ham, Neil Harris, Joanne Hartley, Karlee Lees and Chris Meunier for their invaluable help in reviewing this book.

The author and publisher would like to thank the following individuals and organisations for permission to reproduce photographs:

(Key: b-bottom; c-centre; l-left; r-right; t-top)

123RF.com: 124tl, Anna Ivanova 112tr, Brian Kinney 258tr, Steve Collender 303bl; **Alamy Stock Photo**: age fotostock 249br, AMELIE-BENOIST / BSIP SA 123tr, BrazilPhotos.com 163, BSIP SA 152tr, Cavallini James / BSIP SA 175, 185tl, Cavallini James / BSIP SA 175, 185tl, Cultura Creative (RF) 273bl, 311tl, Daisy Corlett 128cr, David Cairns 121tr, Hero Images Inc. 259tl, Laurent / BSIP SA 214b, Mark & Audrey Gibson / Stock Connection Blue 1, Maximilian / Prisma Bildagentur AG 196br, Mikael Karlsson 297br, Monty Rakusen / Cultura Creative (RF) 203br, Pablo Paul 268bc, 269tl, Radius Images 71, Rafe Swan / Cultura Creative (RF) 66tr, Scenics & Science 182tr, Science History Images / Alamy 184t, StudioSource 98bl, SWNS 137tr, US Coast Guard Photo 139cl; **Brennon Sapp**: 300tr; **Corbis**: Cultura 261, 310tr; **Fotolia.com**: Abhijith3747 182br, CGinspiration 234cr, Couperfield 267bl, Lucvar101 202tl, note_yn 98bc, Puripat1981 98br, sablin 115, Valdas Jarutis 130bc; **Getty Images**: Mario Tama 92tl, T.T. 227tr; **Martyn F. Chillmaid**: 275tr; **Medical Images.com**: Medical Images RM / GIANCARLO ZUCCOTTO 237cl;

Science Photo Library Ltd: 242tr, ASTIER - CHRU LILLE 229, 248br, Biodisc, Visuals Unlimited 187tr, Biophoto Associates 181br/a, Dr Jeremy Burgess 181br/b, 196tl, DR YORGOS NIKAS 124br, Dr. E. Davidson, Visuals Unlimited 188br, Dr. Gopal Murti 182cr, Dr. Klaus Boller 210tr, Eye of Science 185cr, Greg Dimijian 189tr, Jim Varney 275tl, Matteis / look at science 288tr, MAURO FERMARIELLO 300cl, Mike Agliolo 194tl, Steve Gschmeissner 181bl, Wim Van Egmond 203tl; **Shutterstock.com**: Asholove 164, Celebrian 275br, Dreamerb 183bl, egd 121tl, Image Source Trading Ltd 234tl, Sebastian Kaulitzki 277bl, Sirirat 193cl, wavebreakmedia 228; **Supplied by Tetra Scene of Crime Ltd**: 298c; **The Nobel Foundation**: Anthony Hewish's Nobel Lecture in 1974: Copyright © The Nobel Foundation (1974) 144t

Cover images: *Front*: **Shutterstock.com**: kilukilu

All other images © Pearson Education Limited

The publisher would like to thank the following for their kind permission to reproduce their materials:

Figures
Figure 5.3 from *OCR A Level Chemistry A2*, Pearson (Gent, D. and Ritchie, R.) p. 97, Original illustrations © Pearson Education Limited 2008; Figures 5.12, 5.13, 5.14, 5.15, 5.16 from *Applied Science Level 3 Student Book*, Pearson Education Limited (Annets, F., Hartley, J., Hocking, S., Llewellyn, R., Meunier, C., Parmar, C. and Peers, A.) © 2016; Figures 5.18, 5.19, 5.20, 5.21, 5.22 from *OCR A2 Biology Student Book and Exam Cafe CD*, Pearson Education Limited (Hocking, S.) © 2008; Figures 6.9, 6.17, 6.18, 17.2, 17.39, 17.40, 17.41, 17.43, 17.44, 17.45, 17.46, 23.2, 23.20, 23.23, 23.24, 23.34, 23.42 from *BTEC Level 3 National Applied Science Student Book*, Pearson Education Limited (Annets, F., Foale, S, Hartley, J., Hocking, S., Hudson, L., Kelly, T., Llewellyn, R., Musa, I. and Sorensen, J.) Original illustrations © Pearson Education Limited 2010; Figures 17.26, 17.35 from *OCR AS/A level Biology A Student B1*, 2 ed., Pearson Education Limited (Hocking, S., Sochacki, F. and Winterbottom, M.) Original illustrations © Pearson Education Limited 2015.

Text
Case Study 7.01 from 'The shocking picture that shows how a wind farm has disfigured one of Britain's loveliest landscapes' *Daily Mail*, 29/04/2008 (Camber, R. and Derbyshire, D.), Daily Mail & MailOnline; Case Study 7.02 from 'So Ashya's parents were right' *Daily Mail*, 31/01/2016 (Mezzofiore, G. for Mailonline), Daily Mail & MailOnline.

Websites
Pearson Education Limited is not responsible for the content of any external internet sites. It is essential for tutors to preview each website before using it in class so as to ensure that the URL is still accurate, relevant and appropriate. We suggest that tutors bookmark useful websites and consider enabling students to access them through the school/college intranet.

Note from the publisher
Pearson has robust editorial processes, including answer and fact checks, to ensure the accuracy of the content in this publication, and every effort is made to ensure this publication is free of errors. We are, however, only human, and occasionally errors do occur. Pearson is not liable for any misunderstandings that arise as a result of errors in this publication, but it is our priority to ensure that the content is accurate. If you spot an error, please do contact us at resourcescorrections@pearson.com so we can make sure it is corrected.

Contents

5 8 21
7 8 23

How to use this book

Welcome to your BTEC National Applied Science course.

You are joining a course that has a 30-year track record of learner success, with the BTEC National widely recognised within the industry and in higher education as the signature vocational qualification. Over 62 per cent of large companies recruit employees with BTEC qualifications and 100,000 BTEC learners apply to UK universities every year.

A BTEC National in Applied Science qualification will give you the opportunity to develop a range of specialist skills that will prepare you for the world of work, or for continued scientific study at a higher level.

BTEC Applied Science is a vocational course, available at a range of sizes, which is recognised and respected by employers and higher education institutions alike. Its flexible, unit-based structure allows you to choose the areas you wish to study, and focus on the aspects of science that interest you most.

In your BTEC course, you will not only get a solid grounding in scientific theories and concepts, but also develop the practical, investigative skills that underpin this sector. You will have the opportunity to focus on more specialist areas, such as forensics, genetics, material science, molecular biology and cryogenics. In addition to gaining science-specific skills, throughout your BTEC course you will develop more generic skills such as team-working, presentational skills and research strategies. These will ensure that you are ready to meet the demands of the modern workplace.

Scientific developments help to shape our world, and provide a huge range of employment opportunities. The field of genetics and genetic engineering help us to understand human diseases and how we can control food production. Forensics sheds light on how crimes are committed and how accidents can be investigated in a methodical and effective way. Materials science provides an understanding of how materials behave and what uses they can be put to. Biomedical science opens the door to careers in health care and related industries, giving you the choice to follow your interests and realise your ambitions. Most importantly, science is an area that is continually changing, and a BTEC Applied Science course reflects these developments and allows you to keep pace with the exciting innovations that emerge from scientific study.

How your BTEC is structured

Your BTEC National is divided into **mandatory units** (the ones you must do) and **optional units** (the ones you can choose to do). The number of units you need to do and the units you can cover will depend on the type and size of qualification you are doing.

This book covers **units 5, 6, 7, 17, 21 and 23**. If you are doing the **Foundation Diploma** in **Applied Science**, there is an optional unit here for you to help complete your course. If you are studying the **Diploma or Extended Diploma**, this book is designed to be used together with the *Pearson BTEC National Applied Science Student Book 1*, which includes further mandatory and optional units for these larger sizes of qualification. The table below shows how each unit in this book maps to the BTEC National Applied Science qualifications.

Unit title	Mandatory	Optional
Unit 5 Principles and Applications of Science II	Diploma and Extended Diploma	
Unit 6 Investigative Project	Diploma and Extended Diploma	
Unit 7 Contemporary Issues in Science	Extended Diploma	
Unit 17 Microbiology and Microbiological Techniques		Foundation Diploma, Diploma and Extended Diploma
Unit 21 Medical Physics Applications		Diploma and Extended Diploma
Unit 23 Forensic Evidence, Collection and Analysis		Diploma and Extended Diploma

Your learning experience

You may not realise it but you are always learning. Your educational and life experiences are constantly shaping you, your ideas, your thinking, and how you view and engage with the world around you.

You are the person most responsible for your own learning experience so it is really important you understand what you are learning, why you are learning it and why it is important both to your course and your personal development.

Your learning can be seen as a journey which moves through four phases.

Phase 1	Phase 2	Phase 3	Phase 4
You are introduced to a topic or concept; you start to develop an awareness of what learning is required.	You explore the topic or concept through different methods (e.g. research, questioning, analysis, deep thinking, critical evaluation) and form your own understanding.	You apply your knowledge and skills to a task designed to test your understanding.	You reflect on your learning, evaluate your efforts, identify gaps in your knowledge and look for ways to improve.

During each phase, you will use different learning strategies. As you go through your course, these strategies will combine to help you secure the core knowledge and skills you need.

This book has been written using similar learning principles, strategies and tools. It has been designed to support your learning journey, to give you control over your own learning and to equip you with the knowledge, understanding and tools to be successful in your future studies or career.

Features of this book

In this book there are lots of different features. They are there to help you learn about the topics in your course in different ways and understand it from multiple perspectives. Together these features:

▶ explain what your learning is about

▶ help you to build your knowledge

▶ help you understand how to succeed in your assessment

▶ help you to reflect on and evaluate your learning

▶ help you to link your learning to the workplace

In addition, each individual feature has a specific purpose, designed to support important learning strategies. For example, some features will:

▶ get you to question assumptions around what you are learning

▶ make you think beyond what you are reading about

▶ help you make connections across your learning and across units

▶ draw comparisons between your own learning and real-world workplace environments

▶ help you to develop some of the important skills you will need for the workplace, including team work, effective communication and problem solving.

Features that explain what your learning is about

Getting to know your unit

This section introduces the unit and explains how you will be assessed. It gives an overview of what will be covered and will help you to understand *why* you are doing the things you are asked to do in this unit.

Getting started

This appears at the start of every unit and is designed to get you thinking about the unit and what it involves. This feature will also help you to identify what you may already know about some of the topics in the unit and acts as a starting point for understanding the skills and knowledge you will need to develop to complete the unit.

Features that help you to build your knowledge

Research

This asks you to research a topic in greater depth. Using these features will help to expand your understanding of a topic as well as developing your research and investigation skills. All of these will be invaluable for your future progression, both professionally and academically.

Worked example

Our worked examples show the process you need to follow to solve a problem, such as a maths or science equation. This will also help you to develop your understanding and your numeracy and literacy skills.

Theory into practice

In this feature you are asked to consider the workplace or industry implications of a topic or concept from the unit. This will help you to understand the close links between what you are learning in the classroom and the affects it will have on a future career in your chosen sector.

Discussion

Discussion features encourage you to talk to other learners about a topic in greater detail, working together to increase your understanding of the topic and to understand other people's perspectives on an issue. This will also help to build your teamworking skills, which will be invaluable in your future professional and academic career.

Safety tip

This provides advice around health and safety when working on the unit. It will help build your knowledge about best practice in the workplace, as well as make sure that you stay safe.

Key terms

Concise and simple definitions are provided for key words, phrases and concepts, allowing you to have, at a glance, a clear understanding of the key ideas in each unit.

Link

This shows any links between units or within the same unit, helping you to identify where the knowledge you have learned elsewhere will help you to achieve the requirements of the unit. Remember, although your BTEC National is made up of several units, there are common themes that are explored from different perspectives across the whole of your course.

Step-by-step

This practical feature gives step-by-step descriptions of particular processes or tasks in the unit. This will help you to understand the key stages in the process and help you to carry out the process yourself.

Further reading and resources

This contains a list of other resources – such as books, journals, articles or websites – you can use to expand your knowledge of the unit content. This is a good opportunity for you to take responsibility for your own learning, as well as preparing you for research tasks you may need to do academically or professionally.

Features connected to your assessment

Your course is made up of a series of mandatory and optional units. There are two different types of mandatory unit:
▶ externally assessed
▶ internally assessed.

The features that support you in preparing for assessment are below. But first, what is the difference between these two different types of units?

Externally assessed units

These units give you the opportunity to present what you have learned in the unit in a different way. They can be challenging, but will really give you the opportunity to demonstrate your knowledge and understanding, or your skills in a direct way. For these units you will complete a task, set directly by Pearson, in controlled conditions. This could take the form of an exam or it could be another type of task. You may have the opportunity in advance to research and prepare notes around a topic, which can be used when completing the assessment.

Internally assessed units

Internally assessed units involve you completing a series of assignments, set and marked by your tutor. The assignments you complete could allow you to demonstrate your learning in a number of different ways, from a written report to a presentation to a video recording and observation statements of you completing a practical task. Whatever the method, you will need to make sure you have clear evidence of what you have achieved and how you did it.

Assessment practice

These features give you the opportunity to practise some of the skills you will need when you are assessed on your unit. They do not fully reflect the actual assessment tasks, but will help you get ready for doing them.

Plan – Do – Review

You'll also find handy advice on how to plan, complete and evaluate your work after you have completed it. This is designed to get you thinking about the best way to complete your work and to build your skills and experience before doing the actual assessment. These prompt questions are designed to get you started with thinking about how the way you work, as well as understand why you do things.

Getting ready for assessment

For internally assessed units, this is a case study from a BTEC National student, talking about how they planned and carried out their assignment work and what they would do differently if they were to do it again. It will give you advice on preparing for the kind of work you will need to for your internal assessments, including 'Think about it' points for you to consider for your own development.

Getting ready for assessment

This section will help you to prepare for external assessment. It gives practical advice on preparing for and sitting exams or a set task. It provides a series of sample answers for the types of questions you will need to answer in your external assessments, including guidance on the good points of these answers and how these answers could be improved.

Features to help you reflect on and evaluate your learning

II PAUSE POINT

Pause Points appear after a section of each unit and give you the opportunity to review and reflect upon your own learning. The ability to reflect on your own performance is a key skill you'll need to develop and use throughout your life, and will be essential whatever your future plans are.

Hint
Extend

These also give you suggestions to help cement your knowledge and indicate other areas you can look at to expand it.

Reflect

This allows you to reflect on how the knowledge you have gained in this unit may impact your behaviour in a workplace situation. This will help not only to place the topic in a professional context, but also help you to review your own conduct and develop your employability skills.

Features which link your learning with the workplace

Case study

Case studies are used throughout the book to allow you to apply the learning and knowledge from the unit to a scenario from the workplace or the industry. Case studies include questions to help you consider the wider context of a topic. This is an opportunity to see how the unit's content is reflected in the real world, and help you to build familiarity with issues you may find in a real-world workplace.

THINK ▶FUTURE

This is a special case study where someone working in the industry talks about the job role they do and the skills they need. This comes with a *Focusing your skills* section, which gives suggestions for how you can begin to develop the employability skills and experiences that are needed to be successful in a career in your chosen sector. This is an excellent opportunity to help you identify what you could do, inside and outside of your BTEC National studies, to build up your employability skills.

Principles and Applications of Science II

5

Getting to know your unit

Science technicians need to be able to use and apply key science concepts to work efficiently and safely in science and science related organisations. Chemists, for example, need to understand how the uses of chemical products relate to their physical and chemical properties. Scientists in the pharmaceutical and medical industries need to understand how the human body works. They need to know how diseases can be diagnosed and treated. Engineers need to understand the properties and behaviour of different materials to ensure that the materials are fit for purpose. When designing machines, they need to understand how materials will behave under different conditions. They also need to understand how energy is transferred in order to make efficient machines and engines.

How you will be assessed

The external assessment for this unit will be split into three sections.

▶ **Section A** – Chemistry (Properties and uses of substances, Structures, Reactions and properties of commercially important organic compounds, Energy changes in industry)

▶ **Section B** – Biology (Organs and systems, Ventilation and gas exchange to the lungs, Urinary system structure and function, Cell transport mechanisms)

▶ **Section C** – Physics (Thermal physics, Materials, Fluids in motion).

The assessment will contain a range of question types, including multiple choice, calculations, short answer and open response. These question types, by their very nature, generally assess discrete knowledge and understanding of the content in this unit.

You need to be able to apply and synthesise knowledge from this unit. The questions will be contextualised in order for you to show that you can do this.

Throughout this unit you will find assessment practices that will help you prepare for the assessment. Completing each of these will give you an insight into the types of questions that will be asked and, importantly, how to answer them.

Unit 5 has four Assessment Outcomes (AOs). These are:

▶ **AO1**: Demonstrate knowledge of scientific facts, terms, definitions and scientific formulae
  - Command words: describe, draw, explain, identify, name, state

▶ **AO2**: Demonstrate understanding of scientific concepts, procedures, processes and techniques and their application
  - Command words: calculate, describe, draw, explain, give, show, state

▶ **AO3**: Analyse, interpret and evaluate scientific information to make judgements and reach conclusions
  - Command words: analyse, comment, describe, explain, give, state

▶ **AO4**: Make connections, use and integrate different scientific concepts, procedures, processes or techniques
  - Command words: calculate, comment, explain

Command word	Definition – what it is asking you to do
Add/label	This requires you to add labelling to a stimulus material given in the question, for example labelling a diagram or adding units to a table.
Assess	You need to give careful consideration to all the factors or events that apply and identify which are the most important or relevant. You need to make a judgement on the importance of something and come to a conclusion where needed.
Calculate	You will obtain a numerical answer, showing relevant working. If the answer has a unit, you must include it.
Comment on	This requires you to synthesise a number of variables from data/information to form a judgement. More than two factors need to be synthesised.
Compare	This asks you to identify the main factors of two or more items and point out their similarities and differences. You may need to say which are the least important or most important. The word *Contrast* is very similar.
Complete	This requires you to complete a table/diagram.
Criticise	Here you need to inspect a set of data, an experimental plan or a scientific statement and consider the elements. Look at the merits and/or faults of the information presented and back up the judgements that you make.
Deduce	Here you draw/reach conclusion(s) from the information provided.
Derive	Here you combine two or more equations or principles to develop a new equation.
Describe	You need to give a full account of all the information, including all the relevant details of any features, of a topic.
Determine	Your answer must have an element that is quantitative from the stimulus provided, or must show how the answer can be reached quantitatively. To gain maximum marks there must be a quantitative element to the answer.
Devise	You plan or invent a procedure from existing principles/ideas.
Discuss	Here you write about the topic in detail, taking into account different ideas and opinions.
Draw	You produce a diagram either using a ruler or using freehand.
Evaluate	Here you bring all the relevant information you have on a topic together and make a judgment on it (for example on its success or importance). Your judgment should be clearly supported by the information you have gathered.
Explain	You make an idea, situation or problem clear to your reader, by describing it in detail, including any relevant data or facts.
Give a reason why	This is when a statement has been made and you only need to give the reasons why.
Give/state/name	These generally require you to recall one or more pieces of information.
Identify	This usually requires you to select some key information from a given stimulus/resource.
Plot	You need to produce a graph by marking points accurately on a grid from data that is provided and then drawing a line of best fit through these points. You must include a suitable scale and appropriately labelled axes if these are not provided in the question.
Predict	You give an expected result.
Show that	You prove that a numerical figure is as stated in the question. The answer must be to at least one more significant figure than the numerical figure in the question.
Sketch	You produce a freehand drawing. For a graph this would need a line and labelled axes with important features indicated. The axes are not scaled.
State and justify/ identify and justify	You make a selection and justify it.
State what is meant by	Here, you give the meaning of a term but there are different ways in which this meaning can be described.
Write	Here the question asks for an equation.

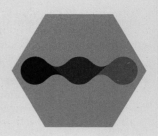

Getting started

This unit builds on prior knowledge. Look at the topics for each section of the exam. Write down two things you already know about each of those topics. Suggest what you are expecting to learn about when studying those topics.

A Properties and uses of substances

Relating properties to uses and production of substances

Scientists need to know the different chemical properties of substances. They can use this knowledge to make new, useful materials.

Metal oxides and metal hydroxides

Alumina is the common name for aluminium oxide. One property of alumina is that it is **amphoteric**. This means that it will act as an **acid** or a **base** depending on the conditions. If an acid is present it will react as a base. If a base is present it will act as an acid.

> **Key terms**
>
> **Amphoteric** – substance that can act as both an acid and a base.
>
> **Acid** – a compound containing hydrogen that dissociates in water to form hydrogen ions.
>
> **Base** – a compound that reacts with an acid to form a salt and water.

When aluminium oxide reacts with an acid it reacts in the same way as other metal oxides, such as magnesium oxide, that are bases.

Aluminium oxide reacts with hot hydrochloric acid to give aluminium chloride and water. The aluminium chloride is soluble so you get aluminium chloride solution.

$$Al_2O_3 + 6HCl \rightarrow 2AlCl_3 + 3H_2O$$

Aluminium oxide will react with bases such as sodium hydroxide to form aluminates. For example:

$$Al_2O_3 + 6NaOH + 3H_2O \rightarrow 2Na_3Al(OH)_6$$
$$\text{(sodium aluminate)}$$

Other aluminates can form depending on the conditions. Hot concentrated sodium hydroxide will give sodium tetrahydroxoaluminate, $2NaAl(OH)_4$. You will not be expected to know all aluminates that may form.

In general, aluminium oxide is chemically inert except under certain conditions, e.g. when a hot acid or base is present. This means it has a lot of uses including filler, paint, sunscreens, and glass. The amphoteric nature of alumina means it can be used as a medium for chromatography as a basic, acidic or neutral medium.

It is also an effective desiccant and can be used at a range of basic and acidic pHs. It is stable over pH range 2–13. This also makes it suitable for environmental clean-up and separation applications.

Group 1 and group 2 metal oxides are basic. If they dissolve in water, then they will react with the water to form metal hydroxides. This forms an **alkali** solution. For example:

$$Na_2O + H_2O \rightarrow 2NaOH$$
$$CaO + H_2O \rightarrow Ca(OH)_2$$

> **Key term**
>
> **Alkali** – a base that dissolves in water to form hydroxide ions.

The oxide ion is a very strongly basic anion due to its very small size and high (2−) charge. The oxide ion in the metal oxide reacts with water to produce hydroxide ions, because a hydroxide ion is the strongest base that can persist in water.

Calcium oxide (quicklime) reacts with water to form calcium hydroxide (lime). This can then be used by farmers to raise the pH of acidic soil.

Magnesium oxide can be used as a desiccant when preserving books in a library. Some papers produce acidic sulfur oxides when oxidised in air. Although magnesium oxide is not a very good desiccant, it is used because due to its basic nature it will neutralise the acidic sulfur oxides produced and so help to preserve the book.

Sodium hydroxide, which is produced when sodium oxide reacts with water, is an important compound in the chemical industry as well as in everyday life. For example, it is used in some processes to make plastics and soaps. It is used in food processing in many ways. Some examples are to peel fruits and to process cocoa and chocolate. It is found in drain cleaner or in oven cleaner. These uses all rely on the basic properties of the compound.

Magnesium hydroxide, $Mg(OH)_2$, is also known as 'milk of magnesia' and is used for treating acid indigestion. It raises the stomach pH due to its basic nature.

Calcium hydroxide, $Ca(OH)_2$, is used in the treatment of acidic effluent. Factories that produce or use sulfuric acid will also produce acidic liquid effluent which may have to be treated before being released to the environment, especially if it has high acid content. Impurities are precipitated and removed. The acidic liquid can be neutralised using calcium hydroxide, lime.

This reaction occurs.

$$H_2SO_4 + Ca(OH)_2 \rightarrow CaSO_4 + 2H_2O$$

The calcium hydroxide is a base and a neutral calcium sulfate is formed. Calcium sulfate is a **salt**. Although other bases are available, calcium hydroxide is cheap and simple to use.

> **Key term**
>
> **Salt** – a compound formed by an acid–base reaction where the hydrogen in the acid has been replaced by a metal (or other positive) ion.

Electrolysis

Electrolysis breaks down compounds into simpler substances. It is used in some industrial processes (see page 9). Ionic substances are decomposed by an electrical charge being passed through them during electrolysis. The ions must be free to move for this to work so the compound must either be molten (reduced to liquid form by heating) or in solution.

> **Key term**
>
> **Electrolysis** – the decomposition of a compound using electricity.

Figure 5.1 shows the electrolysis of molten sodium chloride. Sodium is produced at the cathode (positive electrode) and chlorine is produced at the anode (negative electrode).

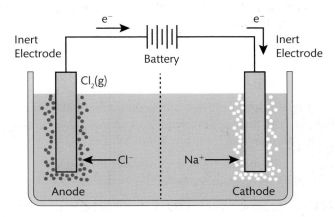

▶ **Figure 5.1** Electrolysis of molten sodium chloride

The only ions present are sodium and chloride ions and so only sodium and chlorine can be produced by this reaction.

Figure 5.2 shows the electrolysis of sodium chloride solution.

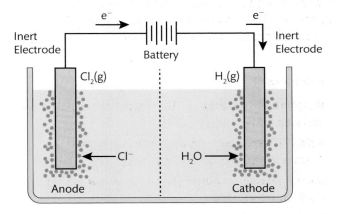

▶ **Figure 5.2** Electrolysis of sodium chloride solution

When the sodium chloride is dissolved in water, hydrogen and hydroxide ions are present as well as the sodium and chloride ions. In this case, the chlorine is still produced at the anode but hydrogen gas is produced at the cathode rather than sodium. It is only possible for one type of ion to discharge at each electrode. The ion which is selected to discharge at an electrode depends on a number of factors.

The position of the ion in the electrochemical series affects its ease of discharge at an electrode. When metals react, they lose electrons to become positive ions. If a metal is placed in water, then the metal atoms will tend to lose electrons to the water and become positive. This in turn attracts the negative electrons back to the metal. This is an example of an equilibrium process.

For example, magnesium in water will have the following equilibrium reaction.

$$Mg^{2+}(aq) + 2e^- \rightleftharpoons Mg(s)$$

A piece of copper in water will behave in the same way.

$$Cu^{2+}(aq) + 2e^- \rightleftharpoons Cu(s)$$

As copper is less reactive than magnesium, it will form ions less easily so the equilibrium for copper in water is further to the left than that for magnesium in water.

For both the metals, there is a potential difference between the negative charge on the metal and the positive charge of the solution around it. The potential difference is bigger for magnesium than it is for copper. This potential difference can be measured as a standard electrode potential and then the metals can be placed in order of the standard electrode potentials ($E°$) in the electrochemical series.

Hydrogen is also included in the electrochemical series (see Table 5.1).

▶ **Table 5.1** Part of the electrochemical series

Equilibrium half equation	$E°$ (volts)
$Li^+(aq) + e^- \rightleftharpoons Li(s)$	−3.03
$K^+(aq) + e^- \rightleftharpoons K(s)$	−2.92
$Na^+(aq) + e^- \rightleftharpoons Na(s)$	−2.71
$Mg^{2+}(aq) + 2e^- \rightleftharpoons Mg(s)$	−2.37
$Al^{3+}(aq) + 3e^- \rightleftharpoons Al(s)$	−1.66
$Zn^{2+}(aq) + 2e^- \rightleftharpoons Zn(s)$	−0.76
$2H^+(aq) + 2e^- \rightleftharpoons H_2(g)$	0
$Cu^{2+}(aq) + 2e^- \rightleftharpoons Cu(s)$	+0.34
$\frac{1}{2}Cl_2(aq) + e^- \rightarrow Cl^-(aq)$	+1.36

The lower the metal is in the electrochemical series, the more likely it is to be discharged. So in a solution of copper sulfate, the copper ions will be discharged to form copper atoms rather than the hydrogen ions. In a solution of sodium chloride, hydrogen gas is given off at the cathode rather than sodium metal.

The concentration of the ions in solution will also have an effect on which ion is discharged. The most concentrated ion tends to be discharged, no matter where the ions are in the electrochemical series. So in a concentrated solution containing the chloride ion, the chloride ion will be discharged rather than the hydroxide ion, even though the hydroxide ion is more easily oxidised.

Assessment practice 5.1

Explain the products of the electrolysis of brine.

Transition metals

Transition metals are the d block elements found between group 2 and 3 on the periodic table. They are used in a variety of industrial processes and transition metals and their compounds have a wide number of uses. Transition metals have incomplete d-sub-shells as a stable ion.

If we look at the transition elements in period 4, across the period, from scandium to zinc, the $3d$-orbitals are being filled. The pattern is regular. (Except for chromium and copper, which do not follow the principle of completely filling the lowest energy levels first.) The sub-shell energy levels in the third and fourth energy levels overlap. The $4s$-sub-shell fills before the $3d$-sub-shell.

When they react, transition element atoms lose electrons to form positive ions. Transition metals lose their $4s$ electrons before their $3d$ electrons. They form ions with more than one stable oxidation state. They can all form compounds with metal ions in the +2 oxidation state.

In solution, transition metal compounds form **complex ions**. A complex ion consists of a central metal ion surrounded by **ligands**. A ligand is a molecule or ion that donates a pair of electrons to the central transition metal ion to form a dative covalent bond.

Figure 5.3 shows the complex ion $[Fe(H_2O)_6]^{2+}$. Each of the six H_2O ligands forms one dative covalent bond with the central metal ion.

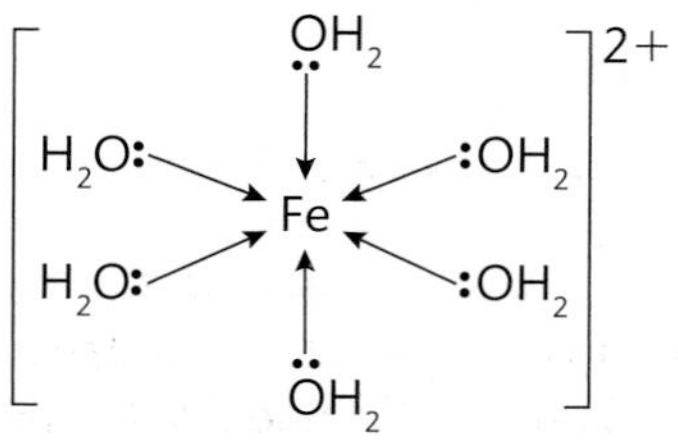

▶ **Figure 5.3** Complex metal ion

Other complex ions include $[Cu(H_2O)_6]^{2+}$, $[Al(H_2O)_6]^{3+}$, and $[CuCl_4]^{2-}$.

Transition metals, oxides of transition metals and transition metal complexes are all used as industrial **catalysts**.

Complex ion – a transition metal ion bonded to one or more ligands by dative covalent bonds.

Ligand – a molecule or ion that can donate a pair of electrons to the transition metal ion to form a dative bond.

Catalyst – a substance that speeds up a reaction. It can take part in the reaction but is left unchanged at the end of the reaction.

Platinum and rhodium are used in catalytic converters in car exhausts. They convert carbon monoxide and nitrogen monoxide emissions into carbon dioxide and nitrogen gas. These gases are not polluting. Manganese dioxide acts as a catalyst in the decomposition of hydrogen peroxide into water and oxygen.

The contact process uses vanadium(V) oxide as a catalyst for the production of sulfuric acid. There are four stages in this process.

1 Sulfur reacts with oxygen to produce sulfur dioxide.

$$S(s) + O_2(g) \rightarrow SO_2(g)$$

Or sulfur dioxide is produced by reacting sulfide ores with excess air.

$$4FeS_2(g) + 11O_2 \rightarrow 2Fe_2O_2(s) + 8SO_2(g)$$

The sulfur dioxide produced is mixed with excess air for use in stage 2.

2 Sulfur dioxide is converted into sulfur trioxide in a reversible reaction.

$$2SO_2(g) + O_2(g) \rightleftharpoons 2SO_3(g)$$

Vanadium(V) oxide catalyses this stage. The following reactions occur.

The overall reaction is:

$$SO_2 + \tfrac{1}{2}O_2 \xrightarrow{\;V_2O_5\;} SO_3$$

Vanadium(V) oxide can act as a catalyst due to the ability of vanadium to change its oxidation state. Sulfur dioxide is oxidised to sulfur trioxide by the vanadium(V) oxide. So the vanadium(V) oxide is reduced to vanadium(IV) oxide.

$$SO_2 + V_2O_5 \rightarrow 2SO_3 + V_2O_4$$

The vanadium(IV) oxide is then oxidised with the oxygen:

$$V_2O_4 + \tfrac{1}{2}O_2 \rightarrow V_2O_5$$

So the vanadium(V) oxide catalyst takes ... the reaction and is changed in the react... remains chemically unchanged by the end of... reaction.

3 The sulfur trioxide is converted into sulfuric acid. Adding sulfur trioxide to water is too uncontrollable to be the method used. First sulfur trioxide is dissolved in concentrated sulfuric acid. This produces fuming sulfuric acid which can then be added safely to water to produce more concentrated sulfuric acid.

$$H_2SO_4(l) + SO_3 \rightarrow H_2S_2O_7(l)$$
$$H_2S_2O_7(l) + H_2O(l) \rightarrow 2H_2SO_4(l)$$

Iron is used as the catalyst in the Haber process for the synthesis of ammonia. Nitrogen reacts with hydrogen to produce ammonia.

$$N_2(g) + 3H_2(g) \rightleftharpoons 2NH_3(g)$$

The iron catalyst speeds up the reaction by lowering the **activation energy** so that the N_2 bonds and H_2 bonds can be broken more readily.

Activation energy – the minimum amount of energy needed by the reactants for collisions to result in a reaction taking place.

The Haber Process

The iron catalyst lowers the energy demands of the Haber Process. This means that the costs of the process are reduced and also helps the environment. Less fuel is needed to be burned to generate the energy needed. This means fossil fuels are conserved and there are fewer carbon dioxide emissions. The Haber Process is an important industrial reaction because it produces ammonia. This is used as the basis for making fertilisers which improve crop yield. This becomes more important each year as we need to feed the world's population, which is increasing every year.

Check your understanding

1 How does the use of iron in the Haber Process lower environmental and financial costs?

List all the different chemical substances discussed so far in this unit.

Hint   Compete a table with the headings metals, metal oxides and metal hydroxides.

Extend   Give one use for each of the substances based on its chemical properties.

Assessment practice 5.2

Describe how vanadium(V) oxide acts as a catalyst in the contact process. Include any relevant equations.

Purification, extraction and manufacture of useful substances

Alumina (aluminium oxide) is a material that retains its strength at high temperatures. It is chemically and physically stable at these high temperatures. It is a **refractory material**. It can be used in linings for furnaces, kilns and reactors. It can also be used in acidic reactions as it does not react.

Key term

Refractory material – material that is physically and chemically stable at very high temperatures, for example, over 3000°C.

Alumina and aluminium extraction

Alumina is extracted from **bauxite** ore using the Bayer process. The bauxite is crushed to form grains. It is then mixed with liquor from the later precipitation stage and crushed again to make a **slurry**. Bauxite can have a high level of silica present so it may have to go through a desilication process to remove this. Hot caustic soda (NaOH) solution is used to dissolve the aluminium-bearing minerals, gibbsite, böhmite and diaspore, in the bauxite, to form a sodium aluminate supersaturated solution (liquor). This is called 'digestion'.

Key terms

Bauxite – aluminium ore.

Slurry – semi-liquid mixture containing fine solid particles.

Conditions for this change due to the composition of the bauxite ores. Those with a high gibbsite content need about 140°C, while those with a high böhmitic content need a higher temperature between 200 and 280°C.

After digestion, the slurry is cooled to around 106°C. It is then clarified to separate the solids from the liquor via sedimentation. Chemical additives, **flocculants**, are added to help the **sedimentation** process. The bauxite residue is transferred to the washing tanks to recover the caustic soda, which is reused in the digestion process.

Key terms

Flocculants – substances that causes particles to clump and so settle out of a liquid.

Sedimentation – small solid particles settling at the bottom of a liquid.

Further separation of the pregnant liquor (the sodium aluminate remains in the solution) from the bauxite residue is performed using a series of filters to ensure that the final product is not contaminated with impurities present in the residue.

Precipitation is used to recover the alumina by crystallisation from the liquor, which is supersaturated in sodium aluminate. The liquor is cooled, resulting in the formation of small crystals of aluminium trihydroxide ($Al(OH)_3$), which then grow and form larger crystals.

The spent liquor is heated through a series of heat exchangers and subsequently cooled in a series of flash tanks. The **condensate** formed in the heaters is re-used in the process, e.g. for washing bauxite residue. The remaining caustic soda is washed and recycled back into the digestion process.

The gibbsite crystals formed in precipitation are classified into size ranges. This is normally done using cyclones or gravity classification tanks.

The filter cakes are fed into calciners where they are heated to temperatures of up to 1100°C to drive off moisture. This produces alumina solids. The following equation describes the **calcination** reaction:

$$2Al(OH)_3 \rightarrow Al_2O_3 + 3H_2O$$

Alumina, a white powder, is the product of this step and the final product of the Bayer Process.

Key terms

Condensate – liquid collected by condensation.

Calcination – heating to high temperature to remove free and chemically bonded water.

Aluminium is extracted from alumina using the Hall-Héroult process. Alumina is dissolved in molten cryolite, Na_3AlF_6 (an aluminium ore), and electrolysis is used to extract the aluminium. Aqueous aluminium oxide is not used because the aluminium is readily oxidised by the hydrogen ions present in the solution. Alumina has a melting point of over 2000°C and so it is not practical to use molten alumina. However, if the alumina is dissolved in molten cryolite, then the melting point is a lot lower (about 1000°C) and so electrolysis is possible. Aluminium fluoride is also added to the mixture and this lowers the melting point even further.

The aluminium sinks to the bottom of the cell and is syphoned off. Carbon dioxide is produced at the anode along with hydrogen fluoride from the cryolite.

Electrolysis of brine

Brine is aqueous sodium chloride. As you saw on page 5, the electrolysis of brine produces sodium hydroxide, hydrogen and chlorine, all of which are industrially useful chemicals. Sodium hydroxide is used in food processing and removing pollutants during paper manufacture. Hydrogen is used in the production of hydrochloric acid and as a fuel. Chlorine is a disinfectant and is also used in making plastics. Figure 5.4 shows electrolysis of brine using a Hoffman voltameter.

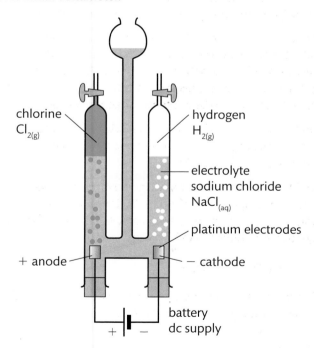

▶ **Figure 5.4** Hoffman voltameter

Inert platinum electrodes are used to prevent reaction with the gases produced. Chlorine is collected at the anode and hydrogen is collected at the cathode. The sodium ions

react with the hydroxide ions in solution to form sodium hydroxide. The chlorine and the hydrogen gases are kept separate as a mixture of these gases will explode on exposure to heat or light.

The chlorine will also react with the sodium hydroxide solution formed to produce a mixture of sodium chloride and sodium hypochlorate. This mixture can be used as bleach. However, specialised cells are need if the desired products are chlorine and sodium hydroxide. There are two types of cell that can be used:

▶ a diaphragm
▶ a membrane cell.

The diaphragm cell has a diaphragm in the centre. This is made of a porous mixture of asbestos and polymers. The brine is added on one side of the diaphragm and the aqueous sodium hydroxide is removed on the other side. The level of liquid on the side where the brine is added is always higher than the side where the aqueous sodium hydroxide is removed, so that the brine does not flow back to mix with the chlorine produced.

The membrane in a membrane cell is made from a polymer that only allows positive ions to pass through it. So the sodium ions can pass through but the chloride ions cannot. So the sodium hydroxide that forms in the right hand compartment does not mix with sodium chloride solution.

The chlorine produced in diaphragm and membrane cells will mix with any oxygen produced at the anode but can be purified by liquefying it under pressure. Under pressure, the oxygen will stay a gas.

Titanium extraction *Come in Post Paper*

Titanium can be obtained by extraction from its ore, TiO_2, called rutile. The titanium cannot be extracted using carbon, as titanium carbide (TiC) will form and the presence of this carbide makes the titanium very brittle. There are two stages in the extraction process. It is a batch process so only small amounts are made at one time. It is also expensive due to the two steps, the conditions and the use of chlorine and magnesium. The magnesium has to be extracted from its ore before being used for this process.

1 Conversion of titanium(IV) oxide to titanium(IV) chloride

$$TiO_2(s) + 2Cl_2(g) + 2C(g) \rightarrow TiCl_4(g) + 2CO(g)$$

2 Titanium chloride is then reduced using magnesium to produce titanium.

$$TiCl_4(g) + 2Mg(l) \rightarrow Ti(s) + 2MgCl_2(s)$$

The magnesium is put into a steel reactor and the titanium chloride is pumped in. The reactor is sealed and heated to 1200°C. The reaction is carried out in an argon atmosphere as the titanium produced would react with any oxygen or water present. Any oxygen or nitrogen present in the reactor would make the metal brittle. The reactor is left sealed for two to three days before the titanium can be removed. A large reactor will only produce about 1 tonne of titanium a day, which is a relatively small amount.

It is important to understand that products are used based on the properties they have. These can be physical or chemical properties. Their production is also dependent on their properties. For example, aluminium is very reactive, so is extracted by electrolysis because it cannot be extracted using the cheaper reduction with carbon method. Titanium can be reduced but magnesium is needed to do this as using carbon would cause the metal to be brittle.

 PAUSE POINT   Draw a mind map of all the production methods discussed in this unit.

> Hint   Link each to purification, extraction and manufacture methods.
> Extend   Link each production method to the properties of the substances.

Structures, reactions and properties of commercially important organic compounds

Organic chemicals are carbon based compounds. They are compounds found in living organisms, e.g. amino acids, fatty acids, etc. They are also used in industry and in everyday products. Clothes, plastics, and food are a few examples of where organic chemicals are used.

There are different families of organic compounds. These families are called **homologous series**. The alkanes found in crude oil are a few of the organic compounds found in the alkane homologous series. Alkenes and alcohols are two more homologous series. All members of a homologous series have the same **functional group** and differ only by the number of CH_2 units present. Organic compounds can be classified as **aliphatic**, **alicyclic** or **aromatic**.

Alkanes and alkenes

There are millions of organic compounds. All scientists use the same system to name them. This is the **IUPAC nomenclature**. Scientists follow the rules of this nomenclature to name compounds. Alkanes are aliphatic **saturated compound** hydrocarbons. They only have single bonds and only contain carbon and hydrogen atoms. Methane is an alkane. The 'ane' part of the name shows it is an alkane. The first part of the name, 'meth' shows how many carbons are present in the chain.

> **Key terms**
>
> **Homologous series** – family of organic chemicals.
>
> **Functional group** – group of atoms responsible for the characteristic reactions of a substance.

> **Key terms**
>
> **Aliphatic** – a hydrocarbon with carbon atoms joined together in straight or branched chains.
>
> **Alicyclic** – a hydrocarbon with carbon atoms joined together in a ring structure.
>
> **Aromatic** – a hydrocarbon containing at least one benzene ring.
>
> **IUPAC nomenclature** – system of the International Union of Pure and Applied Chemistry for naming organic compounds.
>
> **Saturated compound** – contains single bonds only.

The general formula for an alkane is C_nH_{2n+2}. Table 5.2 shows the first ten alkanes, their name, formula and structure.

▶ **Table 5.2** Alkanes

Name	Molecular formula
Methane	CH_4
Ethane	C_2H_6
Propane	C_3H_8
Butane	C_4H_{10}
Pentane	C_5H_{12}
Hexane	C_6H_{14}
Heptane	C_7H_{16}
Octane	C_8H_{18}
Nonane	C_9H_{20}
Decane	$C_{10}H_{22}$

Some alkanes are cyclic. These have rings of carbons. For example, cyclohexane has a ring structure with six carbons. Cyclic alkanes have a general formula of C_nH_{2n}.

Alkenes are organic compounds containing a double bond and have the general formula C_nH_{2n}. They have similar names to alkanes with the same number of carbon atoms but the name ends in '-ene'.

Naming organic compounds

Scientists follow these rules when naming an organic compound.

1   The **stem** is the main part of the name and comes from the longest carbon chain or parent chain in the compound.

2   The **suffix** is the end of the name. This identifies the most important functional group, e.g. 'ane' in alkanes.

3   The **prefix** is the front part of the name and identifies other functional groups. The number of the carbon atom they are attached to is given. You count the

carbons along the chain so that the functional group has the lowest number possible (e.g. butan-1-ol, butan-2-ol, but-1-ene and but-2-ene).

4   If there is more than one functional group, they are written in alphabetical order.

Table 5.3 shows common functional groups.

▶ **Table 5.3** Common functional groups

Functional group	Formula	Prefix	Suffix
Alcohol	—OH	Hydroxy-	-ol
Aldehyde	—CHO		-al
Alkane	C—C		-ane
Alkene	C=C		-ene
Carboxylic acid	—COOH		-oic acid
Ester	—COO—		-oate
Haloalkane	—F, —Cl, —Br, —I	Fluoro-, chloro-, bromo-, iodo-	
Ketone	C—CO—C		-one

Worked example

Name the following organic compounds.

1

Stem – the longest chain contains 3 carbon atoms and so the stem is prop-.

There are two functional groups: i. alcohol —OH, ii. alkyl (methyl-) —CH_3 They are both on the second carbon in the chain.

—OH becomes the suffix. -2-ol

—CH_3 is the prefix 2-methyl-

The name of the compound is 2-methylpropan-2-ol.

2   H_3C—CH—CH_2—CH—CH_2—CH_3
            | |
            Br   Cl

Stem – the longest chain contains 6 carbons so the stem is hex-.

The suffix is –ane, and there are **halo**gens bromine and chlorine present, as this is a **haloalkane**.

There are two functional groups, bromine and chlorine. These are both prefixes. Remember, if there is more than one prefix they are written in alphabetical order. So bromine is first. This means you count the carbons from the end nearest the bromine as above.

The name of this compound is 2-bromo-4-chlorohexane.

Key term

Haloalkane – alkane where at least one hydrogen is replaced by a halogen.

Types of organic formulae

A displayed formula shows the relative positions of all atoms in a molecule and the bonds between them. For example, ethanoic acid has the formula CH_3COOH. Sometimes the functional group is shown as $-COOH$. However, in a displayed formula you must show all the bonds and atoms.

The displayed formula for ethanoic acid is

Organic compounds are not two-dimensional. It is often important to draw the structure of an **organic compound** to show its three-dimensional structure. Methane can

be shown as

This does not show how the hydrogen atoms are arranged in space around the carbon. To see the spatial arrangement, we use wedge and dashed lines.

The solid lines are in line with the plane of the paper. The dashed line shows the bond is angling backwards and the wedge shows the bond is angling forward. It is easier to see the different bond angles when a compound is drawn in this way.

A symmetrical alkene has the same groups attached to each of the carbons in the double bond. Ethene is a symmetrical alkene because both carbons in the double bond are attached to two hydrogens.

But-2-ene can exist in two distinct geometric forms which have different types of symmetry. Both carbons in the double bond have a methyl group and a hydrogen atom attached.

has a mirror plane symmetry.

But exchanging the positions of the methyl group and the hydrogen attached to carbon atom 3 results in another structure which has a centre of two-fold rotational symmetry.

Isomers

These two forms of but-2-ene are called **isomers**. Isomers are two or more compounds that have the same general formula but a different arrangement of atoms in the compound and so have different properties.

Substances with the same molecular formula but different structural formulae are called structural isomers. There are three types of structural isomer; chain isomers, positional isomers and functional group isomers.

Butane and 2-methyl propane both have the molecular formula C_4H_{10}. They are chain isomers.

Butane

2-methyl propane

Chain isomers occur in all carbon compounds with four or more carbons. The longer the chain, the more possible chain isomers there are.

Butane and 2-methyl propane are both alkanes so have similar chemical properties but their boiling points are different.

Positional isomers are molecules with the same carbon chain and functional groups but the functional groups are on different carbons within the chain. For example, butanol has the molecular formula $C_4H_{10}O$. It has two positional isomers where the $-OH$ functional group is either on the first or the second carbon in the chain.

Functional group isomers have the same molecular formula but different functional groups, so they belong to different organic families. For example, propanal is an aldehyde with the molecular formula C_3H_6O. Propanone has the same molecular formula but it is a ketone.

Stereoisomerism is caused by a double or triple bond. Optical isomers are also a form of stereoisomerism. Stereoisomers are compounds that have the same structural formula but the arrangement of their atoms in space is different. Atoms can rotate around a single bond. However, this rotation is restricted around a double or triple bond. For example, ethene C_2H_4 has a double $C=C$ bond. If a methyl group is added to each carbon, replacing a hydrogen atom, stereoisomers form. These are called cis and trans-isomers. Figure 5.5 shows these stereoisomers.

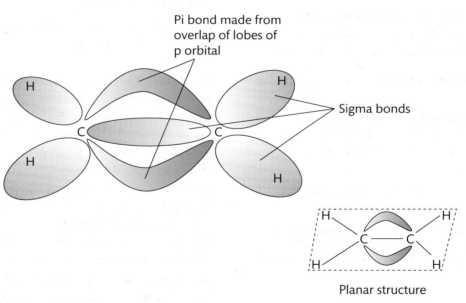

cis–but–2–ene
(Z)–but–2–ene

trans–but–2–ene
(E)–but–2–ene

▶ **Figure 5.5** (*Z*)-but-2-ene and (*E*)-but-2-ene

E/Z and cis/trans are two different methods of classifying geometric isomers. If something is the cis-isomer, then its name is *cis*-but-2-ene; if it is the E isomer, then its name is (*E*)-but-2-ene. E/Z isomerism is different as it applies a priority to each group attached to the double-bonded carbon and compares the position of those with highest priority. Alkenes showing cis/trans isomerism will also show E/Z isomerism, but the opposite is not true. For example, 1-bromo-1-chloropropene will have E/Z isomerism but not cis/trans isomerism.

Sigma and pi bonds

Carbon has four electrons in its outer shell. These form four covalent bonds by pairing with electrons on other atoms. The electron clouds from each electron in a bond overlap. This is called a sigma bond (σ-bond). C—C bonds and C—H bonds are examples of σ-bonds. The overlap of two *s* orbitals or two *p* orbitals forms σ-bonds.

In alkanes such as methane, the carbon 2*s* and 2*p* atomic orbitals make four hybrid orbitals called *sp*³ orbitals. (This notation is a result of hybridisation of the one *s* and three *p* orbitals in the second energy level.) One *sp*³ orbital from each carbon overlap to produce a C—C bond. The others overlap with 1*s* orbitals in each of the hydrogen atoms, forming C—H σ-bonds.

The bonding in the double bond in alkenes is slightly different. A σ-bond forms between the two carbon atoms in the same way as in a single bond. A pi-bond (π-bond) is formed by the electrons in adjacent *p*-orbitals overlapping above and below the carbon atoms. The π-bond can only form if a σ-bond has already formed. The π-bond restricts movement around the double bond. This means the region around the double bond in a molecule is flat. Double bonds are 'planar'. The π-bond is the reactive part of the molecule because there is a high electron density around it. Figure 5.6 shows the bonding in ethene.

Bond length and bond shape

Bond length and bond shape in molecules can be explained by the hybridisation of orbitals (see Table 5.4). A bond is formed by the overlap of atomic orbitals. Hybridisation is when atomic orbitals fuse to form newly hybridised orbitals.

Pi bond made from overlap of lobes of p orbital

Sigma bonds

Planar structure

▶ **Figure 5.6** Bonding in ethene

▶ **Table 5.4** Hybridised orbitals

Hybrid type	Bond angle	Arrangement	Example	C—C bond strength (kJ mol⁻¹)	C—C bond length (nm)
*sp*³	109.5°	tetrahedral	ethane	347	0.147
*sp*²	120°	Trigonal planar	ethene	612	0.135

For example, methane contains four C—H bonds. Carbon has the electronic configuration $1s^2\ 2s^2\ 2p^2$, so only the two p electrons are unshared. This would mean only two bonds are possible. However, we know four bonds are formed. The one s and three p orbitals combine to form hybrid orbitals which have 25% s character and 75% p character. These are sp^3 hybridised orbitals. This makes all four bonds in methane equivalent to each other, giving equal bond angles of 109° and methane a tetrahedral shape.

In ethene, a π-bond is required for the double bond. This uses sp^2 hybridised orbitals. The $2s$ orbitals mix with two of the three available $2p$ orbitals forming three sp^2 orbitals with a remaining p orbital.

The type of bond affects the length and strength of the bond. This is shown in Table 5.5.

▶ **Table 5.5** Bond strength and length

Bond	Number of electrons	Bond strength	Bond length
Single	2	Weakest	Longest
Double	4		
Triple	6	Strongest	Shortest

As the number of electrons shared between two atoms increases, the bond strength increases. The distance between the nuclei decreases and so you get a stronger, shorter bond. So the double C=C bond in ethene is stronger and shorter than the single C—C bond in ethane.

The bond strengths and lengths in the benzene ring of an aromatic hydrocarbon are different than those in a straight chain alkene.

The formula for benzene is C_6H_6. All the bond lengths between each carbon are identical. This is because it has a delocalised electron structure. Each carbon atom in benzene donates on electron from its p-orbital, forming a ring of delocalised electrons above and below the plane of the molecule. The electrons move freely within the ring and do not belong to any one carbon.

The bonds between the carbons in the benzene ring have a length of 0.140 nm which is between that of a C—C bond and a C=C bond. The bond strength is about 518 kJ mol^{-1}, which is between the bond strengths of the C—C bond and the C=C bond.

In order to show the delocalised electrons in benzene when representing it in an equation, you draw a hexane ring with a circle inside to represent the delocalised electrons, as shown in Figure 5.7.

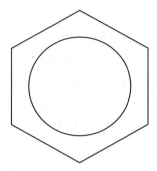

▶ **Figure 5.7** The skeletal structure of benzene

Boiling points

As the chain length of the alkanes increases, so does the boiling point. The relative molecular mass also increases as the chain length increases. The larger the molecule, the more surface area is available for contact with adjacent molecules. This means the number of induced dipole-dipole forces increases (van der Waals forces; see Unit 1). So the longer the chain, the stronger the intermolecular attraction and so more energy is needed to overcome this for the molecule to change state.

Structural isomers of alkanes will have different boiling points. This is because they are branched chains and so have a smaller surface area than their straight chain equivalents. Therefore there is less surface area for contact, and so fewer induced dipole-dipole attractions. So branched-chain isomers will have a lower boiling point than the straight-chained isomer.

Pentane has a boiling point of 31.6°C. 2,2-dimethylpropane only has a boiling point of 9.5°C. They are structural isomers of each other.

 PAUSE POINT Produce a table showing the name formula and structural formula of the organic compounds discussed in this unit.

Hint You have studied alkanes, alkenes and haloalkanes as well as aromatic compounds.

Extend Choose two different organic molecular formulas and draw out all the isomers for each.

Reactions of organic compounds

Alkanes are not very reactive. This is because a large amount of energy is needed to break the covalent bonds in the molecules. (The bonds have high bond enthalpy—see next page). The electronegativities of carbon and hydrogen are very similar and so the C—H bond is not very polar.

Alkanes are often used as fuels as complete combustion releases large amounts of energy. When there is plenty of air, or excess oxygen, all alkanes will completely combust to form carbon dioxide and water.

The word equation for the combustion of alkanes is:

alkane + oxygen → carbon dioxide + water

Methane is the fuel used in a Bunsen burner. The equation for the combustion of methane is:

$$CH_4 + 2O_2 \rightarrow CO_2 + 2H_2O$$

If oxygen supply is limited, for example, when the hole in the Bunsen burner is closed, incomplete combustion will occur.

$$2CH_4 + 3O_2 \rightarrow 2CO + 4H_2O$$

Many organic reactions take several steps to gain the final product. These steps are called the **reaction mechanism**.

Key term

> **Reaction mechanism** – the step by step sequence of reactions that lead to an overall chemical change occurring. It shows the movement of electrons in the process.

Free radical substitution

Alkanes react with halogens to form haloalkanes. For example, chlorine will react with methane to form chloromethane. The reaction needs ultraviolet light or temperatures of about 300°C. **Halogenation** happens by **homolytic fission** of the halogen molecule forming **free radicals**. A hydrogen atom in the alkane is substituted with a halogen atom. If the reaction conditions remain then all the hydrogens can be substituted with a halogen.

Key terms

> **Halogenation** – addition of halogen such as bromine or chlorine to an alkane.
>
> **Homolytic fission** – formation of a free radical by splitting a bond evenly so each free radical has one of the two available electrons.
>
> **Free radical** – atom with a single unpaired electron.

The free radical substitution mechanism has three stages:

1 Initiation – formation of radicals.

2 Propagation – steps that build up the desired product in a side reaction.

3 Termination – two free radicals collide.

These reactions are unpredictable and difficult to control. A mixture of products is formed and these will need to be separated by fractional distillation or chromatography. The reaction is shown in Table 5.6.

▶ **Table 5.6** Free radical substitution

Step	Equation	Description
Initiation	$Cl_2 \rightarrow 2Cl\cdot$	UV light or 300°C
Propagation	$CH_4 + Cl\cdot \rightarrow \cdot CH_3 + HCl$	Step 1 generates an alkyl radical and hydrogen chloride.
	$\cdot CH_3 + Cl_2 \rightarrow CH_3Cl + Cl\cdot$	Step 2 generates the desired product and regenerates the chlorine radical.
Termination	$2Cl\cdot \rightarrow Cl_2$ $2\cdot CH_3 \rightarrow C_2H_6$ $\cdot CH_3 + Cl\cdot \rightarrow CH_3Cl$	A mixture of products is produced by the random collisions between free radicals. The desired CH_3Cl needs to be separated from the rest of the products. Termination is the result of any two radicals colliding.

As there are chlorine radicals present in the mixture, the chloromethane produced could react with these to form other chloroalkanes with chlorine attached to more than one carbon.

Assessment practice 5.3

Write out the three steps for the free radical substitution of ethane, C_2H_6, by bromine, Br.

Electrophilic addition in alkenes

Alkenes are more reactive than alkanes. This is due to the C=C double bond.

The **mean bond enthalpy** for C—C is $+347$ kJ mol^{-1}. So you might expect the mean bond enthalpy for C=C to be double that, $+694$ kJ mol^{-1}. In fact, it is only $+612$ kJ mol^{-1}. This means that the π-bond only uses 265 kJ mol^{-1} of energy to break the bond. The π-bond is weaker than the σ-bond and so breaks more easily and reacts first.

The double bond is an area of high electron density so **electrophiles** such as Br_2 and HBr are attracted to it.

In alkenes, the π-bond breaks and a small molecule is added across the two carbon atoms.

Halogenation is an **addition reaction** in which a halogen is added across the C=C bond. Adding bromine water to an alkene will form a haloalkane and the bromine water will decolourise, because the bromine is removed from the reaction mixture. This is the test for identifying an alkene.

$$CH_2CH_2 + Br_2 \rightarrow CH_2BrCH_2Br$$

> **Key terms**
>
> **Electrophilic addition** – a reaction using an electrophile where two or more molecules bond to become one product.
>
> **Mean bond enthalpy** – the average amount of energy for a mole of a given bond to undergo homolytic fission in the gaseous state.
>
> **Electrophiles** – species that are electron-pair acceptors and are attracted to areas of high electron density.
>
> **Addition reaction** – a reaction where two or more molecules join together to give a single product.

This is a multi-step process. The reaction mechanism is shown in Table 5.7. The arrows show the movement of the electrons.

Haloalkanes can also be formed by an addition reaction between a hydrogen halide such as HBr and an alkene. Hydrogen halides like HCl, HBr and HI are gases at room temperature so they are bubbled trough the alkene to cause a reaction.

$$CH_2CH_2 + HBr \rightarrow CH_3CH_2Br$$

There is a large electronegativity difference between the hydrogen and the halide in a hydrogen halide, so the molecule is polar. The hydrogen end of the molecule, which is δ^+, is attracted to the high electron density of the double bond. Heterolytic fission occurs in the hydrogen halide and the electrons in the π-bond. A bromine ion and a carbocation then form, and react to form a stable product.

Assessment practice 5.4

Copy and complete the following table to show the reaction mechanism for the addition of HBr to ethane.

Stage	Mechanism	Explanation

When a hydrogen halide is added to an unsymmetrical alkene there are two possible products. For example, when hydrogen bromide is reacted with propene both 1-bromopropane and 2-bromopropane are formed. 2-bromopropane is the major product formed.

When H-X (X is a halide) is added to an unsymmetrical alkene then the hydrogen becomes attached to the carbon with the most hydrogen atoms to start with. This occurs because carbocations that have alkyl groups attached are more stable than those with only hydrogen atoms attached.

▶ **Table 5.7** Halogenation of alkenes

Stage	Mechanism	Explanation
Starting reaction	CH$_2$=CH$_2$ Induced dipole Br$^{\delta+}$ – Br$^{\delta-}$	Bromine has non polar molecules. The electron rich C=C bond induces a dipole in the bromine molecule. The δ^+ end of the bromine molecule is attracted to the high electron density of the C=C bond.
Reaction intermediates	CH$_2$=CH$_2$ → CH—CH$_2$ + Br$^{\delta+}$ Br ↓ :Br$^-$ Br$^{\delta-}$	The electrons from the π-bond make a bond with a bromine atom. This causes heterolytic fission of the bromine molecule. The other carbon atom has a positive charge. This positive carbon atom species is very reactive and is called a carbocation.
Forming dibromoethane	CH—CH$_2$ → CHBrCHBr + ↓ Br :Br$^-$	Two electrons from the bromide ion are shared with the carbocation making a second bond and a stable product.

Hydration is an addition reaction between an alkene and steam. This is when water is added. The alkene must be in gaseous form for the reaction to occur. The temperature must be high, about 300°C, as must be the pressure, about 65 atm. Phosphoric acid acts as the catalyst.

$$CH_2CH_2 + H_2O \rightarrow CH_3CH_2OH$$

Alkenes react with concentrated sulfuric acid under cold conditions to form alkyl hydrogensulfates. For example, ethene reacts to form ethyl hydrogen sulfate.

$$CH_2{=}CH_2 + H_2SO_4 \rightarrow CH_3CH_2HSO_4$$

Sulfuric acid acts as an electrophile in this reaction. The hydrogen atoms are attached to the very electronegative oxygen atoms giving the hydrogen atoms a slight positive charge, δ^+. (This is similar to what happens with the hydrogen atoms in the hydrogen halides.) The positive hydrogen is attracted to the high electron density double bond in the alkene. The mechanism is shown in Figure 5.8.

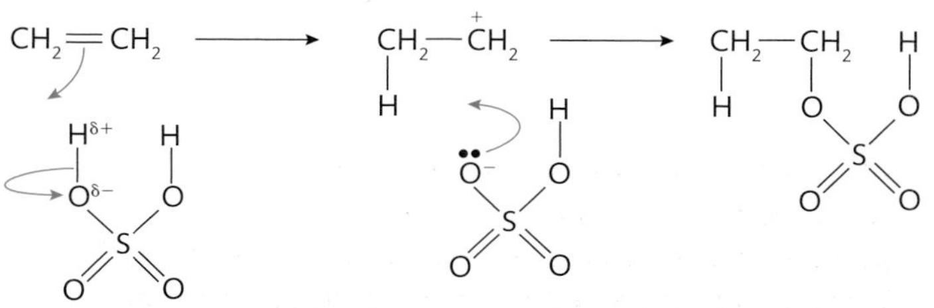

▶ **Figure 5.8** Electrophilic addition of sulfuric acid to ethene

Come in PTT paper

Polymer formation

Polymers are important commercial substances and scientists need to understand the reactions and processes involved in producing them.

One method is the free radical polymerisation of alkenes. This is an addition reaction.

Polythene can be produced by an addition reaction. Thousands of ethene molecules join together to make polythene.

$$n\,NCH_2{=}CH_2 \rightarrow [CH_2{-}CH_2]_n$$

This reaction needs the following conditions: 200°C, 2000 atm, a small amount of oxygen.

▶ **Initiation:** free radicals are produced by the reaction between ethene and oxygen. A variety of free radicals are produced so it is easier to represent them as Ra.

▶ **Propagation:** a free radical hits an ethene molecule to form a longer free radical.

$$Ra\cdot + CH_2{=}CH_2 \rightarrow Ra\,CH_2CH_2\cdot$$
$$Ra\,CH_2CH_2\cdot + CH_2{=}CH_2 \rightarrow Ra\,CH_2CH_2CH_2CH_2\cdot$$
and so on

▶ **Termination:** if two free radicals collide then they will produce a final molecule and no more free radicals are formed.

$$Ra(CH_2)_n\cdot + \cdot(CH_2)_n Ra \rightarrow Ra(CH_2)_n Ra$$

The randomness of this process means that polythene will be made up of many different length chains.

Many hydrocarbons found in crude oil have very long chains. Many of these hydrocarbons have few uses or are in low demand. Hydrocarbons such as ethene which have short chains are much more in demand. A process called **cracking** is used to decompose long chain hydrocarbons into smaller chain hydrocarbons. There are two types of cracking used in the **petrochemical industry**, thermal and catalytic. The hydrocarbon molecules are broken in a random way during the reaction to produce a mixture of smaller hydrocarbons. Alkanes and alkenes are always produced. For example:

$$C_{10}H_{22} \rightarrow C_5H_{12} + C_3H_6 + C_2H_4$$

Pentane, propene and ethene have been produced here. These are not the only possibilities. Another reaction may produce octane and ethene.

$$C_{10}H_{22} \rightarrow C_8H_{18} + C_2H_4$$

Catalytic cracking uses zeolites as the catalyst. Zeolites are complex aluminasilicates and are large lattices of aluminium, silicon and oxygen atoms. The other conditions needed are a temperature of about 500°C and low pressures. The hydrocarbons are mixed with the catalyst which has been ground into a fine powder. It is an example of homogeneous catalysis, as both the reactants and the catalyst are solids.

This method gives a high number of hydrocarbons with between five and ten carbon atoms in the chain. These are useful for petrol. Catalytic cracking also produces high numbers of branched alkanes and aromatic hydrocarbons such as benzene. The hydrocarbons produced are separated by cooling and fractional distillation.

Thermal cracking uses temperatures between 450°C and 750°C. The pressure can be up to 70 atms. Thermal cracking gives high numbers of alkenes. Higher temperatures mean that the chain breaks near to its end. This gives a high proportion of short chain alkenes.

Key terms

Cracking – long chain hydrocarbons are broken down into shorter chain hydrocarbons.

Petrochemical industry – industry that produces materials by refining petroleum.

PAUSE POINT   Describe all the different reaction types in this unit.

> **Hint**   Remember to consider the different conditions for each reaction type.
>
> **Extend**   Write out the reaction mechanism for one of the reaction types.

Energy changes in industry

Chemical reactions always involve an energy change. In industry, most products are made on a very large scale. Industrial chemists want to make these products as safely and cheaply as possible. This means they need to also carry out the reactions as quickly as possible with as little energy input as possible.

Enthalpy

Enthalpy H is the thermal energy stored in a chemical system. It is defined by:

$$H = U + pV$$

where U = internal energy of the system

  p = pressure

  V = volume

You cannot measure the enthalpy of a system but you can measure the energy absorbed or released to the **surroundings** during a reaction. Thermal energy changes are measured by measuring the temperature using the **Kelvin scale**.

> **Key terms**
>
> **Surroundings** – everything that is not part of the system e.g. water bath or beaker or aqueous solution that reactants are dissolved in.
>
> **Kelvin scale** – a temperature scale with absolute zero as zero. The size of one unit (1 K) is the same as one degree Celsius. Water freezes at 273 K.

Enthalpy change, ΔH, is the heat exchange with the surroundings during a reaction. It is the difference between the enthalpy of the products and the enthalpy of the reactants. This is at constant pressure.

$$\Delta H = H_{products} - H_{reactants}$$

It can also be defined as $\Delta H = \Delta U + p \Delta V$.

Enthalpy changes are measured in $kJ\,mol^{-1}$ (kilojoules per mole). In order to compare enthalpy changes they must be measured in standard conditions. These are a pressure of 1×10^5 Pa and a temperature of 298 K. All substances must be in their standard states. These are their most stable states. When enthalpy changes are measured

under standard conditions the enthalpy change is known as $\Delta H°$.

▶ An **exothermic reaction** releases heat and has a negative enthalpy change ($-\Delta H$). The enthalpy of the products is smaller than that of the reactants. Heat is lost from the system to the surroundings.

▶ An **endothermic reaction** absorbs heat and has a positive enthalpy change ($+\Delta H$). The enthalpy of the products is greater than that of the reactants. Heat is absorbed by the system from the surroundings. Figure 5.9 shows exothermic and endothermic enthalpy changes.

> **Key terms**
>
> **Exothermic reaction** – a reaction that releases energy.
>
> **Endothermic reaction** – a reaction that absorbs energy.

The two diagrams in Figure 5.9 are called enthalpy profile diagrams. In an exothermic reaction, excess energy is released into the surroundings so the temperature of the surroundings will rise. In an endothermic reaction, energy is required. This energy comes from the surroundings so the temperature of the surroundings will fall.

In order for a reaction to start, energy must be provided to break bonds. This initial energy needed is called the activation energy. This energy can also be shown on an enthalpy profile diagram, as in Figure 5.10.

The initial activation energy needed for exothermic reactions often comes from an initial flame, for example, from a match. Once the activation energy is overcome, the net output of energy provides the activation energy for the reaction to continue. An example of this is the energy from the match flame you need to light the methane in a Bunsen burner. Once the methane is lit, combustion will carry on happening until the methane runs out, as the combustion gives out enough energy to provide activation energy to break bonds.

A balanced equation that has the enthalpy data included is called a thermodynamic equation. For example, the thermodynamic equation for the reaction of carbon (graphite) with excess oxygen is:

$$C(s) + O_2(g) \rightarrow CO_2(g) \quad \Delta H° = -393.5\ kJ\,mol^{-1}$$

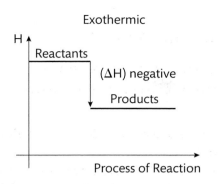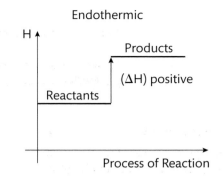

▶ **Figure 5.9** Exothermic and endothermic enthalpy changes

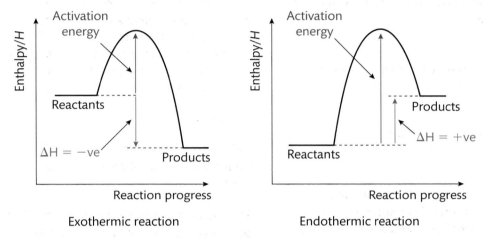

▶ **Figure 5.10** Activation energy for exothermic and endothermic reactions

When writing a thermodynamic equation, you may need to use fractions to balance the equation, to ensure there is only one mole of reactant.

Enthalpy change of formation is the energy change that takes place when one mole of a compound is formed from its constituent elements in their standard state under standard conditions. It has the symbol $\Delta_f H^\circ$. The f stands for formation.

Enthalpy change of combustion is the energy change when one mole of a substance is completely burned in oxygen. It has the symbol $\Delta_c H^\circ$. The c stands for combustion. So the thermodynamic equation for the combustion of ethane is

$$C_2H_6(g) + 3\tfrac{1}{2}O_2(g) \rightarrow$$
$$2CO_2(g) + 3H_2O(l)\ \Delta_c H^\circ = -1500 \text{ kJ mol}^{-1}$$

Hydration enthalpy is the measure of the energy released when attractions form between positive or negative ions and water molecules. There are often only loose attractions between the positive ions and the slightly negative oxygen atoms in the water molecules. There may also be dative covalent bonds. Hydrogen bonds form between the lone pairs of electrons on the negative ions and the slightly positive hydrogen atoms in the water.

The size of the hydration enthalpy depends on the attraction between the ions and the water molecules. The smaller the ion, the stronger the attraction will be and the higher the hydration enthalpy. The higher the charge on the ions, the stronger the attraction will be and so the hydration energy will also increase.

You can find standard values (literature values) for enthalpy changes in data books. You should compare any enthalpy changes you have measured or calculated through experiments to these values. Remember that these will be in kJ mol^{-1} and, unless they state differently, they will be measured under standard conditions.

Measurement of enthalpy changes

It is not possible to measure enthalpy, but you can measure enthalpy change. This can be done by measuring the temperature change on a system.

Calorimetry uses a mathematical relationship to calculate enthalpy change from experimental quantitative data.

The mathematical expression used is:

$$Q = mc\Delta T$$

where Q = the heat exchanged with the surroundings measured in joules.

m = the mass of the substance heated or cooled measured in grams.

c = the **specific heat capacity** of the substance heated or cooled, expressed as $J\,g^{-1}\,K^{-1}$.

ΔT = change in temperature, measured in kelvin.

Worked example

0.5 g of ethanol was used to heat 100 cm³ of water from 21°C to 45°C.

What is the molar enthalpy change of this reaction?

Specific heat capacity of water is $4.18\,Jg^{-1}\,K^{-1}$

Use $Q = mc\Delta T$

Mass – mass of 1 cm³ of water has mass of 1 g so 100 cm³ of water has a mass of 100 g

$c = 4.18$

$\Delta T = 45 - 21$
$= 24$ K as 1°C is equivalent to 1 K

$Q = 100 \times 4.18 \times 24 = 10\,032\,J$

To calculate molar enthalpy change:

Ethanol has relative molecular mass of 46

$n = \dfrac{m}{M_r} = \dfrac{0.5}{46} = 0.011$ so 0.011 moles of ethanol used in this experiment

10 032 J released by 0.011 moles of ethanol so $\dfrac{10032}{0.011} = 912\,000\,J = 912\,kJ\,mol^{-1}$ of energy released

The reaction is exothermic and so the enthalpy change is negative.

$\Delta_c H = -912\,kJ\,mol^{-1}$

However, the reported value is $-1367\,kJ\,mol^{-1}$ so the experimental value is not accurate. This is because it is unlikely that the experiment was carried out under strict standard conditions and incomplete combustion could have occurred. However, the biggest source of error would be the heat lost to the surroundings.

 PAUSE POINT   Define the following enthalpy changes: combustion, formation, hydration.

 Hint   You can use words and equations.

Extend   Draw a step by step guide showing how you would calculate the molar enthalpy change of a burning fuel.

Further reading and resources

www.rsc.org The website of the Royal Society of Chemists.
www.sciencebuddies.org A website giving hands on science projects.
www.virtlab.com A series of hands on experiments and demonstrations in chemistry.
www.iupac.org The website of the International Union of Pure and Applied Chemistry.

Getting started

This unit builds on Unit 1, where you learned about cells and tissues. It is very important that scientists, science technicians, medical personnel, teachers, nursery nurses and others have an understanding of the human body and its organ systems. This knowledge can be used in many ways, including to assess and treat illnesses, and to help prevent disease. It is essential that you understand the body's structure and function in terms of organs and organ systems, and how they all work together to maintain a steady internal environment known as **homeostasis**. List any organs and organ systems you can think of, and briefly note the importance of each system. When you have completed this unit, see if you can add more to your list.

B Organs and systems

The cardiovascular system

In this section you will learn about the structure and function of the heart, the blood vessels and the events of the **cardiac cycle**. You will also investigate the effect of caffeine on the heart rate in *Daphnia*, more commonly known as water fleas.

Many cells in a multicellular organism are not in direct contact with their surroundings. This means that the organism cannot rely on diffusion alone to supply the cells of all its organs with the nutrients and oxygen it needs to survive. The cardiovascular system, also known as the circulatory system, allows blood to circulate around the body to transport and supply the essential nutrients that multicellular organisms need in order to maintain homeostasis and to survive.

The cardiovascular system consists of the heart, blood vessels and blood (see Figure 5.11). It is responsible for not only transporting nutrients and oxygen but also hormones and cellular waste throughout the body.

Blood is a tissue and it always flows in blood vessels. The components that make up human blood are **erythrocytes**, **leukocytes**, **thrombocytes** and plasma.

Key terms

Cardiac cycle – a complete heartbeat from the generation of the beat to the beginning of the next beat.

Erythrocytes – red blood cells, containing haemoglobin, that transport oxygen around the body.

Leukocytes – white blood cells, of which there are many different types.

Thrombocytes (platelets) – component of blood involved in blood clotting.

Blood is responsible for transporting:

▶ oxygen from the lungs to body cells for aerobic respiration

▶ carbon dioxide from respiring cells back to the lungs to be exhaled

▶ nutrients from the intestines to body cells

▶ urea (waste product) from the liver to the kidney to be removed

▶ hormones from the endocrine glands to their target cells

▶ heat from respiring tissue to organs to maintain body temperature or to the skin to be lost.

Blood is also responsible for regulating:

▶ body temperature

▶ pH of body tissues

▶ volume of fluid in circulation.

Blood also protects us because it contains:

▶ platelets that cause clotting, to prevent bleeding and entry of pathogens

▶ leukocytes that defend us against infection.

Link

Go to *Unit 20: Biomedical Science*, Section A, to find more information about the structure and function of erythrocytes, leukocytes and thrombocytes.

Characteristic features of blood vessels

Blood is always in blood vessels, either arteries, arterioles, capillaries, venules or veins. The vessels form a closed transport system which starts and finishes at the heart. We refer to the mammalian circulation system as double circulation because we have two separate circulations. The blood flows from the heart to body tissues and back to the heart. This is known as **systemic circulation**. In a separate

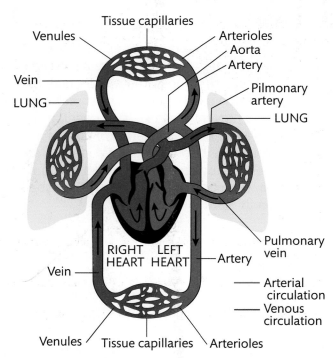

Venules
Tissue capillaries
Arterioles
Aorta
Artery
Vein
LUNG
Pilmonary artery
LUNG
Pulmonary vein
RIGHT HEART   LEFT HEART
Artery
Vein
Arterial circulation
Venous circulation
Venules
Tissue capillaries
Arterioles

▶ **Figure 5.11** The cardiovascular system structure: heart, blood and blood vessels

circulation known as **pulmonary circulation**, blood flows from the heart to the lungs to expel carbon dioxide and take in oxygen and then it returns back to the heart.

> **Key terms**
>
> **Systemic circulation** – parts of the circulatory system concerned with the transport of oxygen to, and carbon dioxide from, the heart to body cells.
>
> **Pulmonary circulation** – parts of the circulatory system concerned with the transport of oxygen to, and carbon dioxide from, the heart to lungs.

Arteries and arterioles

Arteries carry blood away from the heart. The blood leaves the ventricles of the heart and enters thick elastic arteries under high pressure. The artery lumen is small and arteries have thick walls that contain collagen, a fibrous protein, to help maintain the shape and volume of the arteries under pressure. Artery walls also contain elastic tissue to enable them to continuously expand and recoil to keep the blood under pressure. This expansion and recoiling is what you will feel as your pulse where an artery passes near the surface of the skin. Arteries also contain smooth muscle that contracts, enabling the artery lumen to narrow if needed. Arteries are lined with smooth endothelium tissue, reducing the friction as the blood flows through the lumen. As blood flows further from the heart, these large elastic arteries become smaller muscular arteries which

carry the blood to the organs in the body. They contain less elastic tissue and more smooth muscle than larger elastic arteries. Arteries then divide into smaller arterioles. These vessels contain smooth muscle cells wrapped around the endothelium. These eventually become capillaries.

Capillaries

Capillaries allow the exchange of materials between blood and the body's cells via tissue fluid. These are tiny vessels with very thin walls consisting of only one layer of endothelium cells. This thin wall reduces the diffusion distance for the materials being exchanged. The lumen of a capillary is very narrow, with a diameter that is the same size as a red blood cell. The small diameter allows only one erythrocyte through at one time. This ensures that the red blood cell has to squeeze through the capillaries which help it release the oxygen. Capillaries spread throughout tissues, forming capillary networks, and it is here where the plasma of blood, rich with nutrients and oxygen, is forced out through small gaps in the capillary walls. This fluid that is forced out is known as tissue fluid and it also carries away the waste from cellular activity. Capillaries link arterioles to venules.

Venules and veins

Capillaries become slightly larger and form venules, which also have small diameters. Venules join to make veins. Veins have a large lumen and their walls are thinner than arteries. Veins have thinner layers of collagen, smooth muscle and elastic tissue as they do not need to constrict and recoil. Veins have valves to help prevent the back flow of blood as it makes its way back to the heart. The action of the surrounding skeletal muscle can flatten veins. This also helps force the blood back to the heart.

Table 5.8 states the role of the major blood vessels found in the body.

▶ **Table 5.8** Roles of the major blood vessels

Blood vessel	Artery or vein	Role
Vena cava	Vein	To deliver deoxygenated blood from the body into the right atria of the heart.
Pulmonary vein	Vein	To deliver oxygenated blood from the lungs into the left atria of the heart.
Pulmonary artery	Artery	To transport deoxygenated blood away from right ventricle in the heart to the lungs to collect oxygen.
Aorta	Artery	To transport oxygenated blood away from left ventricle in the heart to the rest of the body.
Coronary Artery	Artery	To supply the cardiac muscle with its own supply of oxygen.

⏸ PAUSE POINT   Explain the structure and function of blood and the vessels it travels in.

<table>
<tr><td>Hint</td><td>Close the book and list the components of blood and all the blood vessels.</td></tr>
<tr><td>Extend</td><td>Think about the differences between the components and the vessels and why their structure is important for their function.</td></tr>
</table>

Structure and function of the heart

Your heart is about the size of your fist when it is clenched and it has a mass of about 300 g. It is in the thoracic cavity between the lungs and behind the sternum, enclosed in a fibrous bag made from inelastic connective tissue called the **pericardium**. The heart is a muscular double pump divided into two halves, each of which contain two chambers.

The wall of the heart is made from mainly **myocardium**, which consists of cardiac muscle. Cardiac muscle contracts to make your heart beat. The coronary arteries seen in Figure 5.12 lie on the surface of the heart. They carry oxygenated blood to the heart muscle itself.

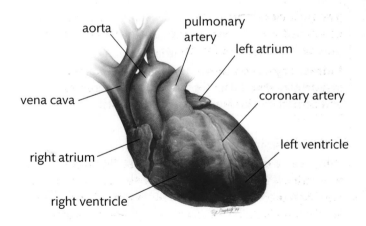

▶ **Figure 5.12** External view of the heart

Key terms

Pericardium – a fibrous membrane that surrounds and protects the heart.

Myocardium – the middle and thickest layer of the heart wall, composed of cardiac muscle.

The heart is divided into four chambers. The two upper chambers are called the **atria**. They have thin walls and

they sit above the two lower chambers called **ventricles**. Ventricles are thicker walled chambers. Deoxygenated blood flows into the right atrium from the **vena cava** (a main vein); at the same time the left atrium receives oxygenated blood from the lungs via the pulmonary vein. From here the blood flows into the lower ventricles through **atrioventricular valves**. The atrioventricular valve between the left atria and left ventricle is known as the bicuspid valve. The tricuspid is the valve between the right atrium and right ventricle. These valves are thin flaps of tissue attached to the ventricles via tendinous chords to stop the valves from turning inside out. When the ventricles are full of blood and ready to contract, the valves close to stop the back flow of blood into the atrium.

The heart is divided into the right and left side and the ventricles are separated by a wall of muscle called the **septum**. This stops the oxygenated and deoxygenated blood coming into contact with each other.

When the ventricles contract, deoxygenated blood flows upwards out of the right ventricle into the pulmonary artery, where the blood is transported to the lungs to collect oxygen.

Oxygenated blood leaving the left ventricle flows into the aorta. This is a major artery that carries the oxygenated blood to a number of arteries to supply the body cells with oxygen. At the base of both of these arteries there are semi-lunar valves that prevent back flow of blood into the ventricle when they relax.

> **Key terms**
>
> **Atria** – two top chambers of the heart.
>
> **Ventricle** – two bottom chambers of the heart.
>
> **Vena cava** – a large vein carrying deoxygenated blood into the heart.
>
> **Atrioventricular valve** – the structure found between the atrial and ventricular chambers of the heart to prevent back flow.
>
> **Septum** – the dividing wall between the right and left sides of the heart.

⏸ **PAUSE POINT**   Label the structure of the heart.

Hint   Use the internet to find an unlabelled diagram of the heart, and see what you can label.

Extend   Draw a flow diagram of the cardiac cycle. Include the functions of all the structures.

The heart muscle is described as myogenic, because it can initiate its own contraction and contracts and relaxes rhythmically, even without stimulation from the nervous system. At the top of the right atrium there is a patch of tissue, commonly known as a pacemaker, called the **sinoatrial node (SAN)**. It is this node that generates the electrical activity and initiates a wave of excitation at regular intervals.

The wave of excitation quickly spreads over the walls of the atrium. As it spreads along the muscle tissue membrane, it causes the muscle cells in both atria to contract simultaneously. This is known as atrial systole. The excitation cannot, however, spread directly to the ventricle because there is non-conducting tissue in the base of the atria. The wave of excitation is picked up by the **atrioventricular node (AVN)** located in the top of the septum separating the two ventricles. The wave of excitation is delayed here, which allows the atria time to complete its contraction and for the blood to flow through the atrioventricular valves into the ventricle before they start to contract.

The AV node stimulates the **bundle of His**, a bundle of conducting tissue made from **Purkinje fibres**. These fibres line the interventricular septum and carry the excitation from the atria to the apex (bottom of the septum in the ventricle). At the apex, the Purkinje fibres spread through the walls of the ventricles and, as the excitation spreads upwards through the muscle tissue, the ventricles contract simultaneously. The ventricles contract from the base upwards and this pushes the blood up into the arteries. This is known as ventricular systole. When the ventricles relax, semi-lunar valves at the base of the aorta and pulmonary artery prevent any blood flowing back into the heart.

> **Key terms**
>
> **Sinoatrial node (SAN)** – a patch of tissue found in the right atrium that generates the electrical activity and initiates a wave of excitation at regular intervals.
>
> **Atrioventricular node (AVN)** – a patch of tissue located in the top of the septum that picks up the wave of excitation from the atria.
>
> **Bundle of His** – a collection of heart muscle cells specialised for electrical conduction.
>
> **Purkinje fibres** – specialised conducting fibres found in the heart.

The cardiac cycle

One whole cardiac cycle takes about 0.8 seconds.

1 Both atria relax and fill with blood from the pulmonary vein and vena cava. This is atrial diastole.

2 The atria contract and force the atrioventricular (AV) valves open. Blood flows into the ventricles and they fill up; this is ventricular diastole.

3 The AV valves close when the pressure in the ventricles rises above the pressure in the atria to prevent the backflow of blood into the atria.

4 The ventricle walls contract and increase pressure in the ventricles. This forces the semi-lunar valves to open and the blood flows into the pulmonary artery and aorta.

5 When the pressure in the aorta and pulmonary artery rises, the semi-lunar valves close to prevent backflow of blood into the ventricles.

Assessment practice 5.5

1 Describe the cardiac cycle. Ensure that you mention the action of the valves.

2 Explain why the sinoatrial node is commonly referred to as a pacemaker.

3 Explain why the ventricles contract from the apex upwards.

Electrocardiograms

An electrocardiogram (ECG) can be used to monitor the electrical activity of the heart. Some of the electrical activity that is generated by the heart spreads through tissue surrounding the heart and to the skin. During an ECG a number of sensors are attached to the skin, they pick up this electrical excitation and convert it into a trace, as shown in Figure 5.13, that can be interpreted by a medical professional.

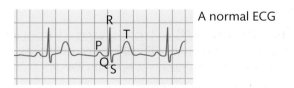
A normal ECG

▶ **Figure 5.13** Electrocardiogram trace showing a normal ECG

A healthy trace consists of a series of waves and has a particular shape. The waves are labelled P, Q, R, S and T. These waves indicate different information:

▶ Wave P shows excitation of the atria, when they begin to contract and therefore represents atrial systole.

▶ Wave QRS indicates excitation of the ventricles, when they begin to contract and therefore represents ventricular systole.

▶ Wave T shows diastole, when the heart chambers are relaxing.

The shape of an ECG can indicate that there is a problem with the heart muscle. It can detect various arrhythmias:

▶ Tachycardia is a condition where the heart rate is very fast and over 100 beats per minute. Here the P waves are evenly spread but are closer together than they should be.

▶ Bradycardia is a condition where the heart rate is very slow and under 60 beats per minute. Here the P waves are evenly spread but further away from each other than they should be.

▶ Ventricular fibrillation is a very serious condition where the contraction of ventricles is not controlled or coordinated. The ventricles fibrillate/quiver and the heart pumps little or no blood, it can cause cardiac arrest. Here there are no identifiable P, QRS or T waves on the ECG, and the heart rate is very fast.

▶ Sinus arrhythmia is a normal variation in the beating of your heart. It occurs when your heart rate cycles with your breathing. When you breathe in, your heart rate speeds up and when you breathe out, your heart rate slows back down. Here there is a normal P wave but they are not evenly spread.

▶ Flat line is where the ECG shows no electrical activity of the heart.

Assessment practice 5.6

1 Describe the shape of an electrocardiogram.

2 Explain why the QRS peak is larger than the P wave.

The effect of caffeine on heart rate

Many different drugs can have an effect on the rate of your heart beat. Caffeine makes your heart 'beat' faster by increasing the electrical activity of the SAN. It also affects the ventricles, leading to an increase in the rate of contraction and relaxation of each heartbeat. As well as beating faster, a larger volume of blood can be pumped out every time the heart beats. Two or three cups of strong coffee or tea contain enough caffeine to increase human heart rate by 5–20 beats/min.

Investigating the effect of caffeine on heart rate in *Daphnia*

Water fleas (*Daphnia*) have a small heart that is very easy to see under a low power light microscope. The heart rate of a daphnia can be up to 300 beats per minute and can be monitored and counted in different conditions. We are going to investigate the effect of caffeine on the heart rate by changing the concentration of caffeine and adding it to the water surrounding the daphnia.

For each step in the investigation, it is important that you understand the purpose of it, and what you need to pay particular attention to, in order that your results are as accurate as possible.

Steps in the investigation	Pay particular attention to . . .	Think about this . . .
1. Take some cotton wool, and place it in the middle of a Petri dish.		
2. Use a pipette to transfer a large daphnia onto the cotton wool fibres.		
3. Add pond water immediately to the Petri dish until the daphnia is just covered by the water.	You should have a good ethical attitude towards the daphnia. Although they are simple organisms that may not 'suffer' in the same way as higher animals, they still deserve respect.	
4. Place the petri dish on the stage of a microscope and observe under low power. The beating heart is located on the dorsal side just above the gut and in front of the brood pouch.		
5. Use a stopwatch to time 30 seconds, and count the number of heart beats in several periods of 30 seconds.	Make sure that you are counting the heart beats, and not the flapping of the gills or movements of the gut. The heart must be observed with transmitted light if it is to be properly visible.	Think about how you might accurately measure the rapid heart rate. For example, to count the beats, make dots on a piece of paper. Once the timer has stopped, count the dots and express heart rate as number of beats per minute (times your dots by 2).
6. Add one drop of water containing caffeine at a concentration of 100 mg/L to 5 cm^3 of pond water in a beaker. Mix well. Draw the pond water off the daphnia with a pipette and replace it with 2 or 3 cm^3 of the water containing caffeine. Record the rate of heart beat again.	Make sure that you measure the volumes accurately, as you will have to repeat this step, and you must keep all the non-variable aspects the same throughout. Scientific investigations produce more accurate results if they are carried out a number of times.	Think about how you might accurately measure heart rate.
7. Repeat this investigation with different concentrations of caffeine, e.g. 10, 1, 0.1, 0.01 and 0.001 mg/L.	Scientific investigations produce more accurate and valid results if they use a large range of data.	In order to find the effect of changing one 'variable' (condition that you can change), you must keep all the other conditions the same in each test you carry out. This is why it is crucial to measure out the volumes accurately each time you repeat the experiment.

Steps in the investigation	Pay particular attention to . . .	Think about this . . .
8. Record your results in an appropriate way and write a report on your investigation.	Your report should inform a reader about how you carried out the investigation, and what you did to ensure accurate results. You should present your results in a way that shows your findings as clearly as possible.	Consider all the ways that you might present your results, e.g. as a table, a graph or a chart. You should aim to make your findings as easy to understand as possible.
9. At the end of the investigation, return the daphnia to the stock culture.	You should have a good ethical attitude towards the daphnia. Although they are simple organisms that may not 'suffer' in the same way as higher animals, they still deserve respect. Animals should be returned to the holding tank after being examined.	

Factors that increase the risk of cardiovascular disease (CVD)

Cardiovascular diseases are diseases that affect the heart and circulatory system. Common diseases are atherosclerosis, coronary heart disease (CHD) and stroke. Factors that increase the risk of CVD are:

▶ **genetics:** if you have a family history of CVD, then this increases your chance of also suffering CVD

▶ **age:** as you get older, the risk of CVD increases

▶ **gender:** statistics indicate that men are at a higher risk of dying of CVD under the age of 50 than women

▶ **diet:** consuming a high level of saturated fat, high intake of salt, and limited healthy fats and vitamins increases the risk of CVD

▶ **high blood pressure:** suffering with high blood pressure can increase your risk of a CVD

▶ **smoking:** smoking cigarettes will increase the risk of suffering with a CVD

▶ **inactivity :** lack of physical activity will increase risk of suffering with a CVD.

Case study

Coronary heart disease

Dr Amy Bucknell regularly treats people with CVD, in particular people who suffer with coronary heart disease. This is when the blood supply to the heart becomes restricted as a result of the hardening and narrowing of the coronary arteries. One of the common medicines she prescribes are statins. They are usually in tablet form and taken once a day. In most cases they are continued for life. Statins help lower the level of low-density lipoprotein (LDL) cholesterol in the blood. Statins can't cure CVD conditions, but benefits of taking them are that they help prevent the patient's condition from getting worse and statistics tells us that 1 in 50 people taking statins for 5 years will avoid a major heart attack or stroke. One of the important parts of Dr Bucknell's job is to ensure that she discusses the

risks of taking statins. Common side effects that affect 1 in 10 patients are nosebleeds, sore throat, headaches, sickness, constipation, increased blood sugar levels and a risk of diabetes. Dr Bucknell also explains that there are further side effects that affect 1 in 100 patients, and these are more severe, such as memory problems, inflammation of the liver and pancreas.

Check your knowledge

1 What may cause arteries to become narrow and restricted?

2 Why are statins an appropriate form of medicine for coronary heart disease?

3 Compare and contrast the benefits and the risks of taking statins.

Benefits and risks of treatments for CVD

Antihypertensives

Antihypertensives are a class of drugs that are used to treat hypertension. Hypertension is commonly referred to as high blood pressure. There are many different types of antihypertensives, which lower blood pressure in different ways. The most widely used drugs are thiazide diuretics, calcium channel blockers, and beta blockers. Table 5.9 lists the advantages and disadvantages of common antihypertensive medication.

▶ **Table 5.9** Antihypertensive medication advantages and disadvantages

Antihypertensive	Advantage	Disadvantage
Thiazide diuretics	Lower blood pressure and they are an option for those unable to take or tolerate calcium channel blockers.	They are not suitable during pregnancy and can raise potassium and blood sugar levels so regular blood and urine tests required. They can also cause impotence.
Calcium channel blockers	Lower blood pressure and they are effective in black people and those aged 55 or over.	Not suitable for people with a history of heart disease, liver disease or circulation problems and they can have side effects such as flushed face, headaches, swollen ankles, dizziness, tiredness and skin rashes.
Beta blockers	Lower blood pressure and are an option for people who do not respond to other medication. They are safer to use during pregnancy than other medications.	Cause many side effects such as tiredness, cold hands and feet, slow heartbeat, diarrhoea and nausea. Can also cause sleep disturbances, nightmares and impotence.

Transplantation and immunosuppressants

A heart transplant is a major operation that comes with many risks and complications, but it may be the final option to ensure survival of the patient. These risks include the following.

▶ **Rejection of the donor heart:** this occurs when the immune system mistakes the new heart as foreign and attacks it. It can happen immediately after surgery or even years after.

▶ **Infection:** immunosuppressant drugs, taken to reduce the chances of rejection of the donor organ, weaken the immune system and make the patient more vulnerable to infections such as bacterial infections, for example pneumonia, and fungal infections.

▶ **Failure to pump properly:** the donated heart may not work as it may not start beating, or it may stop beating soon after surgery.

▶ **Narrowing of the arteries that are connected to the heart:** this is potentially serious as it can restrict the supply of blood to the heart. This could potentially trigger a heart attack.

After a patient has undergone a heart transplant, they need to take immunosuppressants. These can also cause:

▶ kidney problems
▶ high blood pressure
▶ diabetes
▶ a higher risk of cancer.

Ventilation and gas exchange in the lungs

In this section, you will learn about the structure and function of the lungs and the **ventilation** system. You will also investigate the effect of exercise on the ventilation system using a spirometer to collect data. Respiration is the release of energy from organic molecules found in food; in order for this to occur oxygen is needed.

> **Key term**
>
> **Ventilation** – the exchange of air between the lungs and the surroundings.

Structure of the human lung and ventilation system

The ventilation system is also known as the respiratory system (see Figure 5.14). It allows oxygen to move into the body and carbon dioxide to be removed from the body as waste. The respiratory system consists of the:

▶ trachea
▶ bronchi
▶ bronchioles
▶ alveoli
▶ intercostal muscles
▶ diaphragm.

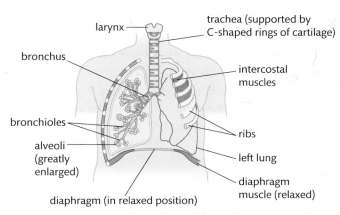

larynx

trachea (supported by C-shaped rings of cartilage)

bronchus

intercostal muscles

bronchioles

ribs

alveoli (greatly enlarged)

left lung

diaphragm muscle (relaxed)

diaphragm (in relaxed position)

▶ **Figure 5.14** The ventilation system structure

Trachea and bronchi

Air is inhaled through the nose and mouth and travels down the trachea, bronchi and bronchioles. These are airways that allow passage of air into the lungs and out again. The trachea and bronchi are similar in structure, they are just different in size. They have thick walls lined with several layers of tissue, and both are supported by walls of cartilage. The trachea has incomplete rings of cartilage, in the shape of a 'C'. The C-shaped rings of cartilage are also present in the bronchi, but they are less regular. Cartilage holds the airways open and prevents them collapsing when there is low pressure. The C-shaped rings of the trachea allow the oesophagus to sit and function alongside it. On the inside of the cartilage there are different layers of tissue.

▶ Glandular tissue secretes mucus to trap any pathogens and reduce the risk of infection. Further mucus is produced by the goblet cells of the epithelium tissue that lines the airways.

▶ Smooth muscle can contract to restrict airflow onto the lungs.

▶ Elastic fibres recoil as the smooth muscle relaxes, helping the airway to widen.

The airways are lined with epithelium tissue, which consists of ciliated epithelium cells and goblet cells. Ciliated epithelium move in a synchronised pattern to waft mucus up and out of the airways.

Link

Go to *Unit 1: Principles and Applications of Science 1* to find more information about tissue structure and function.

Bronchioles

Air continues its journey and travels through the bronchioles. These are much narrower than the bronchi. Bronchiole walls are made mostly of smooth muscle and elastic fibres. Larger bronchioles consist of some cartilage

and the smaller bronchioles have no cartilage. The smallest bronchioles have clusters of alveoli at the end.

Alveoli

Finally, when the air reaches tiny sacs called alveoli, gas exchange takes place between the air in the alveoli and the blood in the capillaries that surround the alveoli. Gases pass both ways across the walls of the alveoli to ensure efficient provision of oxygen for respiration, whilst also ensuring that carbon dioxide is removed for efficient production of adenosine triphosphate (ATP) during cellular respiration. The structure of the lungs and the alveoli ensures effective and efficient gas exchange through the following features.

▶ **Large surface:** there are millions of alveoli, providing more space for the gas molecules to pass through.

▶ **Short diffusion pathway:** the walls of the alveoli are made of squamous epithelial cells and are only one cell thick. This reduces the distance these molecules have to diffuse through.

▶ **Capillary network:** each alveolus is close to a capillary which has a well which is also one cell thick to give a short diffusion pathway to molecules.

▶ **Diffusion gradients:** oxygen diffuses from a high concentration inside the alveoli down a **concentration gradient** to an area of lower concentration in the blood capillary, where it joins with haemoglobin in the erythrocytes. Carbon dioxide diffuses from a high concentration inside the blood capillary down a concentration gradient to an area of lower concentration in the alveoli. If it stayed in the blood it could be toxic because it lowers the pH.

▶ **Moisture:** a layer of moisture lines the alveoli. Gases can only diffuse across the membrane if dissolved. Therefore this moisture allows gases to dissolve in order to cross.

▶ **Surfactant:** this is a chemical produced by the lungs to stop the alveoli from collapsing by reducing the surface tension of water.

Key term

Concentration gradient – the difference in the concentration of a substance between two regions.

Ventilation of the lungs

Breathing in is called inspiration. The following steps explain what the body does in order to breathe in oxygen.

1 The intercostal muscles between the ribs contract and raise the rib cage up and out.

2 The diaphragm muscle contracts and the diaphragm flattens and moves down.

3 Both these actions increase the volume in the thoracic cavity.

4 This reduces the air pressure inside the thoracic cavity.

5 As a result, air moves down the trachea, bronchi, bronchioles and into the alveoli from the higher atmospheric air pressure to the lower air pressure of the thoracic cavity.

6 Oxygen diffuses through the alveolar membrane into the blood capillaries and carbon dioxide diffuses from the blood capillaries across the alveolar membrane into the alveoli.

Breathing out is called expiration. This is done in order to push air with a high concentration of carbon dioxide out of the body.

1 External intercostal muscles relax and the rib cage moves down and inwards.

2 The muscles of the diaphragm relax and the diaphragm moves up in a dome shape.

3 Both these actions reduce the volume in the thoracic cavity.

4 This increases the air pressure inside the thoracic cavity.

5 As a result, air is pushed out.

6 In the case of expiration during exercise, the internal intercostal muscles contract to reduce the volume of the thoracic cavity further so that a larger volume of air can be breathed out to get rid of the extra carbon dioxide (CO_2) that has been produced by a higher rate of respiration in muscle tissue.

Role of pleural membranes

The role of the pleural membranes is to protect the lungs because lung tissue is delicate and can be easily damaged. The pleural membranes enclose a fluid-filled space surrounding the lungs which provides lubrication. Our lungs are constantly expanding and contracting so the pleural membranes and fluid enable the lungs to move easily, minimising friction from other organs.

⏸ PAUSE POINT   Explain the structure and function of the lungs.

> **Hint**   Breathe in deeply and think about the passage of air.
> **Extend**   How are the alveoli specialised for efficient and effective gas exchange?

Spirometer readings of lung volumes

One way to investigate pulmonary ventilation (breathing) is by using a spirometer. A spirometer consists of a chamber filled with medical grade oxygen that floats on a tank of water. A disposable mouthpiece is connected to a tube. This is connected to the tank, and the patient breathes in and out (see Figure 5.15). Breathing in removes oxygen from the chamber so it moves down, while breathing out pushes carbon dioxide in so the chamber moves up. The movement of the chamber up and down is recorded using a datalogger and produces a trace.

Soda lime is attached to the tube to absorb carbon dioxide that is breathed out. This means that the total volume of gas in the spirometer will gradually decrease. The volume of carbon dioxide breathed out is the same as the volume of oxygen breathed in. Therefore, as carbon dioxide is removed, this total decrease equals the volume of oxygen used up by the person breathing in and out. The trace will show a slope and this can be used to measure the volume of oxygen used in a specific period, such as when exercising.

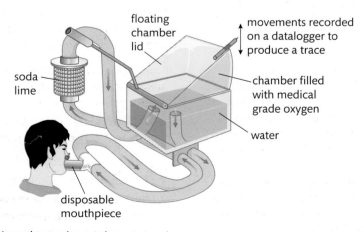

▶ **Figure 5.15** Person breathing in and out using a spirometer

Worked example

You can calculate oxygen uptake per minute for a subject at rest.

If a *y*-axis shows a reduction of 0.5 dm³ between point A and B:

Step 1: Use the *y*-axis for the reduction in oxygen volume which is 0.5 dm³.

Step 2: Look at the *x*-axis for the time taken: 65 s.

Step 3: So in 65 s, 0.5 dm³ of oxygen is used up.

Step 4: $0.5/65$ dm³s⁻¹.

Step 5: Times it by 60 to work it out per minute $0.5 \times 60/65$ dm³min⁻¹.

Step 6: 0.46 dm³min⁻¹.

The air in the nose, trachea and bronchi is not involved with gas exchange and is known as dead space.

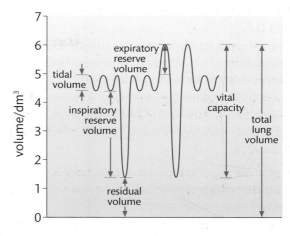

▶ **Figure 5.16** A spirometer trace showing tidal volume, residual volume, inspiratory reserve volume, expiratory reserve volume, vital capacity and total lung capacity

Methods used to measure lung function for respiratory conditions

Medical personnel may need to measure the function of people's lungs on occasion to determine if there is an underlying medical condition that is stopping them from functioning properly.

Peak expiratory flow

A peak flow measures the speed of air flowing out of a person when they breathe out as fast and as much as they can. A peak flow reading can indicate how well the lungs are functioning. A peak flow is commonly used to determine how well an asthmatic person's lungs are working. The test is carried out three times and the best reading is recorded.

Forced vital capacity

Forced vital capacity (FVC) is a lung function test that can be measured using a spirometer. It is used to diagnose obstructive lung diseases such as asthma and chronic obstructive pulmonary disease (COPD).

To achieve data for forced vital capacity, the patient breathes out as forcefully and rapidly as possible into a spirometer. The forced vital capacity is the total volume of air exhaled; it is normally equal to the vital capacity. It can help determine the amount of obstruction that a person has in their airways.

The effects of exercise

When you are at rest, air moves in and out of your lungs about 12 times per minute. Each breath renews the air in your lungs and expels carbon dioxide from your body.

Types of lung volume

There are different elements of lung volume, as shown in Table 5.10. An example of a spirometer trace showing these elements is shown in Figure 5.16.

▶ **Table 5.10** Elements of lung volume

Element	Description
Tidal volume (TV)	The volume of air that is taken in and breathed out during each breath when you are at rest. At rest TV is approximately 0.5 dm³.
Inspiratory reserve volume (IRV)	The volume of air that can be breathed in when you take a big breath, over and above normal tidal volume. This is approximately 2.5–3 dm³.
Residual volume	The volume of air that always remains in your lungs even after you have breathed out. This is approximately 1.5 dm³.
Expiratory reserve volume (ERV)	How much air can be breathed out after you have taken a big breath in, over and above normal tidal volume. This is approximately 1 dm³.
Vital capacity	The maximum volume of air that can be moved in and out of your lungs in one breath. This varies depending on size, age and sex of the person; it is approximately 5 dm³.
Total lung capacity	This can be calculated by adding together the person's vital capacity and the residual volume. This will provide you with data to suggest the total lung volume of a person.

When you exercise or become frightened, your breathing becomes quicker and deeper to supply your body with more oxygen and remove more carbon dioxide. This is known as breathing rate.

The frequency at which we take breaths in must increase to meet the demand for oxygen. This therefore increases the **respiratory minute ventilation**, which is the volume of air breathed in or out per minute. If a person were to carry out exercise and then breathe in and out through a spirometer, we would notice a difference in tidal volume.

> **Key term**
>
> **Respiratory minute ventilation** – volume of air breathed in or out per minute.

Exercise causes tidal volume to increase, to meet the oxygen demands of respiring muscles and to accommodate the exhalation of the increased production of carbon dioxide. During exercise, your body consumes large amounts of oxygen. The harder you exercise, the more oxygen your body consumes.

Assessment practice 5.7

1 Explain how a spirometer works.

2 Explain the data that can be analysed from a spirometer.

3 Use the spirometer trace to determine the oxygen uptake per minute.

Urinary system structure and function

In this section you will learn about the structure and function of the urinary system, which is also known as the renal system. You will learn about the purpose of the urinary system in excretion and **osmoregulation** and how the kidneys are involved in water, electrolyte and acid base balance. The section also covers problems with kidney function and how kidney disease can be treated.

> **Key term**
>
> **Osmoregulation** – the control of water and salt levels in the body which prevents problems with osmosis.

The function of the urinary system

The urinary system consists of the kidney, ureters, bladder and urethra. Most people have two kidneys. They are positioned each side of the spine at the back of the abdominal cavity just below your waist. The ultrastructure of the kidney will be discussed later in this unit. Figure 5.17 shows the structure of the urinary system.

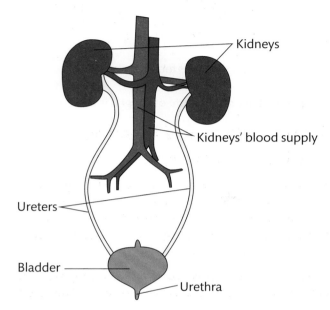

▶ **Figure 5.17** Structure of the urinary system

Ureter

The ureters are muscular tubes made of smooth muscle fibres. They transport urine from the kidneys to the bladder.

Bladder

The urinary bladder is a hollow muscular organ that collects urine from the kidneys before disposal by urination. The urine enters the bladder through the ureter and urine leaves the bladder by the urethra.

Renal artery and vein

There are two types of blood vessel attached to the kidneys. The renal arteries deliver an oxygen-rich blood supply to the cells in each kidney. Once the blood has been processed here it leaves the kidney via the renal veins and is transported in the inferior vena cava back to the heart.

The kidney

The kidney filters waste products from the blood before turning them to urine. The kidney consists of three very easily identifiable regions (see Figure 5.18). It is surrounded by a strong capsule and consists of the cortex, medulla and in the centre the renal pelvis which leads to the ureter.

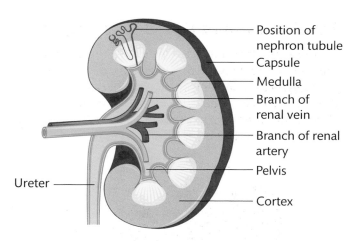

▶ **Figure 5.18** The structure of the kidney

The structure and function of a kidney nephron

The nephron is the functional unit of the kidney. Urine is produced here. Nephrons are microscopic structures that make up the bulk of the kidney. Each kidney has approximately one million nephrons, and each is close to many blood capillaries (see Figure 5.19).

Each nephron starts in the cortex of the kidney. Here, the blood capillaries from the renal artery form a knot known as the glomerulus. This sits inside a cup-shaped structure called the Bowman's capsule. Ultrafiltration is the process whereby fluid from the blood is pushed into the Bowman's capsule so that the process of selective reabsorption can take place as the fluid flows along the nephron. Substances that the body needs to conserve are reabsorbed back into the blood capillaries and anything not reabsorbed ends up as urine to be expelled from the body.

The nephron is divided into four parts:

▶ proximal convoluted tubule
▶ loop of Henle
▶ distal convoluted tubule
▶ collecting duct.

Ultrafiltration and the glomerulus

The glomerulus receives blood from the **afferent arteriole** and the blood leaves through the efferent arteriole. The afferent arterioles are wider than the efferent arterioles; this difference in diameter increases the pressure in the blood capillaries of the glomerulus and pushes fluid out of the capillaries and into the Bowman's capsule where there is a lower pressure (see Figure 5.20). The fluid pushed out of the blood capillaries contains water, amino acids, glucose, urea and inorganic ions for example sodium, chloride and potassium. Blood cells and proteins are left in the capillary as they are too large

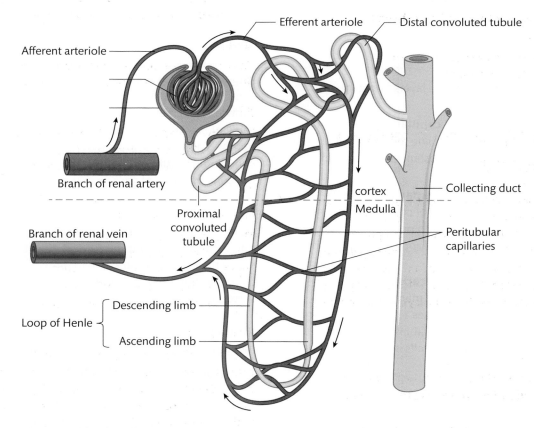

▶ **Figure 5.19** A nephron and blood capillaries

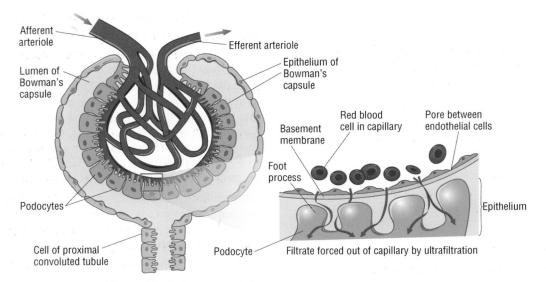

Afferent arteriole
Efferent arteriole
Lumen of Bowman's capsule
Epithelium of Bowman's capsule
Red blood cell in capillary
Pore between endothelial cells
Basement membrane
Foot process
Podocytes
Cell of proximal convoluted tubule
Podocyte
Filtrate forced out of capillary by ultrafiltration
Epithelium

▶ **Figure 5.20** Glomerulus and Bowman's capsule

to pass through the small gaps in the capillary wall. This means the **water potential** in the blood capillaries is very low after ultrafiltration. This is important to help reabsorb water later in the process.

> **Key terms**
>
> **Afferent arterioles** – a group of blood vessels that supply the nephrons in many excretory systems.
>
> **Water potential** – a measure of the ability of water molecules to move freely in solution.

The fluid must pass through three layers to get into the Bowman's capsule.

- **Endothelium of the blood capillary:** there are small gaps in between the cells that line the blood capillary so that blood plasma and the substances dissolved in it can pass through.
- **Basement membrane of Bowman's capsule:** this consists of a very thin network of collagen fibres and glycoproteins. It acts as a filter to prevent the movement of larger substances from the blood capillary into the Bowman's capsule.
- **Epithelial cells of Bowman's capsule:** these cells are called podocytes and have many finger-like protections called foot processes to ensure there are gaps between the cells. This enables the fluid from the capillary to pass between the cells into the Bowman's capsule.

Selective reabsorption and the nephron

As the fluid moves along the nephron, all the glucose and amino acids, and some salts and water that have been pushed out of the blood capillaries are reabsorbed back into the blood. 85% of the filtrate is reabsorbed in the proximal convoluted tubule. The cells that line the proximal convoluted tubule are specialised for this reabsorption.

The proximal convoluted tubule cell surface membrane that is in contact with the fluid in the tubule is highly folded into microvilli to increase the **surface area**. This increases the space available for reabsorption to take place. The membrane also contains co-transporter proteins that transport proteins and glucose along with sodium ions from the convoluted tubule into the cells lining the tubule by **facilitated diffusion**. The opposite surface of the membrane that is situated next to the blood capillaries is highly folded and contains sodium-potassium pumps. These pump sodium out of the cells lining the convoluted tubule and potassium ions in. This reduces the concentration of sodium in the cells' cytoplasm. This means that sodium ions from the filtrate are transported into the cells, by facilitated diffusion, from the tubule along with glucose and amino acids also from the filtrate. This increases the concentration of glucose and proteins inside the cell so they are able to diffuse from a high concentration inside the cell into the tissue fluid that surrounds the cells and back into the blood capillaries. The reabsorption of salts, glucose and proteins reduces the water potential in the cells that line the tubule and increases the water potential in the tubule. This means water will also move into the cells down a water potential gradient from an area of high water potential in the tubule to an area of low water potential in the cells lining the tubule. From the cells it can be reabsorbed back into the blood capillaries.

Link

Go to *Unit 9: Human Regulation and Reproduction* to find more information about the sodium potassium pump.

Key terms

Surface area – a measure of the total space occupied by the surface.

Facilitated diffusion – the movement of molecules down their concentration gradient, across a membrane, with the help of carrier proteins. Energy is not required.

The loop of Henle

The loop of Henle creates a low water potential in the surrounding medulla tissue, to ensure that water can be reabsorbed from the nephron and back into the blood capillaries. The loop of Henle consists of:

▶ the descending limb, which descends from the cortex into the medulla

▶ the ascending limb, which ascends from the medulla into the cortex.

The arrangement of the loop of Henle creates a hairpin counter current. The effect of this arrangement is to increase the efficiency of salt transfer from the ascending limb to the descending limb. This causes a high concentration of salt (sodium chloride) to build up in the surrounding tissue, so that it diffuses out of the loop of Henle into the medulla tissue. This reduces the water potential of the medulla tissue so water moves out of the collecting duct and can therefore be reabsorbed. The steps below show how this is achieved.

Step-by-step: loop of Henle

7 Steps

1 At the base of the ascending tubule, sodium and chloride ions diffuse out of the tubule into the tissue fluid; this reduces the water potential of the surrounding tissue.

▼

2 As fluid moves down the descending limb the water potential inside the tubule becomes lower.

▼

3 As the fluid ascends up the ascending limb towards the cortex the water potential inside the tubule increases.

▼

4 Higher up the tubule sodium and chloride ions are actively transported out into the surrounding tissue fluid. This reduces the water potential of the surrounding tissue.

▼

5 The wall near the top of the ascending limb is impermeable to water, so water can't leave the tubule. This means that the fluid in the ascending limb loses salts but not water.

▼

6 A consequence of the movement of salts from the ascending limb is that water moves out of the descending limb by **osmosis** into the surrounding tissue fluid where the water potential has become lower.

▼

7 Also, sodium and chloride ions diffuse into the descending limb from the surrounding tissue, as the concentration of these ions is higher in the tissue than in the tubule, which decreases the water potential in the tubule.

The water potential in the medulla becomes more negative the deeper the nephron goes into the medulla. The removal of salts from the ascending limb means that the urine at the top of the ascending limb is very dilute so the water in this urine can be reabsorbed in the distal tubules and collecting ducts. The amount of water we reabsorb depends on the body; this is referred to as osmoregulation, the regulation of water.

Key terms

Osmosis – the net movement of water molecules from an area of high water potential to an area of low water potential, down a water potential gradient and across a partially permeable membrane.

Osmoregulation - the control of water and salt levels in the body which prevents problems with **osmosis**.

Distal convoluted tubule and collecting duct

The fluid leaves the ascending loop of Henle and passes along the distal convoluted tubule, where **active transport** is used to adjust the concentration of salts. The fluid, which is still very dilute and contains a high concentration of water, flows from here into the collecting duct. The collecting duct descends into the medulla to the renal pelvis. The fluid in the tubule passes down the collecting duct and, because of the low water potential in the surrounding tissue, water moves out of the collecting duct by osmosis into the surrounding tissue and enters the blood capillaries to be carried away and used in the body. The amount of water that is reabsorbed depends on the permeability of the walls of the collecting duct. This is controlled by antidiuretic hormone (ADH).

> **Key term**
>
> **Active transport** – the movement of molecules against a concentration gradient, across a membrane.

⏸ PAUSE POINT Explain ultrafiltration and selective reabsorption.

Hint Think about the ultrastructure of the kidney and the functional unit.

Extend Explain the change in water potential throughout the medulla of the kidney and how this affects reabsorption of substances.

Osmoregulation and the role of anti-diuretic hormone (ADH)

In order to take water in, we drink and eat. We also produce water in respiration. Water is lost from our body in urine, sweat, faeces and water vapour when breathing out. The correct water balance between cells and the fluid that surrounds them must be controlled. **Osmoregulation** is the control of water and salt levels in the body which prevents problems with osmosis. You may notice that sometimes you have very small amounts of concentrated urine but on other days you have large amounts of dilute urine.

Depending on your body's needs, the walls of the collecting duct can be made more or less permeable. For example, if it is a hot day, you need more water, so the walls of the collecting duct are made more permeable so that more water can be reabsorbed into the blood to keep levels of water high. This will mean you will produce a smaller volume of urine. On a cooler day, when you do not need as much water, the walls of the collecting duct are made less permeable so that less water is reabsorbed and you lose more water in your urine.

Antidiuretic hormone (ADH) is manufactured by special cells called neurosecretory cells. These are found in the hypothalamus in the brain. ADH is released into the blood and binds to receptors on the wall of the collecting duct of the nephron. This causes a chain of enzyme-controlled reactions to occur inside the cells. The outcome of these reactions is that water permeable channels called aquaporins are inserted into the collecting duct cell surface membrane (see Figure 5.21). These channels facilitate the movement of water and more channels mean that more water can be reabsorbed, by osmosis, from the collecting duct across the membrane and into the blood capillaries so less water leaves the body in urine. Less ADH in the blood means that there are fewer aquaporins, because the cell surface membrane folds inwards to create vesicles that contain the water permeable channels. This leaves behind fewer channels, so the collecting duct walls are less permeable to water. Less water is reabsorbed and therefore more water leaves the body in the urine.

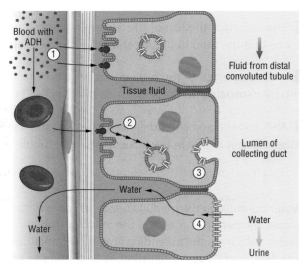

1. ADH detected by cell surface receptors
2. Enzyme-controlled reactions
3. Vesicles containing water-permeable channels (aquaporins) fuse to membrane
4. More water can be reabsorbed

▶ **Figure 5.21** Effect of ADH on the collecting duct wall

Step-by-step: Osmoregulation   8 Steps

1 As blood flows through the hypothalamus in the brain, osmoreceptors in the hypothalamus monitor the water potential of the blood.

▼

2 If the water potential in the blood is low, then the osmoreceptor cells lose water by osmosis and they shrink.

▼

3 This stimulates neurosecretory cells in the hypothalamus to produce ADH. ADH is made in the cell body of these special neurons.

▼

4 ADH flows down the axon of the neurosecretory cell to the terminal bulb in the posterior pituitary gland, where it is stored until it is needed.

▼

5 When the neurosecretory cells are stimulated, they send action potentials down their axons and the ADH is released into the blood capillaries running through the posterior pituitary gland.

▼

6 ADH is transported around the body and acts on the cells in the collecting ducts.

▼

7 When the water potential rises, the osmoreceptors in the hypothalamus detect this change, and less ADH is produced.

▼

8 ADH has a **half-life** of about 20 minutes, so ADH present in the blood is broken down and the collecting ducts will not be stimulated any further.

Key term

Half-life – the time it takes for the concentration to reduce to half of its original value.

Assessment practice 5.8

1   How do neurosecretory cells differ from normal nerve cells?

2   Explain how the kidney controls water potential in the blood.

3   Why is it important that ADH is broken down and does not remain in the blood?

Kidney disease

Medical problems such as kidney failure can occur due to many different reasons. Some of the more common causes are diabetes mellitus, hypertension and infection. If the kidneys fail completely, the body is not able to control the levels of water in the body. It is unable to remove excess water and waste products such as urea, and this can lead very quickly to a fatality.

Haemodialysis

Lillian has kidney failure and has to undergo haemodialysis at a clinic three times a week. Each session is several hours. Dialysis removes waste and excess fluid and salt from Lillian's blood. Blood from Lillian's vein is passed through a dialysis machine that contains an artificial dialysis membrane. The membrane is partially permeable and allows the exchange of substances between Lillian's blood and dialysis fluid. The dialysis fluid contains the correct concentrations of salts, water, urea and other substances that are required in Lillian's blood plasma. If there are high concentrations of any of these substances in Lillian's blood, then they will diffuse out of the blood across the membrane into the dialysis fluid. If Lillian's blood has low concentrations of any of the substances, then they will diffuse across the membrane from the dialysis fluid into her blood. This helps to achieve the correct concentrations of substances in Lillian's blood.

Check your knowledge

1 What is the function of dialysis fluid?

2 Why must the dialysis membrane be partially permeable?

3 How may this impact on Lillian's everyday life?

4 What are the benefits for Lillian?

Peritoneal dialysis

Mark is researching peritoneal dialysis as an alternative to Lillian having to attend a clinic three times a week. He has found that peritoneal dialysis uses a membrane in your body called the peritoneal membrane. It requires a surgeon to implant a permanent tube in the abdomen so that the dialysis fluid can be poured into the abdomen space to allow exchange of substances across the peritoneal membrane. After several hours the solution is drained. Lillian would have to perform 4–6 exchanges every day, but it would mean she could do it in the comfort of her own home, when it is suitable for her.

Check your knowledge

1 What are the main differences between haemodialysis and peritoneal dialysis?

2 What are the advantages to Lillian carrying out peritoneal dialysis?

3 What are the disadvantages of changing to peritoneal dialysis?

Kidney transplant

A kidney transplant involves major surgery, where the patient is under anaesthetic and the surgeon implants the donor kidney into the lower abdomen and attaches it to the blood supply and bladder. Patients are given immunosuppressant drugs to help prevent rejection of the foreign organ.

Assessment practice 5.9

1 Explain why haemodialysis fluid must be sterile and at 37°C.

2 Create a table of advantages and disadvantages for dialysis and kidney transplant.

Cell transport mechanisms

For all the processes mentioned in this unit (e.g. reabsorption of glucose, water and gas exchange in the lungs), cell membranes must have a specific structure to enable the movement of substances from one place to another.

Fluid mosaic model

The term **fluid mosaic model**, proposed by Singer and Nicholson, is used to describe the arrangement of biological membranes. Fluid mosaic membranes consist of the following layers.

- The phospholipid bilayer, which forms the basic structure.
- Protein molecules, which are present within the phospholipid bilayer.
- Extrinsic proteins, which are embedded on the surfaces of the membrane.
- Intrinsic proteins, which completely span the bilayer.

Phospholipid bilayer

Phospholipid molecules consist of a phosphate head and two fatty acid tails. The phosphate head is described as **hydrophilic** (water-loving) due to the charges distributed across the molecule while the two fatty acid chains are **hydrophobic** (water-hating).

> **Key terms**
>
> **Fluid mosaic model** – description of the cell membrane structure, a phospholipid bilayer with proteins floating in it.
>
> **Hydrophilic** – associates with water molecules easily.
>
> **Hydrophobic** – repels water.

When a phospholipid molecule is mixed with water the heads stick into the water, while the fatty acid tails stick up out of the water. When phospholipids are completely surrounded by water, they form a bilayer.

All biological membranes are made from phospholipid bilayers, and the hydrophobic layer formed by the fatty acid tails creates a barrier that helps separate the cell contents from the outside. All membranes are permeable to water because it can easily diffuse through the bilayer. It is also very important for survival that cells get a supply of nutrients that they need and that any waste produced is removed from the cell. These molecules usually enter or leave the cell across the membrane.

Methods used to transport molecules through cell membranes

Diffusion

Diffusion is the movement of molecules from an area of high concentration to an area of lower concentration down a concentration gradient. Molecules possess kinetic energy which keeps them moving and they are passively transported across biological membranes. Diffusion is therefore known as a passive process as the molecules only rely on their kinetic energy and a concentration gradient for movement, they do not use energy from the cell.

Lipid-based molecules

Fat-soluble molecules such as steroid hormones can simply pass through the phospholipid membrane because the bilayer consists of fatty acid tails. They diffuse down a concentration gradient through the membrane and into the cell.

> **Key term**
>
> **Fat soluble** – dissolves in fats.

Small molecules

Oxygen and carbon dioxide molecules are small enough to just pass through the spaces in the phospholipid bilayer and can be transported across phospholipid bilayers by diffusion.

Facilitated diffusion

Larger and charged molecules

Large molecules, such as glucose, and small charged particles, such as sodium **ions,** are not able to pass through the phospholipid bilayer. They need help to cross the membrane. This is known as facilitated diffusion. There are two types of proteins present in the membrane that facilitate diffusion.

- **Channel proteins** form pores in the membrane which are shaped to allow particular molecules/ions, for example sodium and calcium ions, to pass through. Many are 'gated', which means they can be open and closed.
- **Carrier proteins** are shaped for a specific molecule, for example glucose or amino acids. When the molecule binds to the protein, the protein changes shape to allow the molecule to pass across the membrane.

> **Key term**
>
> **Ions** – particles that carry a positive or negative charge.

Active transport

Carrier proteins in the membrane can act as pumps to carry large and charged molecules across the membrane. The shapes of the proteins are complementary to the molecules they carry, which they transport one way across the membrane. As a molecule moves through the protein, its shape changes. This means that as a molecule exits it cannot enter again, as the protein shape is no longer complementary. These protein pumps use metabolic energy in the form of ATP (adenosine triphosphate) to move molecules across the membrane and they can carry molecules in the opposite direction to the concentration gradient from a low concentration to a high concentration. This process of active transport is much faster than diffusion.

Endocytosis and exocytosis

Sometimes large quantities of materials need to be moved into cells, by **endocytosis**, or out, by **exocytosis**.

Bulk transport requires energy in the form of ATP. The energy is used to move the membrane around to form and move vesicles around the cell. Vesicles are used to carry the bulk material, for example insulin, to be transported and they easily fuse with membranes and can separate from membranes by 'pinching off'.

> **Key terms**
>
> **Endocytosis** – movement of bulk material into a cell.
>
> **Exocytosis** – movement of bulk material out of a cell.

Osmosis

Osmosis is diffusion of water molecules only. Water potential is the pressure exerted by water molecules.

If there is a higher concentration of water molecules they will exert a higher pressure and therefore have a higher water potential. You must always refer to osmosis as the movement of water molecules from a region of high water potential to a region of low water potential, down a water potential gradient across a partially permeable membrane. Osmosis will occur until the concentration of water molecules is even either side of the membrane, meaning the water potential is the same on both sides.

Surface area to volume ratio

The larger the surface area to volume ratio, the more effective transport is. Single celled organisms have a large surface area compared to their volume and they can rely on diffusion alone to meet their needs. Larger multi-celled organisms cannot meet their nutrient need by diffusion alone, hence the need for transport systems and specialised surface areas where there is a large surface for diffusion to take place.

Assessment practice 5.10

1   Describe the structure of a biological membrane.

2   Explain the methods of transport across a partially permeable membrane.

3   Produce a table to compare and contrast the different methods of transport.

> **Further reading and resources**
>
> Annets, F., Foale, S., Hartley, J., Hocking, S., Hudson, L., Kelly, T., Llewellyn, R., Musa, I. and Sorenson, J. (2010) *Applied Science Level 3*. Pearson.
>
> Boyle, M. and Senior, K. (2008) *Biology* (3rd edition). Collins.

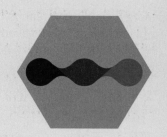

Getting started

Thermodynamics is about the links between heat, energy and power. Although it started with engineers designing and improving steam engines, it now helps shape our understanding in almost every area of science – from the chemical reactions that provide the energy for life to the mystery of space and time and the origins and future of the universe.

What do you already know about heat and energy? Try writing down all the definitions and important equations you can think of. When you have completed this unit, come back and see what you can add to your summary.

C Thermal physics, materials and fluids

Thermal physics in domestic and industrial applications

This subject builds up in small steps. When you can put it all together, it is surprisingly powerful and answers important questions. So try to make sure you grasp each definition and law along the way.

Measurements

You will need to be able to recognise and use the SI (Système Internationale) units for the following important quantities. Each unit starts with a capital letter because it is the name of a scientist who did important work.

Energy: unit Joule (J)

Energy is closely related to work (see below) and they share the same unit.

Power, symbol P: unit Watt (W)

We commonly deal with large quantities of power, so you also need to be familiar with the following multiples:

▸ kilowatt (kW) = 1000 Watt
▸ megawatt (MW) = 10^6 Watt
▸ gigawatt (GW) = 10^9 Watt.

1 Watt = 1 Joule per second (J s^{-1}).

(Do not confuse with the kilowatt-hour (kWh), which is a unit of energy, not power. As there are 3600 seconds in an hour and 1000 watts in a kW, 1 kWh = 3,600,000 J.)

Temperature, symbol T: unit Kelvin (K)

In this subject it is often important to use the **absolute temperature** in equations. So you need to know how to convert from degrees Celsius (°C) – also known as

°Centigrade – to Kelvin (K). (Note that there is no degree sign before Kelvin.)

When you see the symbol T in an equation you should always use the absolute temperature in Kelvin.

> **Key terms**
>
> **Absolute temperature** – or '**thermodynamic temperature**' is measured on a scale starting at **absolute zero.** (Symbol T, SI unit Kelvin (K)). Convert temperatures in °C to Kelvin by adding 273.15 K, which is the freezing point of water on the Kelvin scale.
>
> **Thermodynamic temperature** – (symbol T) the modern way of defining absolute temperature, which does not rely on the existence of an ideal gas, but instead uses the amounts of heat transferred in idealised engine cycles.
>
> **Absolute zero** – the lowest temperature an object can be cooled to, where all thermal energy has been removed from it, there is minimal particle movement and it is in its lowest possible energy state.

Pressure, symbol p: unit Pascal (Pa)

Notice that the letter 'p' is used for a lot of things in thermodynamics; so it is important to be careful and consistent about how you write them. Capital P means power. Try to always use a lower case p for pressure. The unit symbol Pa has the extra 'a' to distinguish it from power and pressure.

1 Pascal = 1 Newton per square metre (N m^{-2})

Work done

When objects interact, the forces between them can lead to energy being transferred. Sometimes the energy is stored in a useful form, e.g. in a spring or in a gravitational or electric field (this is called potential energy). Sometimes energy is due to the speed of a moving object (this is called kinetic energy). Alternatively, the thermal energy (heat) content of an object may be increased.

When you measure **work**, you are focussing on the amount of energy transferred during the process.

> **Key term**
>
> **Work** – the work done in a process is the amount of mechanical energy transferred. (Symbol: W, SI unit: Joule (J)).
>
> (Be careful not to confuse W, for the energy transfer *quantity* 'work', with W meaning the power *unit* 'Watt'.)

Work done by a force

The most obvious way of doing work is to exert a force in order to move something, e.g. pushing a child on a swing, or lifting a bag of shopping onto a table. Calculating the work done in both those examples is straightforward:

work done = force × perpendicular distance moved in the direction of the force ($W = F \times \Delta s$)

The symbol Δ means 'change in'. So the 'change in position', $\Delta s = s_2 - s_1$, where s_1 is the initial value of the position (displacement) and s_2 is its final value. Note that we always start with the final value and take away the initial value.

Forces occur in opposing pairs: so the force you exert to lift the shopping is opposed to the weight of the shopping itself. When you lift the shopping and it moves up in the same direction, you do positive work on it. But if you let it down again onto the floor, the work you do is negative – or to put it another way, the shopping then does work on your arm. Some old clocks operate in that way, with weights that slowly fall and drive the mechanism.

But sometimes the force exerted is in a different direction from the movement. For example, when you push something up a slope, the force you are working against is gravity. So, for calculating work done, it is only the vertical height gained that counts as the distance moved – the part of the movement that is along the line that the force acts.

Work done by a gas expanding

Another important way that work is done is when a gas expands to take up a larger volume. This is what happens in the cylinders of a steam engine or motor car engine (petrol or diesel). Hot, high pressure vapour or gas pushes against a piston and moves it. To calculate the force exerted by the gas you would need to multiply the gas pressure by the area of the piston head, A. So $F = p A$.

When the piston moves, the volume change for the gas is that same area, A, multiplied by the distance moved by the piston. So $\Delta V = A \Delta s$.

Putting those two facts together, you can now calculate the work done by the expanding gas:

work done = pressure × volume change
($W = p \times \Delta V$)

Note that this definition of work saves you having to worry about directions. A gas exerts its pressure in all directions on every surface of the container that defines its volume.

Law of conservation of energy

Because forces always occur in pairs that are equal in size but opposite in direction (Newton's First Law of Motion), the energy *transferred to* an object when work is *done on it* is always equal in size to the amount of *energy lost* by the second object that is *doing the work*. So overall the total energy of the pair of objects remains unchanged. As this is true for every pair of objects in the universe, you can assume that the total energy of the whole universe must also be constant, whatever goes on within it.

This is known as the Law (or principle) of Conservation of Energy. It is one of the most fundamental laws of physics.

Thermodynamics is largely about simplifying things so that they can be studied and measured. You cannot study the whole universe all at once. So scientists define a **system** by drawing an imaginary boundary around what they are studying. Then they can talk about the system and its **surroundings**, i.e. the rest of the universe. (You will often study two systems and measure the interactions between them.)

> **Key terms**
>
> **System** – the part of the universe whose properties you are investigating. It is enclosed by a boundary defined by you, the experimenter.
>
> **Surroundings** – the rest of the universe, outside the system boundary.

The energy contained in a system can be in many different forms, including those shown in Table 5.11.

▶ **Table 5.11** Types of energy and their nature

Type of energy	Nature of the energy
mechanical energy	either potential energy, due to its mass and position in a gravitational field, or kinetic energy, due to an object's mass and speed of motion
electrical energy	can be associated with static electric charge and potential, or with moving charges, current and magnetism
chemical energy	intrinsic to the microscopic structure of the material and chemical bond energies
nuclear energy	due to the binding of subatomic particles (protons and neutrons) in the nuclei of atoms – changes in that are what cause radioactive decay, and also where nuclear power comes from
thermal energy	due to the microscopic vibrations and movements of atoms and molecules in a material – movements that you cannot directly measure, but that give rise to a measurable quantity, temperature

Link

See *Unit 1: Principles and Applications of Science 1*, Learning Aim A, for more on bond energies.

The principle of energy conservation means that when, in some process, you observe a reduction in one kind of energy, you will always find an equivalent increase in other forms of energy. Energy is transferred, but never lost. (Nevertheless, sometimes when energy is transferred to the surroundings in a form that is not 'useful', people do talk about energy losses. You will think more about 'useful' and 'wasted' energy when calculating efficiency.)

Heat and temperature

When two systems interact, there may be a flow of **heat** between them due to a **temperature** difference. This is a transfer of thermal energy from one system to the other. Flow of heat between two bodies in thermal contact (i.e. they are not completely insulated from one another) will continue until they reach **thermal equilibrium** and have the same temperature.

Temperature is such a commonly experienced quantity that scientists did not realise they need to define it until after they had developed the first two laws of thermodynamics. So the definition of temperature is often called the Zeroth Law of Thermodynamics. That simply states that there exists a property called temperature such that any two bodies at the same temperature will be in equilibrium with one another, and indeed also in

equilibrium with a third object – a **thermometer** – used to measure each of their temperatures – see Figure 5.22.

Key terms

Heat – the quantity of thermal energy transferred during a process.

Temperature – the physical quantity that determines the rate at which heat will flow – from a 'hot' body (system) to a 'colder' (i.e. lower temperature) one. It is directly proportional to the average kinetic energy of the molecules.

Thermal equilibrium – exists when two systems are in thermal contact, but there is no net transfer of heat because they are at the same temperature.

Thermometer – a device with a readily measurable property that varies directly with temperature.

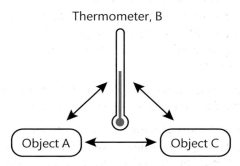

▶ **Figure 5.22** Thermometry – systems in thermal equilibrium

At a microscopic level, temperature is linked to the average kinetic energy of vibration and of motion of the atoms and molecules. You can picture that particles continually bump against one another and so exchange thermal energy, with the result that the energy gradually gets spread more and more evenly. At equilibrium, the rate of energy transfer in one direction will exactly match that transferred back in the other direction. So overall there will be no net transfer – no heat flow.

Calculating efficiency

Not all forms of energy are as useful and controllable as others. Mechanical energy is what you can most readily control, and it can be converted into electrical energy using a generator. Electrical energy can be converted into many other energy forms. But it is difficult to produce anything other than thermal energy from nuclear or chemical energy – fuel cells being one exception.

Every other energy form can eventually be completely converted into thermal energy. When you bounce a ball, stir a liquid, hammer a nail . . . the so-called 'lost' mechanical energy always turns up as an increase in temperature in those things. It is not actually lost. An electric kettle or a microwave oven is always 100% efficient at converting electrical energy into thermal energy, even though some of the heat from the process may be 'lost' in heating electrical cables or the air in the surroundings rather than in heating your food.

But the same is not true in reverse. You cannot take the warmth from your cup of tea or coffee and turn it back 100% into electricity or into mechanical energy. Thermal energy can only be made to do useful work if you have a temperature difference to create a heat flow; and at the end of the process a lot of the energy will still be thermal – just at a lower temperature.

As so much of the energy our society uses comes from heat – from the high temperature of a nuclear reactor or from burning a chemical fuel – improving as far as is possible the efficiency of engines that use heat is very important. For example, steam turbines used in power stations and the internal combustion engines of motor cars are **heat engines**.

The idea of **energy efficiency** is all about how much of the energy you use actually ends up in the useful form you are trying to create.

You can therefore define the **thermal efficiency** of a heat engine quite straightforwardly as:

$$\text{efficiency } (\eta) = \frac{\text{work output, } W}{\text{heat input, } Q_{in}}$$

This is because the purpose of the engine is to turn a source of thermal energy into useful mechanical work.

You can understand why this efficiency is always much less than 100% by looking at the diagrams of heat engines in Figure 5.23.

From these diagrams, you can see that heat engines always have a heat input, Q_{in}, at a high temperature, T_H, but they cannot avoid also having a heat output, Q_{out}, at a lower temperature, T_C. It is that temperature difference that makes the heat flow, and so drives the process in the engine – usually the expansion of a gas or vapour in a turbine or cylinder.

Using the principle of energy conservation:

total energy input = total energy output

So, in the case of a well-designed heat engine that produces useful work, with nothing wasted as friction or noise:

$Q_{in} = W + Q_{out}$ and rearranging: $W = Q_{in} - Q_{out}$

Therefore, the equation for the efficiency of the heat engine becomes:

$$\text{efficiency } (\eta) = \frac{W}{Q_{in}} = \frac{Q_{in} - Q_{out}}{Q_{in}} = 1 - \frac{Q_{out}}{Q_{in}}$$

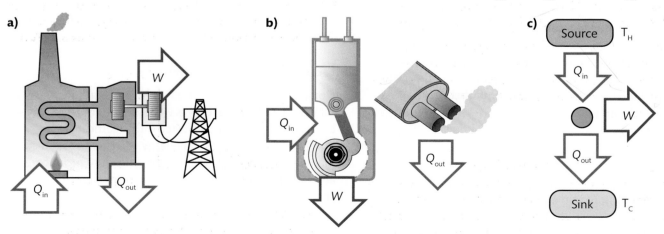

▶ **Figure 5.23** Energy transfers in a) a steam turbine b) a car engine c) a generalised heat engine

Nineteenth-century engineers, working to improve steam engines, began to realise that the efficiency they could achieve seemed to be limited by the temperatures involved, T_H and T_C. But to understand why, you will first need to develop your understanding of a few more concepts of thermal physics, which these engineers were the first to discover.

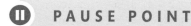

PAUSE POINT   You have covered many of the key ideas and terms used in thermal physics. Close the book and test how many you can recall and explain.

   Hint   Try explaining how thermometry works, and what a heat engine does.

   Extend   Why is electricity such a useful form of energy? Why is it so expensive to produce?

Ideal gas equation

Solids, liquids and gases are all composed of atoms and molecules. In a gas, the molecules are separated from one another and are constantly in motion. This understanding of the nature of a gas is called kinetic theory ('kinetic' means 'moving') and it explains why the pressure, p, of a gas varies according to its temperature, T, and the number of molecules, N, that are contained in a volume, V.

If you make the approximations that, in an **ideal gas** (i.e. one that fits this simplified theory):

▶ the molecules are mostly so far apart that they do not exert any forces on one another, except when they actually collide

▶ their collisions are always elastic – i.e. they exchange kinetic energy, but none is lost overall

▶ the size of the molecules themselves is negligibly small compared with the volume the gas is occupying

then, from this theory, you can derive an **equation of state** for the gas:

$$pV = NkT$$

where k is the Boltzmann constant, which is defined statistically from the average kinetic energy of the gas molecules at a given temperature – average molecular kinetic energy $= \frac{1}{2}kT$ in each of the directions, x, y and z.

This ideal gas equation is useful, because at relatively low pressures and large volumes, and well above their condensation temperature (the temperature at which they become liquid), real gases show behaviour that is very close to the 'ideal'.

Key term

Ideal gas – a theoretical model of a gas, where the molecules are assumed to be point particles that take up no volume and that exert no forces on one another in between their elastic collisions. This simplified model makes it easier to calculate their behaviour, based on Newton's laws of motion.

Key term

Equation of state – an equation connecting the measurable quantities that define the physical state of a system; for example, the pressure, volume, temperature and amounts of material (numbers of molecules) present in the system.

It is also very helpful for developing the idea of a thermometer to measure temperature, because measurements of gas pressure and volume can give us a direct proportionality with absolute temperature, T. (See Figures 5.24 and 5.25.)

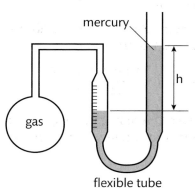

▶ **Figure 5.24** A constant volume gas thermometer; the gas expands on heating, but the mercury manometer is adjusted to bring it back to its original volume, and then a pressure reading is taken.

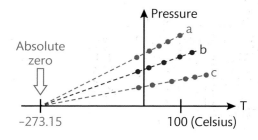

▶ **Figure 5.25** Extrapolating the readings taken from a constant volume gas thermometer indicates an intercept on the temperature axis at approximately $-273\,°C$. This is how the existence of an absolute zero for temperature was first predicted.

Worked example: Work done by gas escaping from a balloon – the rocket principle

An inflated balloon has a volume of 5 litres and contains air at 120 kPa, which is higher than the surrounding atmospheric pressure of 100 kPa. If the air is allowed to escape rapidly, the balloon flies off like a rocket because the expanding gas is doing work. But there is no piston.

To estimate the work done by the gas, first calculate the volume change:

The temperature of the air at the end of the process will be the same as at the start so, using the ideal gas equation of state, $p_1V_1 = p_2V_2$. Thus the final volume of the gas after expansion is given by:

$$V_2 = V_1 \times p_1/p_2 = 5 \text{ litre} \times 120/100 = 6 \text{ litre}$$

So ΔV = initial volume − final volume = $V_2 - V_1$ = 6 litre − 5 litre = 1 litre = $1 \times 10^{-3} \text{m}^3$

The escaping air pushes against the external atmosphere at 100 kPa.

So Work done, $W = p \Delta V = 100 \text{ kPa} \times 1 \times 10^{-3} \text{m}^3 = 100 \text{J}$

The balloon itself has quite a small mass – just a few grams – so when 100 J is converted to kinetic energy using the equation $\frac{1}{2}mv^2$, it will equate to quite a high velocity.

Changes of state

When the temperature of a gas cools, and its pressure and volume reduce accordingly, the average spacing between the molecules becomes much less so that they exert significant attractive forces on one another. A point comes where those forces are sufficient to bind the molecules into a disorganised but quite closely packed structure – a liquid. This is **condensation**, and it happens at definite temperature and pressure values – the boiling point.

The molecules in a liquid vibrate and even flow round one another, but they do not have enough energy to break free of the bonding forces and to travel randomly over long distances as molecules do in a gas. The reverse process, when a liquid is heated and, progressively, more and more of its molecules gain enough kinetic energy to break free of the liquid surface, is called **vaporisation**.

Vaporisation requires a lot of energy input and, correspondingly, during condensation an equivalent amount of energy is given up by the condensing gas. This energy is called the **latent heat** of vaporisation – 'latent' because it appears to be used up without doing anything: there is no temperature rise during the process of vaporisation and no temperature fall during condensation. All the latent heat energy goes into the process of changing the state of the material from liquid to gas, or back again.

When a system contains more than one substance – e.g. when it also contains air – then the **saturated vapour** pressure in equilibrium with the liquid is only part of the total air pressure. Boiling occurs when the temperature increases to a point where the saturated vapour pressure (SVP) has risen to equal the atmospheric pressure. The boiling **vapour** then drives out the rest of the air, filling all the space. So rapidly boiling off a liquid in a confined space can easily cause asphyxiation and death.

A similar energy change occurs when temperature falls and pressure increases sufficiently for the material to form a more ordered and fixed structure – **freezing** to form a solid. The reverse process of solid melting to a liquid is called **fusion**.

Both the freezing point and the boiling point happen at different temperatures depending on the pressure of the system. Look at Figure 5.26 to see the regions of temperature and pressure in which the material exists as a mixture of two **phases** – solid and liquid or liquid and vapour. Because of these two-phase regions it is called a phase diagram.

Examining the phase diagram, you will see that there is one specific temperature and pressure at which three phases are in equilibrium – the solid state, liquid state and vapour state of the material: in the case of water these are ice, liquid water and water vapour (steam). This is called the **triple point** of water, and its temperature −273.16 K (0 °C) is the upper fixed point defining the Kelvin thermodynamic temperature scale, the lower fixed point being 0 K = absolute zero.

At the upper end of the liquid-vapour two-phase region is a maximum temperature and pressure called the **critical point**, above which there is just one phase, a relatively hot, dense supercritical gas.

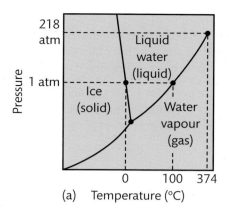
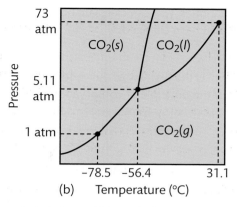

▶ **Figure 5.26** Temperature–pressure phase diagrams (a) for water and (b) for carbon dioxide.

Key terms

Latent heat – energy transferred that has the effect of changing the physical state of a substance without changing its temperature.

Vapour – is the name given to a gas that is at a sufficiently low temperature that it will change state to a liquid or solid if its pressure is increased sufficiently. In other words, a gas below its critical temperature (see below).

Vaporisation – is the change of physical state from liquid to gas. The reverse process is **condensation**.

Saturated vapour – is vapour that is in equilibrium with a liquid phase of the same substance. Any further compression causes more liquid to condense. The vapour cannot exert a higher pressure at that temperature than its saturated vapour pressure (SVP).

Fusion – (or melting) is the change of physical state from solid to liquid. The reverse process is **freezing**.

Phase – a separate part of system which is not uniform throughout. For example, a mixture of ice and water is a two-phase system, but a solution of salt in water is a one-phase system.

Triple point – the temperature and pressure at which all three phases: solid, liquid and vapour, can exist in equilibrium.

Critical point – the temperature and pressure at which the density of the liquid and vapour phases become identical. Above that critical temperature, only a supercritical gas phase can exist.

Theory into practice

Measuring temperature: Using a container of gas as a thermometer and measuring its changes of volume or of pressure is cumbersome. Moreover, there is no such thing as a perfectly ideal gas. As gases and vapours cool towards their condensation points, the distances between molecules become smaller and the forces exerted between them increase. So they begin to deviate from the Ideal Gas Equation.

So, in practice we define practical temperature scales that use a series of fixed temperature points defined by physical properties of a material – e.g. the triple point of water – and some good practical thermometers to divide the scale in between those points. Look up the International Temperature Scale of 1990 (IPS-90) to see the fixed points and thermometers specified for really accurate temperature measurement.

Internal energy and the first law of thermodynamics

If you take an object, like a tennis ball, you can lift it or throw it or hit it with a racquet to give it more mechanical energy – i.e. potential energy and kinetic energy. But that is only a small fraction of the total energy that the ball has. Inside it, and invisible to you, all the atoms and molecules are in thermal motion. Moreover, those particles are bound together by chemical bonds, and the nuclei of the atoms contain vast quantities of nuclear binding energy.

All the hidden energy inside a system that you are studying is called its internal energy, denoted by the letter U.

Of course, it is impossible to measure the total value of U in any direct way. What you can measure is the amount of energy transferred into or out of the system during any process – transfers of heat, Q, or work, W. From those definite measured quantities, you can calculate how much the internal energy must have changed:

$$\Delta U = Q - W$$

This equation is known as the first law of thermodynamics. It is just a statement of the principle of conservation of energy as it applies to thermal systems – systems that use and transfer heat.

In the equation:

▶ Q has a *positive* sign because we count heat input to the system as positive and heat transferred out as negative

▶ W has a *negative* sign because we count work as positive when the system does work for us and so gives up some of its internal energy.

The symbol Δ means 'change in', i.e. (final value) – (initial value). So ΔU means 'change in internal energy'.

Internal energy, U, is a function of the state of a system. This means that if the pressure, temperature and volume of each component in a system is known, then there must be a fixed amount of internal energy, U_1, in the system, associated with those values. And so, when there is a change in those variables that define the state of the system, there will be a corresponding change, ΔU, to a new fixed amount of internal energy, U_2. You can calculate the change, ΔU, from the measured values of Q and W, but you never know the absolute values of either U_1 or U_2.

(Note that Q and W are measured quantities of energy transferred, not changes in anything. So, correctly, you should not use the symbol Δ with them. However, doing so is a common mistake that you will even see in some books. So if you see that, just ignore the Δ.)

Natural processes and the second law of thermodynamics

Some things naturally tend to happen, and others never do. For example, a hot cup of tea or coffee gradually cools down, losing heat and slightly warming its surroundings. But you never see a warm drink absorbing heat from its cooler surroundings and heating up.

Similarly, if you put sugar into your hot drink it will gradually dissolve and mix naturally, though you might stir it to hurry the process along. But, once mixed, you never see all the sugar separating out again as crystals. It takes a lot of effort and energy input to concentrate a sugar solution and to get it to crystallise, and even then some of the sugar remains in solution.

If you watch a film or video running backwards, some things appear normal – they are reversible in time; but

many other things are obviously wrong. A collapsed wall will not rebuild itself naturally. A spilt drink will not collect itself up and flow back into the cup. You can easily spot that time has been reversed in the film played backwards.

These everyday observations are the basis of the second law of thermodynamics, which can be stated in several different but equivalent forms.

▶ A natural process can never be reversed in its entirety.

▶ It is impossible to completely change heat into work.

▶ Heat will not flow from a colder body to a hotter one without an input of work.

▶ You cannot reverse the direction of time.

The second and third of the statements above apply the Second Law to heat engines and to heat pumps respectively, and you will be learning more about them in a later section.

But everything that happens around you, and indeed in your own body, is a result of **natural processes**. Nothing is ever fully reversible. Nevertheless, all the other laws and equations of physics – including the First Law of Thermodynamics (energy conservation) – are written as if time could be reversed. So it is only by using the Second Law alongside these other laws that you can determine which way things will go. This means the Second Law is particularly important for:

▶ engineers designing engines and machinery

▶ chemists working out which reactions will happen, and where the position of equilibrium will lie

▶ biologists understanding what drives life processes

▶ cosmologists exploring the past and predicting the future of the universe.

Key terms

Natural process – any real process that is not driven by an input of work or heat from a source outside the system being studied, but just occurs naturally.

Entropy – 'potential for change'. One way to calculate the entropy of a system would be to count how much randomness it contains – i.e. the number of equivalent energy states and positions its particles could be arranged in. The more ordered and compact and high in temperature something is the lower is its entropy. Things naturally become more randomly mixed up, more spread out and cooler – i.e. their entropy increases. So, at its start, the universe must have had a very low entropy indeed.

Link

Entropy can also be defined by using the concept of ideal heat engine cycles.

Ⅱ PAUSE POINT Close your book and write down definitions of the First and Second Laws of Thermodynamics. Think about how those two laws work together to explain what happens in the world around you. What drives the direction of industrial processes and the biochemical reactions in living organisms?

Hint One system (e.g. our sun) increasing in entropy may release energy to drive another system (e.g. plant leaves) which then creates some more ordered, lower entropy structures. But what happens overall to the order and entropy of the universe?

Extend Highly unusual things – coincidences – happen all the time. But can *you* repeat them? What does the Second Law mean when it states that certain things are 'impossible'?

Reversible processes (isothermal and adiabatic)

Having just said that no process is ever fully reversible in time, the *ideal* of a fully **reversible process** is a very useful idea.

There are two sets of conditions which, if fully achieved, would make a process reversible:

Heat transfer at constant temperature – an isothermal process

When two bodies are at thermal equilibrium, heat could move in very small amounts either way across a boundary between them. So that would be a reversible process.

But in reality, to make the heat flow there always has to be a small temperature difference. So an **isothermal** process is an ideal that you can get close to but never quite achieve. Nevertheless, you can use the *idea* of a reversible isothermal process to think about the limits to achieving greater efficiency.

Slow heat transfer processes with excellent thermal contact, achieved in a large heat exchanger, are close to being isothermal.

No heat transfer at all – an adiabatic process

If no heat is transferred, and work is done with no conversion of energy into heat, then all the work will result in increased potential or kinetic energy. That energy could then be used to reverse in its entirety the work previously done – a fully reversible process, which could go on being repeated for ever and so create perpetual motion.

But it is never possible to thermally isolate a system completely. If its temperature rises above, or falls below, the temperature of its surroundings, it will transfer some heat. Even the best insulation gives some heat leakage. Also, any practical machine for doing work always suffers from some kind of frictional or viscous energy conversion into heat. Once again, it is the *idea* of a reversible

adiabatic process that you will find useful – an ideal to strive for.

Very quick compression or expansion in a machine with low friction losses is close to being adiabatic.

Key terms

Reversible process – one which could be fully reversed in time, following the same path and exactly reversing the quantities of heat transferred and of work done. This is an ideal concept that can never be fully achieved in practice.

Isothermal – happens at one fixed temperature. To be reversible, an isothermal process would have to have a negligibly small temperature gradient across the system boundaries, so would be extremely slow.

Adiabatic – means 'no transfer'. It describes a process in a system that is not just well-insulated but is totally thermally isolated, so there is zero heat transfer. Also all work done in the process must be friction free and not create any extra thermal energy.

Idealised engine cycles

Now that you have the idea of two kinds of reversible process, you can go on to imagine a perfectly reversible engine for converting heat from a high temperature heat source into work. This *ideal* engine would be more efficient than any real engine you could build. It tells you the limits of what could be possible.

The ideal cycle is named after its inventor, Carnot, and comprises four processes for a working fluid – e.g. a gas or a liquid-vapour mixture:

1–2 adiabatic compression – zero heat transfer.

2–3 isothermal expansion – heat absorbed by the system, Q_{in}

3–4 adiabatic expansion – zero heat transfer

4–1 isothermal compression – heat absorbed by the system, $-Q_{out}$

Completing the four processes returns the working fluid to its original state, but work has been done, which can be measured from the areas under the pressure-volume curve ($W = \Sigma\, p\, \delta V$ where δV signifies a minute change in volume):

▶ over the two expansion processes: work done *by* the system counts positive, W_{out}

▶ over the two compression processes: work done *on* the system counts negative, $-W_{in}$.

So the net work done in the cycle is $W = W_{out} - W_{in}$. That is equivalent to the area on the p-V diagram (see Figure 5.27) enclosed by the cycle of processes.

Maximum thermal efficiency – the efficiency of an ideal Carnot cycle engine

You have already seen that the efficiency of a heat engine producing useful work from a heat input is given by:

$$\eta = \frac{W}{Q_{in}} = \frac{Q_{in} - Q_{out}}{Q_{in}} = 1 - \frac{Q_{out}}{Q_{in}}$$

Now you can apply this general formula to calculate the efficiency of the idealised 'Carnot' engine cycle.

What is the ratio of the heat input to the heat output, (Q_{out}/Q_{in})? How is it affected by the temperatures, T_H and T_C, at which those two heat transfers occur?

Well, imagine that the working fluid for your Carnot cycle were an ideal gas, obeying the equation: $pV = NkT$.

During the two isothermal processes:

▶ $\Delta U = 0$, i.e. the amount of thermal energy in the gas stays constant because the temperature is constant. So, applying the First Law equation, $\Delta U = Q - W$, you can see that: the heat transferred, $Q = W$, the work done, which equals the area under the p-V curve

▶ pV is constant for an ideal gas. So the process follows a p-V curve where p is proportional to $1/V$.

During the two adiabatic processes:

▶ $Q = 0$. So, again applying the First Law equation, you get: $\Delta U = -W$, which again equals the area under the p-V curve.

▶ For an ideal gas, **heat capacity** at constant volume, $C_V = \Delta U/\Delta T$, is a constant value because all the energy supplied goes into raising the average kinetic energy of the gas molecules, and thereby also raising the gas temperature. So that equation, combined with the ideal gas equation, fixes the shape of the p-V curve and makes it possible to calculate the volume changes in the adiabatic processes.

Doing the maths to sum up the areas under those curves for an ideal gas gives a simple relationship between the volumes at the four points of intersection between the four process curves. That, in turn, leads to a surprisingly straightforward relationship between the temperature of each isothermal in a reversible cycle and the amount of heat transferred reversibly in that process:

$$\left(\frac{Q_{in}}{Q_{out}}\right)_{rev} = \frac{T_H}{T_C}$$

This important result does three things:

1 It means you can rewrite the formula for **maximum theoretical efficiency** of a heat engine using absolute temperatures instead of amounts of heat:

$$\eta_{rev} = \frac{T_H - T_C}{T_C} = 1 - \frac{T_H}{T_C}$$

2 It gives a new way to define absolute temperature – by saying that thermodynamic temperature, T, varies in proportion to the amounts of heat transferred in an ideal engine cycle. That is a definition that does not depend on the existence of an ideal gas. The Carnot ideal cycle could equally well be performed with a real gas or with any other working fluid. In fact, the cycle of isothermal and adiabatic processes can also be performed with electrical or magnetic systems, rather than using a fluid of any kind.

3 It gives the idea that the ratio $(Q/T)_{rev}$ has a fundamental meaning – that heat transferred at a higher temperature is intrinsically more useful, and that this usefulness can be measured using a quantity derived from that ratio, which is called entropy.

PAUSE POINT 'Laws' apply generally and can be used in real situations. 'Ideals' are simplified ideas and equations that only almost work under specified conditions. Use examples from thermodynamics to help explain this difference between laws and ideals.

> **Hint**   Examples of ideals include: Gas equation of state; Adiabatic changes; Isothermal changes. Explain why each of these is never exactly true.

> **Extend**   Also explain why each of those 'ideals' is nevertheless a useful idea.

Heat engines, refrigerators and heat pumps

Real heat engine cycles

Engines are an important part of modern life and their efficiency is a matter of concern, firstly because fuels in general are a limited resource and secondly because the burning of fossil fuels releases carbon dioxide, CO_2, and hence contributes to global warming and climate change.

Steam turbine engines

The majority of electrical power worldwide is produced by generators that are turned by a steam turbine. Water is heated in a 'boiler' heat exchanger to produce steam, and the steam is then expanded through a turbine, with the low pressure steam being cooled and condensed back into water. Cooling the steam quickly in a heat exchanger reduces its pressure to below atmospheric, and the partial vacuum drawn by this cooling increases the pressure difference that drives the turbine. Because liquid water is nearly incompressible, only a small amount of energy is needed to pump it back up to the pressure of the boiler inlet.

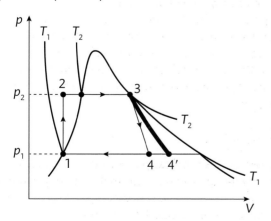

▶ **Figure 5.27** The Rankine cycle of a steam turbine engine shown on a water/steam *p-V* diagram

This practical engine cycle is named after its Scottish inventor, Rankine. The water (working fluid) operates in a closed cycle, as shown in Figure 5.27, with condensed

steam being returned as liquid water to the boiler. But the heat rejected from the condensing steam and transferred in the condenser heat exchanger to a supply of cooling river or sea water, makes that cooling water hot. It is cooling that down again so that it is ready to be returned to the river or sea which causes the clouds of steam and water droplets that are seen escaping from massive cooling towers.

The high specific latent heat of vaporisation of water, $2.26\,MJ\,kg^{-1}$, has the advantage that large quantities of heat energy are transferred by moving only a relatively small mass of water, keeping machinery sizes compact. But it also limits the temperature at which most of the heat is input, which in turn limits the thermal efficiency obtainable – typically up to only about 42% efficiency. Even this figure is only achieved when the steam continues to be heated after it has boiled, which is called 'superheating'. This makes the superheated steam 'dry' (i.e. there is no trace of liquid water) and at a higher temperature than the boiling point.

Nuclear power stations use saturated steam with no superheating (straight from the boiler and 'wet' because it is in contact with liquid water). So they have even lower thermal efficiencies of around only 30%. But efficiency is less of a concern for them than safety of containment of radioactive hazards. They are not emitting CO_2 into the atmosphere to add to global warming, and the supply of heat from their fuel is plentiful.

To keep the heat exchange processes efficient – i.e. near reversible – the size of the heat exchangers used with steam engines needs to be very large indeed, so steam turbine systems are only practical for stationary engines or for powering large ships; they are much too large and heavy for use in road vehicles.

Internal combustion engines – petrol or gas, diesel and gas turbine

The working fluid in these engines is mostly air. Air is sucked or pumped in and then compressed, with some fuel added. Combustion inside the engine heats the gas very rapidly indeed, and then that mixture of hot air and combustion products does work as it expands. Finally, the gases are exhausted out to the atmosphere (see Figure 5.28).

Piston engines, used in motor vehicles, are small and lightweight, but they are mechanically complicated, creating friction losses, and it is difficult to achieve fully effective combustion in such a short space of time. On the other hand, the high temperatures at which the heat is supplied give them the potential for higher thermal efficiencies. The reason diesel engines give the higher efficiencies is because the air can be more highly compressed, giving a higher initial temperature, at which point fuel is injected and ignites immediately. Petrol is much more flammable, so compression ratios in petrol engines have to be kept fairly low to avoid problems with 'pre-ignition'.

Higher thermal efficiency makes diesel engines burn less fuel and produce less CO_2 than petrol engines. But incomplete combustion in diesel engines creates other problems of air pollution in cities.

Liquefied natural gas is a tricky fuel to handle in vehicles because of the high pressure containment needed, but it can combine clean complete combustion with fairly high thermal efficiency.

Gas turbine engines also use internal combustion (i.e. burning the fuel inside the engine), but the ignition and the hot gas expansion happen in a turbine, instead of a piston engine. Turbines are mechanically efficient, directly producing the rotational motion needed for electricity generation, and, with exhaust gas recycling included, they can become the most efficient and clean burning engines of all. But to achieve this, they must have large heat exchangers and they need to be running continuously at close to full

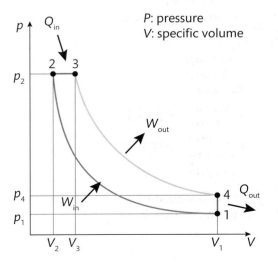

▶ **Figure 5.28** The Diesel cycle, shown on an air pressure-volume diagram

power. So they are widely used in power stations that provide the main electricity supply load – i.e. the base level of power demand that is used all day and all night long. Other types of power station, which can be switched off and on easily or run on reduced load, are needed to provide for the surges in electricity demand at peak load times like early mornings and evenings, when everyone wants to turn lights on, cook and heat their homes.

Muscle power

The chemical processes operating in biological systems are complex, but it is still possible to investigate the thermal efficiency of muscle tissues, measuring their energy conversion into useful work. Results show that, once again, their efficiency depends strongly on the temperature at which a muscle is working. Of course, the temperature differences are very small compared with the numerical size of the absolute temperature, so the maximum theoretical efficiency of an ideal Carnot engine under those conditions would be very low, and the actual practical efficiency values are even smaller.

Being warm-blooded makes a huge difference to the power that an animal's muscles can deliver. Equally, taking away the heat that is generated by 'burning' foods like sugars and fats inside body cells is also essential to keeping a body working efficiently. That's why humans perspire and dogs pant. Precisely because living organisms are so diverse and complex, simplifying a subject of interest and looking at it as a 'system', with inputs and outputs of energy and materials, often a powerful approach in biology.

Reverse cycles

You have been thinking a lot about idealised reversible cycles. So, what happens if you do reverse an engine cycle and run the machinery and the heat transfer processes in the opposite direction?

The reverse carnot cycle
1–2 Adiabatic compression raises the working fluid to a higher pressure and the temperature, T_H.
2–3 Isothermal compression then progressively rejects heat, Q_{out}, to the surroundings at that high temperature.
3–4 Adiabatic expansion next allows the still hot but compressed fluid to reduce pressure and cool rapidly to a new lower temperature, T_C.
4–1 Isothermal expansion finally has the fluid absorbing heat, Q_{in}, at the low temperature and gradually expanding until it is in its original condition and ready to start the cycle again.

This is a machine for moving heat from a colder body to a hotter body, as shown in Figure 5.29, but it requires an input of work, W_{in}, to perform the adiabatic compression, which is larger than the amount of work, W_{out}, that could be recovered in the adiabatic expansion process. The net work done in the cycle is $W = W_{in} - W_{out}$.

Everything is back to front compared with the equations for the normal engine cycle. The equation for W is reversed because you are now focussing on the work done *on* the system, i.e. an input of work that would have a negative sign.

ΔU, for the whole cycle must be zero, so using the First Law equation: $\Delta U = Q - W$, simply gives $Q = W$.

This means that Q for the cycle must also be negative – i.e. there is a net quantity of heat given out. In other words Q_{out} is bigger than Q_{in}.

Q_{in} is now associated with the low temperature, T_C, and Q_{out} with the high temperature, T_H. There is more heat given out at the high temperature than is absorbed at the low temperature.

Vapour compression refrigerators and heat pumps

The main components of a refrigerator or heat pump are shown in Figure 5.30. For a real cycle to move heat from cold to hot, you would usually choose a liquid and condensable vapour system as the working fluid, rather than just a gas. That is because latent heat of vaporisation is helpful to the cost-effectiveness of these reverse cycle machines in several ways.

▶ Condensation and evaporation in the two heat exchangers are naturally isothermal processes, and hence closer to reversible, because the latent heat stops the temperature rising or falling during those changes of state.

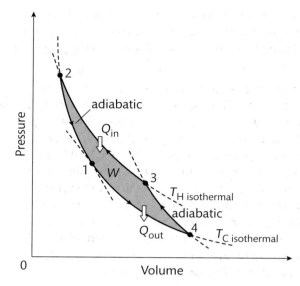

▶ **Figure 5.29** A reverse Carnot cycle removes heat from a low temperature reservoir, has an overall input of work equal to the area enclosed by the cycle, and rejects all that energy as heat into the higher temperature reservoir.

▶ That in turn keeps the temperature difference between the two heat exchange processes as small as possible. While a small temperature difference limits the efficiency of an engine, it maximises the coefficient of performance (CoP) of a refrigerator or heat pump – see below.

▶ A high specific latent heat of vaporisation minimises the quantity of fluid that has to be moved round the system, and so keeps the size of the compressor equipment smaller and its price lower.

▶ Liquids are nearly incompressible – i.e. they have virtually no volume change when the pressure on them

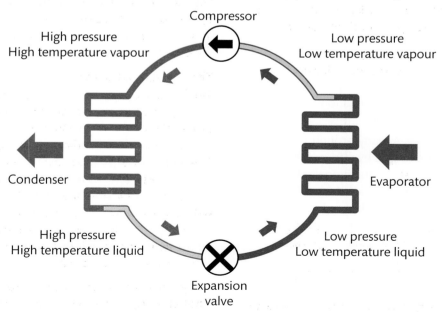

▶ **Figure 5.30** The main components of a refrigerator or a heat pump

is dropped. So the work done, $W = p \Delta V$, is very small indeed. This means it is not necessary to manufacture and use a turbine to recover that work. Just allowing liquid to drop at pressure through a simple valve, or in small domestic systems, using only a length of capillary tube, comes close to achieving the adiabatic expansion part of the cycle with only a very small amount of

potential work output being wasted. At the new lower pressure, the liquid immediately begins to boil and in doing so begins to absorb latent heat. But by that time it is already in the low temperature heat exchanger – known as the evaporator – undertaking the isothermal expansion part of the cycle.

 PAUSE POINT   Check your understanding of the systems - engines, heat pumps and refrigerators, and even muscles – that you have just studied. They all operate a cycle of processes. Explain why that is the best way to think about them.

   Think about the working fluid in each, where it comes from and where it goes to.

Heat pumps produce more heat energy output than the electrical energy input they use. Explain why that doesn't violate either the First Law or the Second Law of Thermodynamics.

Maximum theoretical coefficient of performance (CoP)

In the response to global warming and climate change, heat pumps and refrigerators are becoming increasingly important. Heat pumps can help us save energy on heating bills. But, on the other hand, refrigerators, freezers and air conditioning need a lot of electric power to keep them running. You therefore need to know how to rate these systems' performance and hence choose the most effective solution.

The term 'coefficient of performance' is used, rather than 'efficiency' because the value is generally greater than 1. These machines move more thermal energy than the amount of energy (usually electrical) that they consume doing the work of compression. That is why heat pumps are so popular – they can be big energy savers.

$$\text{Coefficient of performance} = \frac{\text{Heat usefully moved}}{\text{Work input required}}$$

CoP for a heat pump

In a heat pump, the useful heat is the amount, Q_{out}, rejected at the high temperature, T_H. So:

$$\text{CoP (heat pump)} = \frac{Q_{out}}{W} = \frac{Q_{out}}{Q_{out} - Q_{in}}$$

And so the maximum theoretical CoP for an ideal Carnot cycle heat pump would be:

$$\text{CoP (heat pump)}_{max} = \frac{T_H}{T_H - T_C}$$

This is the exact inverse of the maximum theoretical efficiency of a heat engine.

Exam tip

So, in an exam you can find the maximum theoretical efficiency equation on the formula sheet, turn the formula upside down and use it to answer heat pump CoP questions.

If the temperature difference $(T_H - T_C)$ is kept small, the CoP can become quite a large number, meaning a large multiplication factor for the heat delivered compared with the electrical energy consumed – so it can be a big energy saver.

But real heat pumps are not fully reversible; their compressors have frictional and viscous losses, their heat exchangers are not really isothermal but have temperature differentials to drive the heat flow, and the work done by the expanding liquid as it passes through the valve or capillary tube is not recovered. So the CoP of a real heat pump will always be lower than the theoretical maximum.

CoP for a refrigerator

In a heat pump, the useful heat is the amount, Q_{in}, absorbed at the high temperature, T_C. So:

$$\text{CoP (refrigerator)} = \frac{Q_{in}}{W} = \frac{Q_{in}}{Q_{out} - Q_{in}}$$

And so the maximum theoretical CoP for an ideal Carnot cycle heat pump would be:

$$\text{CoP (refrigerator)}_{max} = \frac{T_C}{T_H - T_C}$$

By comparing the two equations for CoP, you can see that:

$$\text{CoP(refrigerator)} = \text{CoP(heat pump)} - 1$$

Exam tip

Again, this makes it easy to answer CoP questions. For refrigerators just work out the heat pump CoP and then subtract 1.

So, for example, a chilled food cabinet refrigeration system in a super market, operating with a CoP value of 2.5, if also adapted to deliver warm air into the shopping space, could be viewed as a heat pump with a CoP value of 3.5.

Maximising CoP

Of the various departures from an ideal cycle in a real system, the easiest to tackle is the size of the heat differentials in the heat exchangers. Simply by using an oversized condenser and an oversized evaporator, the temperature differences can be kept low between the refrigerant (i.e. the working fluid) and the heat exchange fluid (usually either air or water) in those heat exchangers – to just one or two kelvin for example. Another way of keeping temperature differentials low is better circulation of the heat exchange fluid. But care has to be taken not to expend too much energy driving the circulation of the air or water, as that adds to the cost of running the plant and so will bring the overall effective CoP down.

Keeping to a minimum the temperature difference that the heat pump or refrigerator has to work across is the key to increasing the maximum theoretical CoP. So, if you are installing a heat pump system, you should look for a low temperature heat source that will give you the highest possible working temperature through the periods when your heat pump will do most of its work. Similarly, a refrigeration system needs a reliable coolant for its condenser during the hottest times of the year.

For that reason, simple air-to-refrigerant heat exchangers are often not the best option. Air temperature rises in summer, when refrigerators have to do most of their work. And in winter, when heat pumps need to be most effective, the outside air temperature can fall very low. Water therefore often makes a better heat exchange fluid. Water towers are often used to provide cool water to the condenser of an air-conditioning or large refrigeration plant, even during hot summer days. Some of the water is allowed to evaporate as it falls through an air stream in the cooling tower, and the loss of latent heat of evaporation cools it before it is fed into the water-cooled condenser of the refrigeration system.

Underground coils carrying water can provide a steady heat reservoir at around 8°C all the year round. This is good for a heat pump as a heat source during the winter months, when outside air can be much colder than that. But it can also be good for cooling the condenser of an air-conditioning system during the summer. With appropriate design and the use of valves, an air-conditioning system can be reversed to become a heat pump, providing cost effective winter heating to the same building.

If a refrigerator has to use an air-cooled condenser, e.g. for a domestic fridge or freezer, then the siting of the equipment is important. Too often they are placed very close to a wall and under a work-top, where there is limited natural air circulation. The ambient air temperature around the condenser can then become much higher than desirable for efficient operation. The consequent raised temperature and pressure of refrigerant inside the condenser can lead to cut-out or complete failure of the equipment on a hot summer day.

Assessment practice 5.11

Figure 5.31 shows the results of an experiment in which a 1.5 kW kettle was filled with 1.2 kg of water, placed on a digital balance, and a temperature probe monitored the water temperature from the point at which the kettle was switched on.

a) Explain why there is a change in both graphs after 5.5 minutes.

b) Calculate from the data on the graphs a value for the specific latent heat of vaporisation of water

c) In a second experiment, where only 0.6 kg of water was put into the same kettle at the same starting temperature, the change point in the two graphs occurred after only 3.2 min. Use these two results to calculate a value for the specific heat capacity of water.

d) Suggest why both these results may be a little lower than the accepted literature values.

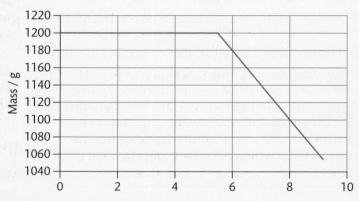

▶ **Figure 5.31** Heating water in a jug kettle

Assessment practice 5.12

A power station steam turbine which drives a 150 MW generator is fed by a boiler that burns fuel to give a heat input of 998 kJ every hour producing superheated steam at 565°C. The low pressure steam output from the turbine is condensed at 30°C. The generator is 99% efficient at converting mechanical work into electrical energy.

a) Calculate the maximum theoretical thermal efficiency of an ideal engine working between these temperatures.

b) Calculate the actual thermal efficiency of the steam turbine and explain what factors make it less than ideal.

c) Calculate the quantity of heat that the condenser unit must remove each hour and the mass of condensate that will be produced in the same period. (The latent heat of vaporisation of water is 2.26 MJ kg^{-1}.)

Assessment practice 5.13

A supermarket operates chilled food cabinets with an evaporator temperature of 2°C and a water-cooled condenser at 15°C.

a) Calculate the maximum theoretical coefficient of performance (CoP) for a refrigerator operating between these temperatures.

b) A suggestion has been made that the condenser heat output could be used to provide heating to the shopping space, using a heat exchanger operating at 30°C. Calculate the impact of doing this on the refrigeration CoP.

c) Explain how, alternatively, a two-stage system, with a heat pump taking its heat input from the condenser of the refrigeration system, might offer more control and flexibility across the seasons.

Materials in domestic and industrial applications

'Materials Science' is about understanding the properties of the *solid* materials you use to construct things, so that you can make them cost-effective, safe and reliable.

Elasticity and Hooke's Law

Solids have a shape and size that do not change unless a sufficient force is applied. Elastic behaviour is when a solid material is able to regain its original dimensions after the applied force is removed. If a shape or size change becomes permanent it is called a **plastic deformation**.

Solid materials are generally elastic – i.e. they can bounce back to shape – up to a certain limit, known as their **elastic limit**. After that point the material will either fail (break) or it will undergo plastic shape change.

Tensile experiments – i.e. testing materials by stretching them – were carried out originally by Sir Robert Hooke, and have been performed regularly since then by materials scientists. These experiments show that, for metals and many other solids over much of their elastic region, the relationship between applied force, F, and extension (increase in length), Δx, is a linear one. That is:

$$F = k\,\Delta x$$

The constant of proportionality, k, depends not only on the material used, but also on the original dimensions of the sample tested. So, while you may choose to use the equation in that form for a specific object – e.g. a spring of a certain size and specification, and to measure k, the object's spring constant; in general it is much more useful to write the law in terms of tensile stress and strain:

$$\tau = E\,\sigma$$

where:

▶ tensile stress, τ, is defined as: (force applied)/(cross-sectional area of the sample) $\tau = F/A$

▶ tensile strain, σ, is defined as: (extension)/(original length of the sample) $\sigma = \Delta x/L$

▶ and the constant of proportionality, E, is called the elastic modulus, or commonly Young's modulus.

Young's modulus is a measurable property of a given material, with a reliable value that can be used in calculations to predict the actual **strength** of components manufactured from it.

The relationships above are known as Hooke's Law. This is not a general law of physics, but is a very useful rule that is obeyed closely by many solid materials for a limited range of applied stresses – up to what is called their limit of proportionality.

For many materials, the limit of proportionality and the elastic limit may be quite close together, but they are not identical – the physically determined elastic limit and the mathematically derived limit of proportionality have different meanings and in general they will have different values.

> **Key terms**
>
> **Plastic deformation** – occurs under stress levels that are sufficient to make the solid material begin to flow, rather like a liquid. When the stress is removed, a change in an object's shape and size remains. This is called a **permanent set**.
>
> **Elastic limit** – the point on the stress-strain curve, beyond which a material begins to suffer plastic deformation, and so will not completely regain its original dimensions when the stress is removed.
>
> **Strength** – also called **ultimate tensile stress**, is the maximum stress that the material can bear. This occurs just before the material fails and fractures.

Stress-strain curves

Materials science involves a practical, experimental study of materials. Each material that has a slightly different chemical composition or manufacturing history – for example, a heat treatment or a shaping process – needs to have its tensile properties tested and recorded, so that its behaviour under load can be predicted and used in the design process. The results are most often presented graphically, as a stress-strain curve, and so you should become familiar with the shapes of these and how to interpret them.

Every stress-strain curve tells its own story about the internal structure of the material, including how tiny imperfections can grow or move under the influence of tensile stresses. These changes can lead to a hardening of the material, called work hardening, so that plastic deformation becomes much harder to produce. Their growth can lead to sudden brittle failure as cracks spread right across the width of the sample under test.

Ductile materials

The structure of solid metals is made up of tiny crystals – they are microcrystalline. The regular arrays of identical atoms make it relatively easy for layers of those atoms to slip across one another and find a new rest position that still keeps the crystal structure intact. This means that most metals show some **ductility** – i.e. under tensile stress they can be drawn out into a new shape, which they then take on permanently. Ductile materials can be drawn out into rods, wires or tubes, which makes them ideal for manufacturing.

Metals can generally also be formed by processes that use compressive force – hammering, rolling, stamping with a die. This is described by the term **malleability**. While all ductile materials are generally also malleable, there are some materials – e.g. lead – that are very malleable but which fail when put under quite small tensile stresses.

A lack of ductility or malleability is called **brittleness** – an example is glass.

> **Key terms**
>
> **Ductility** – the ability of a material to be formed by drawing into new shapes, primarily by means of tensile forces. Ductile materials are generally also malleable.
>
> **Malleability** – the ability to be shaped by means of compressive forces such as occur in rolling, hammering or stamping. Not all malleable materials are also ductile.
>
> **Brittleness** – the tendency of a material to fracture under stress. Brittle materials cannot undergo plastic flow and so are not suitable for manufacturing processes such as drawing, rolling, hammering or stamping.

Stress-strain curves for ductile materials have the basic characteristics shown in Figure 5.32. The stress values and the extents of strain vary markedly between different materials.

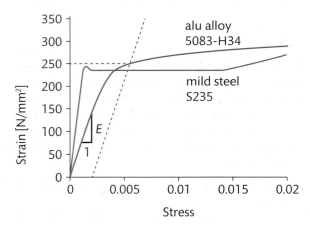

▶ **Figure 5.32** Stress-strain curves for an aluminium alloy and a mild steel

Creating alloys – mixtures of metals of varying compositions – is an important way of obtaining materials with properties suitable for specific applications. Aluminium and iron, perhaps the most widely used constructional metals, are almost never used in a pure form, but always alloyed with other elements. In iron (and steel, which is a

class of iron alloys), the proportion of carbon included is especially important to both its strength and its ductility.

Iron and steel, along with just a few other metals, have a particular crystal structure that allows slippage to occur along planes in several different directions. The special ductility which this gives them is demonstrated on the stress-strain curve by a marked drop in stress as the **yield point** is passed. This important property links with iron and steel exhibiting a degree of protection against **fatigue** failure (see below) that other metals like aluminium do not share.

> **Key terms**
>
> **Yield point** – the point where the start of plastic flow causes a change of slope on the stress-strain curve. Iron and steel and a few other metals show a clearly defined yield with a drop of stress, while in many other materials the exact position of the yield point is hard to spot.
>
> **Fatigue** – the embrittlement and failure of a material that can occur with relatively low levels of stress if these are repeatedly applied and then relaxed over many cycles.

Creep and fatigue

In a simple tensile test, the stress is gradually increased and the strain is recorded, generating the familiar stress-strain curve. It is a process that takes just a few minutes and is done once. But when materials are used in practical situations, the stress pattern is not always like that.

Typically, a component may have to carry a sustained stress for a long period of time. Design engineers use the yield stress as an indication of the strength of a material in tension. But a component that carries a stress well below its yield strength for a long period may begin to show a very slow deformation, which is called **creep**. This phenomenon becomes much more prevalent when components are working at a high temperature, even though still below their melting point. Random thermal energy of the atoms' microscopic vibrations combines with the applied stress to cause slippage and results in a plastic deformation.

> **Key term**
>
> **Creep** – sometimes called 'cold flow', occurs when a material under stress deforms gradually over time. It is more severe in materials that are subjected to heat for long periods.

Alternatively, some components have to withstand regular cycles of stresses increasing and then relaxing, only to increase again. Any part that is subject to mechanical vibrations will be affected in this way. Once again, the maximum stress allowed for in the component's design, including the extra stress due to the vibrations, may be well below the yield strength of the material. Nevertheless, these components often suffer catastrophic breakages known as fatigue failure.

The mechanism of fatigue at a microscopic level is a gradual growth of initially very small cracks. If the structure of the material were completely uniform, then the stress distribution across a component would also be smooth and even. But in reality the tiny crystals making up a metal object are of all sorts of shapes and sizes, and lie in random orientations. So the boundaries between them contain many irregularities. Impurities in a material add to the lack of regularity, as do tiny cracks that can occur due to contraction during rapid cooling and solidification. These imperfections mean that the internal distribution of stress is far from even, and stress concentrations occur, particularly around the tips of cracks. These potentially strength-limiting cracks can be seen if a component is examined using X-rays. The sharper the crack, the greater the stress concentration.

Under a cycling stress pattern, cracks grow progressively, and as they do the stress concentrations around them also increase. Finally, a critical point is reached where the concentrated stress is sufficient to cause a sudden brittle failure. In other words, a material that was originally strong and ductile has changed its character because of the long-term cyclic stress pattern that it has endured.

Case study

An example of where creep can cause a critical failure is in the vanes of a gas turbine engine. The high centrifugal forces and extremely high temperatures in parts of the engine could result in the blades gradually extending with creep until eventually they touch the engine casing at high velocity and rupture. But, before that happens, multiple creep-induced slippages along grain boundaries tend to form tiny voids, which initiate cracks. Cracks, once started, grow until failure occurs. The mechanism of the crack growth could involve a combination of further creep with some fatigue due to stress cycling as engine turbines speed up and slow down.

Check your understanding

1 Why are there high centrifugal forces in a gas turbine engine?
2 Why do creep-induced slippages happen along grain boundaries?

For most metals, if the cycling continues long enough, even quite low stress levels compared with the ultimate tensile strength will still eventually cause fatigue failure. So fatigue testing is carried out, where a component is deliberately tested under a cycling stress to determine a safe lifetime after which it must be replaced. This is an important part of the engineering design process, especially in critical applications like aeronautics.

Iron, steel and titanium alloys have the unusual property of having a well-defined fatigue limit – a stress level below which they can endure any number of cycles without developing fatigue failure. That means that a well-engineered component made in these materials can have an unlimited safe working life.

Theory into practice

In some cases ductility and malleability are wanted – mostly to aid in manufacturing items to particular shapes. For instance, metal drink cans are deep drawn from very ductile flat sheets of aluminium alloy. But in many other cases, resistance to deformation – hardness – is what is wanted – e.g. in the blades of knives and the tips of tools. The hardness of a metal can be increased by working it – stretching, hammering etc. – or by putting it through a heat treatment with rapid heating and cooling. Conversely, hardened objects can be softened again by slower cycles of heat treatment.

Reflect

Diamond is one of the hardest of materials, so one practical test for hardness uses a diamond cone and presses it into the surface of a test piece of material to see what size of impression it will make with a given force. Hardening increases the ultimate tensile strength of a metal, but it also makes it more liable to brittle failure.

In what laboratory applications might you want a hard material?

Elastomers

Objects made from rubber or synthetic polymers with rubber-like properties are able to exhibit large amounts of strain for a relatively small level of stress, but they still have the elastic property of springing back to their original dimensions. The elastic properties of metals are basically due to the strong chemical bonds between adjacent atoms being stretched under the applied stress, but having the ability to pull back afterwards. But the stretchiness of rubber is due to the very long molecules from which it is made. Under an applied stress, these can be stretched out nearly straight; but when the stress is removed, thermal motion quickly curls them back up into much shorter loops and knots.

If stress is applied gradually, and the material is warm, the same thermal motion can lead to the unravelling of some of the knots, so that a larger extension occurs. Afterwards, when the applied stress is removed, there may be a degree of permanent set. But if time is allowed for recovery, the random motion of the molecules gradually shortens and knots them up again, so recovering all or most of the original shape.

So, as a result, the stress-strain relationship for elastomer materials is heavily dependent on the rate at which the strain is induced. Moreover, the stress-strain curve for a reducing stress level follows a different path from that for a rising stress. This is called **elastic hysteresis**, and the shape traced out on the stress-strain graph is called a hysteresis loop (see Figure 5.33). The area enclosed by the hysteresis loop represents an amount of work done in stretching the material which is converted into thermal energy, and so is not recoverable as work in the relaxation phase of the cycle.

> **Key term**
>
> **Elastic hysteresis** – occurs in materials like rubber, where internal friction between large molecules dissipates energy producing heat. Loading and unloading of a sample each produces a different stress-strain curve, creating a hysteresis loop, the area of which represents the energy absorbed in the cycle.

The ability of rubber and materials like it to absorb mechanical energy and transform it into heat is very useful, and is why rubber is widely used for absorbing mechanical shocks and for damping vibrations.

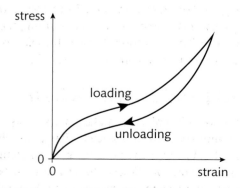

▶ **Figure 5.33** Stress-strain for rubber, showing a hysteresis loop. The loading and unloading curves are different.

Brittle materials

The majority of non-metallic solids are brittle in nature, the main exception being some of the long-chain polymer solids known as 'plastics'. All solids, including metals, become brittle if their temperature is lowered sufficiently to prevent any plastic flow occurring.

Brittle materials still have an elastic region in which applied stress produces a small strain, from which the material springs back when the stress stops. But they tend to be very stiff – i.e. they have large values for Young's modulus – and so stress rises quite sharply with only a small amount of strain (see Figure 5.34).

The microscopic structure of these materials doesn't allow for any plastic flow. Some of them are glassy, meaning that their atoms are rather randomly packed together with no long-range ordering – no crystalline regularity. Others, like concrete, may be composites made up of hard, strong particles held together by a binding medium. Composites that contain fibres, like wood, carbon fibre or fibreglass, may have a good tensile strength, but those like concrete that contain just granular material tend to be strong in compression, but very weak in tension. Concrete structures must either hold together by their weight, causing constant compression, or be reinforced with steel to take any possible tension loads.

The lack of any capacity for plastic deformation means that, as stress rises, the next event is a brittle failure. The stress concentration around the tips of small internal cracks or surface scratches in the material reaches a critical level, and the cracks propagate rapidly right across the structure, causing a sudden failure.

Some brittle materials can be very strong, but the lack of any yield point to give a warning of impending failure

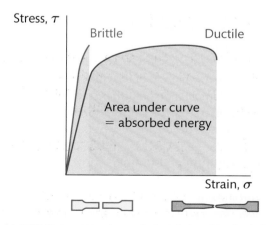

▶ **Figure 5.34** Comparing stress-strain relationships for brittle and ductile materials

means that loading on them needs to be carefully calculated and controlled. The high strength of glass in fibreglass stems from the fact that the glass fibres used are coated with a polymer resin immediately after manufacture and before they develop any surface scratches. It is those scratches that quickly weaken normal unprotected glass.

⏸ PAUSE POINT   Close the book and try writing down what you now know about measuring and classifying the properties of materials.

> **Hint**   Think about materials you use at work and at home. Which need to be strong, hard, flexible, resilient . . . ? Which properties are particularly important for each?

> **Extend**   Look up in reference books or on the internet some tables of properties and some stress-strain curves for metals and non-metals. Which are the strongest, the stiffest, the most ductile?

Assessment practice 5.14

a) Sketch the tensile stress-strain relationships for a brittle material and for a ductile material, marking and labelling key features.

b) Diamond and steel are both strong materials. Referring to your sketches, explain why they are suitable for use in very different applications.

Assessment practice 5.15

a) State the meanings of yield strength and ultimate tensile strength.

b) Explain why engineers designing metal structures usually focus on the yield strength.

c) Creep and fatigue can both lead to failure at stress levels well below the yield strength. Describe under what conditions each of these may occur.

Fluids in motion

What makes fluids (i.e. liquids and gases) useful in systems is their ability to move – i.e. to flow, and in doing so to transmit pressure, to transfer heat or to simply deliver quantities of substance to a new location.

Fluid flow patterns – streamline and turbulent

Fluid flow fundamentally involves layers of molecules sliding over one another. For slow, gentle flow, motion occurs just in the direction of the applied stress, and is called streamline or laminar. This ordered motion does absorb some energy because of the viscous drag – a kind of internal friction – between the layers, but it is the most energy efficient kind of flow.

In streamline flow, the lines of flow are all parallel, and they can easily be made visible by introducing a dye, or by watching the motion of particles carried along by the liquid. Where the liquid is in contact with a solid surface, the speed of that layer is very close to that of the solid – stationary if it is the bank of a river, or moving if it is the hull of a boat or one of the rotating surfaces of an engine bearing. As you move through the liquid, across the direction of flow, the speeds of the layers progressively change. So, for example, in a river the fastest flow streams are found in the middle, well away from the banks. This means there is a velocity gradient perpendicular to the streams of flow.

Turbulent flow begins to appear as the velocity gradient is increased and the viscous drag forces also grow. The drag forces have a natural tendency to rotate parts of the fluid, and any obstacle or sharp corner that interrupts the smooth flow can set off turbulence, which will then spread to neighbouring parts. The rotational motion of turbulent flow absorbs much more energy, and so the effective drag force also becomes much larger. For that reason, except where stirring and mixing is needed, you will normally want to avoid turbulence as much as possible and to design for streamline flow.

Viscosity and Newton's Law

Dependent on their molecular make-up, some liquids offer much more viscous resistance to flow than others. So it is useful to have a measure of viscosity to compare them. This behaviour was first studied by Sir Isaac Newton, who defined a coefficient of dynamic viscosity, η, which is fundamental to exploring fluid flow.

Because the sliding of the layers of liquid over one another generates a resistance force, the kind of stress involved is a shear stress, equal to the size of that viscous drag force divided by the area of contact between the sliding layers. The 'strain' in this dynamic process is not a change in length or a distance moved, but rather the rate of shear strain, which is another name for the velocity gradient across the streams of flow.

Using those concepts, Newton's definition of viscosity is:

$$\text{coefficient of dynamic viscosity, } \eta = \frac{\text{shear stress, } \tau}{\text{rate of shear strain}}$$

$$\tau = \frac{F}{A} = \frac{\eta \, \Delta u}{\Delta y}$$

where:

▸ F is the viscous drag force generated across a layer of fluid of area A

▸ u is fluid velocity (in direction x) and y is distance perpendicular to the flow, so that $\Delta u/\Delta y$ is the velocity gradient or rate of shear strain (see Figure 5.35).

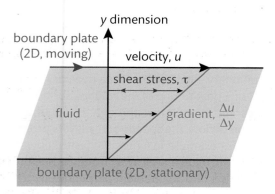

▸ **Figure 5.35** Defining viscosity

For the majority of simple fluids, the dynamic viscosity has a constant value at any given temperature. These liquids or gases are said to be Newtonian fluids, or to 'obey Newton's law of fluid flow'. As with Hooke's law for solids, Newton's law of fluid flow is not a general law of physics, but a useful rule obeyed by many fluids under streamline flow conditions.

An increase in temperature will generally lower the viscosity of a liquid, as thermal energy assists molecules in sliding over one another. But for gases, extra thermal energy increases the rate of collisions between the molecules, so gas viscosities increase when there is a rise temperature. You might notice this if you play badminton on a warm day.

Theory into practice

Terminal velocity: When an object falls under gravity through a viscous fluid – liquid or gas – it will reach a top speed, known as its terminal velocity. George Stokes studied this for small spherical objects falling relatively slowly so that the fluid flow around them was streamlined and he developed an equation, known as Stokes' Law, for the terminal velocity, v:

$$v = 2g \, r^2 \, (\rho_1 - \rho_2)/9 \, \eta$$

where g is the acceleration due to gravity, r is the sphere radius, ρ_1 and ρ_2 are the densities of the sphere and of the fluid medium, and η is the viscosity of the fluid.

This gives a simple practical method for measuring and comparing viscosity – simply dropping steel balls of known diameter into a long tube containing a fluid and measuring the terminal velocity. The equation is also useful when small droplets of liquid are falling through a gas (e.g. rain) or when small bubbles rise through a liquid.

Where more complex shapes and/or higher velocities result in turbulent flow – e.g. around a car, will the viscous drag be higher or lower than for streamline flow? What would this mean for the terminal velocity?

Investigation 5.2

Why not use Stokes Law and the steel ball-bearing method to devise your own investigation into how viscosities vary – e.g. a) between different kinds of oil; or b) as the temperature of a liquid is varied. For b) you would need to think about how you could ensure all the liquid was at the right temperature during each measurement.

Non-Newtonian fluid flow

Many rather more complex fluids have viscosities that change, either immediately or gradually, when they are sheared, e.g. by stirring or shaking. So these are called non-Newtonian. They are generally composed of large molecules in solution or of small solid particles or immiscible droplets of liquid or even gas bubbles in a liquid suspension. The interactions between the carrier liquid and what it contains result in the unusual rheological (i.e. viscous flow) properties of the combination fluid, and these can have various useful applications.

Non-Newtonian fluids may be classified into a number of types, as shown in Table 5.12.

▶ **Table 5.12** Non-Newtonian fluids

Fluid	Description	Examples
Shear thickening or 'dilatant' liquids	*Viscosity increases* immediately with a rise in the rate of shear strain	• Corn flour mixed with a small amount of water • Damp sand that firms when you walk on it
Shear thinning or 'pseudoplastic' liquids	*Viscosity decreases* immediately with a rise in the rate of shear strain	• Nail varnish • Most modern wall paints that brush out and then immediately set before they dry • Whipped cream • Quicksand – i.e. sand containing a higher proportion of water, which gives way and continues to thin as a victim thrashes around • Blood, which flows effectively under pressure through the vessels, but immediately thickens if it seeps out (a process that is followed by full coagulation, aided by platelet changes and clotting factor)
Rheopectic liquids	Gradually become *more viscous over time* when shaken, agitated, or otherwise stressed	• Synovial fluid, which provides both lubrication and shock protection in body joints by increasing its viscosity as it is being stressed • Some printer inks
Thixotropic liquids	Gradually become less viscous over time when shaken, agitated, or otherwise stressed	• Yoghurt • Ketchup that contains xanthan gum • Gelatin or pectin gels • 'One coat non-drip' paints that require a light brushing action and set fairly quickly, but not immediately, to give a single smooth, thick coating
Bingham plastics	Behave as a solid at low stresses but flow as a viscous fluid at higher stresses	• Toothpaste and similar pastes that extrude from a tube as a plug • Mayonnaise • Mud and slurries, which are important in drilling technology

Rate of fluid flow and pressure

A useful result for streamline flow of a Newtonian liquid down a pipe is Poiseuille's equation:

$$\frac{\Delta V}{\Delta t} = \frac{\pi r^4}{8\eta} \frac{\Delta p}{\Delta L}$$

That is, volume flow rate $\Delta V/\Delta t$ is directly proportional to the pressure gradient across the length of the pipe, $\Delta p/\Delta L$. The flow rate is very heavily dependent on the pipe internal radius, r, varying with the fourth power of r.

Exam tip

You don't need to learn this equation, but you should remember the meanings of 'flow rate', 'pressure drop' and 'pressure gradient', and have a practical grasp of how they work together.

If the flow becomes turbulent, the same equation can be used, but the viscosity, η, appears to increase. This is not a real change in viscosity, but just the effect of the extra energy absorbed because of the turbulence.

Gases are compressible, so the volume of a gas entering the pipe will be smaller than the volume on exit, where the measurement of volume flow rate is usually made. Nevertheless, the same formula for flow rate can be applied for gases if a correction factor is applied, multiplying it by (average pressure/outlet pressure).

So this formula can have wide application – including in medical fields such as the study of blood flow rates and air flow in lung alveoli, or applied to flow through a hypodermic needle.

Mass flow rate continuity

A simple but important observation for piped flow systems is that 'what goes in must come out'. This is true for mass flow rate across any pair of boundaries when a system is in steady flow. It can help in analysing what happens at pipe junctions where there are alternative flow paths, and it means that pressure drops and flow rates can be calculated for a complex network of pipes with different flow resistances using similar rules to those used for voltage, current and resistance in electrical circuits.

Bernoulli's principle

Another important result comes from applying the principle of conservation of energy to fluid flow. It is called Bernoulli's principle, and can be stated as follows:

> At any point in a tube through which fluid is flowing: the sum of the pressure energy, the potential energy and the kinetic energy is constant.

In fact, Bernoulli's principle also applies to any smoothly flowing fluid, whether in a tube or not, and it has some very important applications.

Aerofoil lift

Aircraft wings, helicopter blades and sails for yachts are all designed with a curved profile that splits the airflow around them and creates two streams of flow (see Figure 5.36) with different lengths and hence different speeds. The faster air, travelling the longer distance, cannot exert as much pressure as the slower air-stream. So there is a net pressure difference across the aerofoil that gives it a 'lift' force. (Note that the 'lift' created by a curved sail on a sailing boat is a horizontal force, pulling the boat forward.)

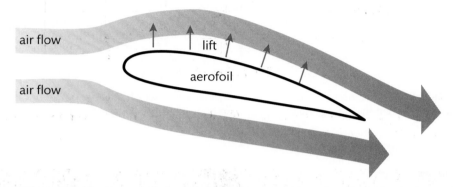

▶ **Figure 5.36** Aerofoil lift is created by splitting an airstream so that it travels further, and therefore faster, on one side than on the other.

Pitot tube velocity meters

Commonly used to measure the air speed of aircraft, but also in laboratory flow meters, a Pitot tube (see Figure 5.37) compares the pressure exerted by the moving fluid with a static pressure. The difference can be interpreted as an indication of the fluid velocity, using Bernoulli's equation.

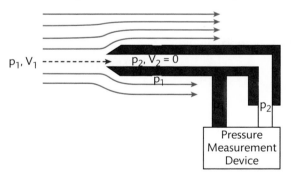

▶ **Figure 5.37** A Pitot tube velocity sensor senses the pressure of static air in its central tube, but also the lower pressure of moving air through annular holes.

❚❚ PAUSE POINT

Close the book and make some notes on what you now understand about viscosity and fluid flow. What applications can you think of in your laboratory or your home?

(Hint)

What laboratory equipment do you use to move fluids? – pipettes, syringes, vacuum pumps . . . How does the size of pipe or tube used affect the flow?

(Extend)

Look around your kitchen or in your medicine or cosmetics cupboards for examples of non-Newtonian fluids. Test how they react to stirring, squeezing, shaking etc. and try to classify them according to type.

Assessment practice 5.16

The graph shows air pressure against time in a bell jar that is being evacuated by a rotary vacuum pump.

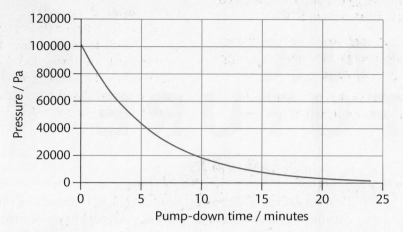

a) Explain the shape of the curve.

b) State which will improve the pump-down speed most: (i) halving the length of the connecting pipe, or (ii) doubling the internal diameter of the connecting pipe. Explain why.

Assessment practice 5.17

a) Describe the difference between streamline and turbulent flow patterns.

b) Give an example where turbulence is beneficial.

c) Explain why in most design applications streamline flow is preferred.

THINK ▶FUTURE

Cemile Alkis

Research and development scientist in a medical laboratory

I started work at 22 just after I had left college. I worked in a chemical plant cataloguing materials and ensuring all the reactants we needed were of a high quality. I then moved jobs to work in a medical lab that develops and makes hip replacements. When I first started, I did a similar range of jobs to that in my first company, from cataloguing resources to testing materials. I have now worked with the company for five years and after two promotions I am the buyer for the company and I work with a range of organisations that produce our raw materials.

I spend a lot of time at their factory ensuring the quality of the production methods they use. I use my knowledge of extraction and purification techniques when doing this. My knowledge of the extraction process for titanium has been invaluable.

The scientists at my company give me a brief for the amount and quality of metal they need. I sometimes have to source raw materials from a range of companies.

I have to have good communication skills as I need to talk to everyone at the factory from the sales department to workers in the plant. I also need good maths skills as I have to negotiate good deals for my company. I do quite a lot of driving between my lab and the factories. I have to write month end reports to show what deals I have made and how I have met targets.

I often have to use online video conferencing to talk to my boss or people in the lab or factory. I nearly always have my laptop with me.

No two days are the same for me, which I really like. I get to use my science skills but also my other, personal skills. This challenges me and keeps the job interesting.

Focusing your skills

Think about the role of a buyer. Consider the following:
- What types of people will you work with and how will you support them?
- What role will you play in helping them achieve their goals?

- What experience would you need to carry out this role?
- What skills do you currently have? What skills do you think may need further development?

Getting ready for assessment

This section has been written to help you to do your best when you take the assessment. Read through it carefully and ask your tutor if there is anything you are still not sure about.

About the test

The test is in three sections (Biology, Chemistry and Physics).

Remember that all the questions are compulsory and you should attempt to answer each one. Consider the question fully and remember to use the key words to describe, explain and analyse. For longer questions, you will need to include a number of explanations in your response; plan your answer and write in detail.

Preparing for the test

To improve your chances on the test, you will need to make sure you have revised all the key assessment outcomes that are likely to appear. The assessment outcomes were introduced to you at the start of this unit.

To help plan your revision, it is very useful to know what type of learner you are. Which of the following sounds like it would be most helpful to you?

Type of learner	Visual	Auditory	Kinaesthetic
What it means	Need to see something or picture it, to learn it	Need to hear something to learn it	Learn better when physical activity is involved – learn by doing
How it can help prepare for the test	<ul><li>Colour code information on your notes</li><li>Make short flash cards (so you can picture the notes)</li><li>Use diagrams, mind-maps and flowcharts</li><li>Use post-it notes to leave visible reminders for yourself</li></ul>	<ul><li>Read information aloud, then repeat it in your own words</li><li>Use word games or mnemonics to help</li><li>Use different ways of saying things – different stresses or voices for different things</li><li>Record short revision notes to listen to on your phone or computer</li></ul>	<ul><li>Revise your notes while walking – use different locations for different subjects</li><li>Try and connect actions with particular parts of a sequence you need to learn</li><li>Record your notes and listen to them while doing chores, exercising, etc. and associate the tasks with the learning</li></ul>

Remember!

Do not start revision too late! Cramming information is very stressful and does not work.

Useful tips

- **Plan a revision timetable** – schedule each topic you need to revise and try and spend a small time more often on each of them. Coming back to each topic several times will help you to reinforce the key facts in your memory.
- **Take regular breaks** – short bursts of 30–40 minutes revision are more effective than long hours. Remember that most people's concentration lapses after an hour and they need a break.
- **Allow yourself rest** – do not fill all your time with revision. You could schedule one evening off a week, or book in a 'revision holiday' of a few days.
- **Take care of yourself** – stay healthy, rested and eating properly. This will help you to perform at your best. The less stressed you are, the easier you will find it to learn.

Sitting the test

Listen to, and read carefully, any instructions you are given. Lots of marks are often lost because people do not read questions properly and then do not complete their answers correctly.

Most questions contain command words. Understanding what these words mean will help you understand what the question is asking you to do. These were also introduced at the start of this unit.

Remember the number of marks can relate to the number of answers you may be expected to give. If a question asks for two examples, do not only give one! Similarly, do not offer more information than the question needs: if there are two marks for two examples, do not give four examples.

Planning your time is an important part of succeeding on a test. Work out what you need to answer and then organise your time. You should spend more time on longer questions. Set yourself a timetable for working through the test and then stick to it. Do not spend ages on a short 1 or 2 mark question and then find you only have a few minutes for a longer 4 or 6 mark questions. It is useful when reading through a question to write down notes on a blank page. This way you can write down all the key words and information required and use these to structure an answer.

If you are writing an answer to a longer question, try and plan your answers before you start writing. Have a clear idea of the point your answer is making, and then make sure this point comes across in everything you write, so that it is all focused on answering the question you have been set.

If you finish early, use the time to re-read your answers and make any corrections. This could really help make your answers even better and could make a big difference in your final mark.

Hints and tips for tests

- Revise all the key areas likely to be covered. Draw up a checklist to make sure you do not forget anything!
- Know the time of the test and arrive early and prepared.
- Ensure that you have eaten before the test and that you feel relaxed and fresh.
- Read each question carefully before you answer it to make sure you understand what you have to do.

- Make notes as you read through the question and use these to structure your answer.
- Try answering all the simpler questions first then come back to the harder questions. This should give you more time for the harder questions.
- Remember you cannot lose marks for a wrong answer, but you cannot gain any marks for a blank space!

Q. Describe how transition metals are able to form ions with different oxidation numbers. (3)

Transition metals can lose two electrons from the 4s-orbital giving the +2 oxidation state. The 4s electrons are in the highest energy level and so are lost first. The 3d and 4s energy levels have similar energies so 3d electrons can also be lost.

This is a 3-mark describe question. The examiner is looking for 3 points. This answer has given 4 points, all of which are correct and so gains the marks. The examiner will not negatively mark the question but if you write more information that is incorrect and disagrees with what you have written you may lose a mark. For example, if this learner had gone on to write that 4s energy levels had much higher energy than 3d energy levels, this would be wrong and would negate the last mark given.

Q. Ethane contains a single C—C bond. Ethene contains a C=C double bond. Compare the bonds between the carbons in ethane and ethene. (6)

Question number	Answer	Mark
	Indicative content **Similarities** ▶ Both covalent bonds ▶ Electrons in the bonds overlap ▶ Both contain σ-bonds ▶ Hybrid sp^3 orbitals (ethane has sp^2) **Differences** ▶ Double bond contains π-bond ▶ p-orbitals overlap above and below the carbon atoms ▶ Movement restricted around double bond ▶ Region around double bond is flat ▶ Double bond more reactive than single bond/single bond more stable than double bond ▶ Single bond stronger than double ▶ Double bond shorter than single	(6)
0	No rewardable content.	0
Pass level	Some simple statements that are not linked.	1–2
Merit level	Some similarities and differences.	3–4
Distinction level	Detailed similarities and difference.	5–6

Ans 1. Double bonds are strong and very reactive. They make ethene a flat shape.

This is a 6-mark levelled question. It is worth 2 pass marks, 2 merit marks and 2 distinction marks. You gain marks for showing understanding rather than there being one mark per point. The more detailed and in-depth your discussion, the more likely you are to gain 6 marks. You would be expected to use all your knowledge of single and double bonds. You should discuss the electrons and orbitals, as well as strength, length and reactivity.

This would be a pass-level answer. The candidate has given some differences in a simple way. He has not compared double bonds to single bonds. This is worth 2 marks.

Ans 2. Both single and double bonds are covalent. The single bond is formed by a hybridised sp3 bond. In ethene there is also a pi bond. Both bonds contain a sigma bond and then in the double bond the p orbitals overlap. This forms a stronger bond but a shorter one. There are more electrons in a double bond so it is more reactive.

This would be a distinction-level answer. The learner has given similarities and differences. The ideas are mostly quite detailed and are linked. The learner does not have to give all indicative content to gain 6 marks.

Investigative Project 6

Getting to know your unit

Science helps us to make sense of our world. The fundamental laws and understanding used in our everyday lives have been developed through centuries of questioning and subsequent investigation by those who have an inherent need, both to gain an understanding of the world around them, and also to push the boundaries in scientific discovery and advancement. This unit will enable you to establish your understanding and skills in scientific investigation by carrying out an investigative project. You will learn how to choose a project based on your interests and literature search. You will be able to apply knowledge gained in other units to manage your project, outline the plan to be followed and produce an evaluation supported by your new scientific skills.

How you will be assessed

This unit will be assessed using a series of internally assessed tasks within assignments set by your tutor. Throughout this unit you will find activities that will help you work towards your assessment. Simply completing these activities will not mean that you have achieved a particular grade, but you will have carried out useful research or preparation that will be relevant when it comes to completing your assignments.

In all the tasks in your assignments, it is important to check that you have met all of the Pass grading criteria. You can do this as you work your way through the assignments.

If you are hoping to gain a Merit or Distinction, you should also make sure that you present the information in your assignments in the manner required by the relevant assessment criterion. For example, Merit criteria require you to analyse and demonstrate skilful application of procedures whilst Distinction criteria require you to evaluate your practice.

The assignments set by your tutor will consist of a number of tasks designed to meet the criteria in the table. This is likely to consist of a written report but may also include activities such as:

▶ demonstrating correct and appropriate practical techniques confirmed by observational record and/or witness statement

▶ presenting findings to your peers and reviewing the procedures and applications of your work during class discussion

▶ analysing and reviewing your own performance in a critique which highlights your strengths and weaknesses.

Assessment criteria

This table shows what you must do in order to achieve a **Pass**, **Merit** or **Distinction** grade, and where you can find activities to help you.

Pass	Merit	Distinction
Learning aim A Undertake a literature search and review to produce an investigative project proposal		
A.P1 Carry out a literature search and review into a chosen scientific area **Assessment practice 6.1**	**A.M1** Analyse a literature search and discuss its relevance to inform the investigative project proposal **Assessment practice 6.1**	**A.D1** Evaluate the different methods of investigation considered for the investigative project proposal, justifying the hypothesis chosen **Assessment practice 6.1**
A.P2 Produce an appropriate project proposal for an investigative project proposal, to include hypothesis **Assessment practice 6.1**	**A.M2** Produce a project proposal for a scientific investigation, to include hypothesis and potential limitations **Assessment practice 6.1**	
Learning aim B Produce a plan for an investigative project based on the proposal		
B.P3 Produce a realistic working plan for the project, including health and safety and risk assessments **Assessment practice 6.2**	**B.M3** Produce a realistic working plan for the project, including health and safety and risk assessments and contingency planning **Assessment practice 6.2**	**B.D2** Analyse the effectiveness of the working plan, justifying changes made **Assessment practice 6.2**
Learning aim C Undertake the project, analysing and presenting the results		
C.P4 Demonstrate practical skills to assemble relevant apparatus/equipment and materials, and carry out the project using safe working practices **Assessment practice 6.3**	**C.M4** Justify the choice of experimental and data analysis techniques used as a means of increasing accuracy, reliability and validity **Assessment practice 6.3**	**CD.D3** Evaluate the conclusions of the investigative project and its practical aspects, discussing limitations, improvements and recommendations for further study **Assessment practice 6.3, 6.4**
C.P5 Accurately collect, analyse and present the results obtained **Assessment practice 6.3**		
Learning aim D Review the investigative project using correct scientific principles		
D.P6 Produce a report using findings, scientific terminology and protocol appropriately and drawing conclusions **Assessment practice 6.4**	**D.M5** Produce a report using findings, correct scientific terminology, protocol and formatting and drawing valid conclusions **Assessment practice 6.4**	

Getting started

Undertaking a scientific investigative project is an important part of the duties of an industrial technician and many science-related workplace positions. Completion of a science project of your choice will develop your understanding of the processes involved and help to develop your skills in science for future education or employment. Research an investigation into a chemistry-, biology- or physics-related subject from an article in a scientific journal. Identify, if you can, the following aspects within the report: literature review, project proposal, outline plan, health and safety considerations, method used, data collection, data analysis and presentation, structure of the report, referencing and evaluation with further project proposals.

A Undertake a literature search and review to produce an investigative project proposal

This section outlines the essential process of deciding on a scientific investigation proposal by carrying out a comprehensive **literature review** to help you make an informed decision. You will be guided through the various types of study and sources of information suitable for your proposal and how you should record your review. The reason for your investigative proposal is an important point which will be addressed since it helps to ensure that you are able to complete the investigation comprehensively, following useful and correct scientific processes and maintaining your full interest throughout.

Key term

Literature review – a search and evaluation of the available information in your given subject or chosen topic area.

Literature review

To determine on which aspect of science to focus your investigation, you need to research areas in which you may have a personal interest or areas which have been provided by your tutor. Science consists of three general sciences; biology, chemistry and physics, and the breadth of study can be quite daunting. Being able to narrow down the field of potential investigation is vital if the project proposal is to stand a chance of 'getting off the ground'. If your tutor has given you a short and specific list to consider, your job will be much easier.

Identification of criteria

There are numerous areas worthy of investigation in all scientific disciplines. Your final decision will obviously be made by consideration of the equipment available, the degree of difficulty, the results which can be obtained and the time scale involved. Table 6.1 lists some examples which may help narrow the field.

▶ **Table 6.1** Examples of potential investigations

Physics	Chemistry	Biology
• Resistivity of various metal wires • Finding the factors which affect the length of a ski jump • Determination of terminal velocity of a ball through fluid using light gates and viscosity tube • Investigating stress and strain of different materials • Resistance change with temperature for thermistors • Improving a pinhole camera with lenses • Determination of relationship between aileron angle and lifting force for aircraft wing models	• Electrical charge calculations from electrolysis of copper(II) sulfate, silver nitrate, lead (II) bromide • Factors determining rates of reaction • Enthalpy changes in combustion of alcohols or flammable liquids • Determination of concentrations of unknown samples • Vitamin C content in fruit juices • Oxidation of ethanol to ethanoic acid • Electrolysis of aqueous salt solutions • Testing water hardness	• The effect of penicillin on bacterial growth • Effect of exercise on reaction time • Testing for reducing sugars • Factors affecting the rate of photosynthesis • The actions of enzymes in digestion • Plant responses to stimuli • Factors affecting plant growth • Effect of polluted water on seed development • Effects of caffeine on water fleas (daphnia) • Factors affecting transpiration in plants

It is important to try to develop a critical review when researching literature. To review *critically* means that you should analyse both the merits and faults of the information that you are reviewing. You should not allow your review to become a general list of research titles with some additional facts and figures, although your knowledge in the topic chosen may not yet be sufficient to produce a substantial critical review. A literature review is much more than a general list and, if carried out effectively, will help to set the foundation of your project proposal and subsequent investigative project completion.

'Literature' in this context refers to anything which provides information related to your subject. It includes:

▶ journals
▶ science articles
▶ government reports
▶ Internet sites
▶ newspaper reports
▶ textbooks
▶ dissertations.

The whole point of a literature review is to give you an overview of your field of study, outlining the current knowledge and theories which may apply and the questions which may still need to be answered. This will also include studies and experiments which have already taken place in the same scientific area. Most laboratory investigative projects which you will carry out should have known outcomes, tested over many years and so your literature review will be useful as supportive information but may also provide an idea of what your **hypothesis** would be or help you to formulate an effective method.

> **Key term**
>
> **Hypothesis** – an explanation, with some evidence, to be further tested by investigation.

The number of sources of literature and therefore **references** used to help with your study will depend on the chosen topic and the level of the qualification. In this case, for example, a useful number at Level 3 study may be anywhere from 5 to 15 references, but your tutor may also provide you with a suitable number.

> **Key term**
>
> **References** – a list of sources at the end of a report used to help provide information for an activity.

'Out with the old – in with the new.'

In science, this statement is not always necessarily useful to follow, since much of what is known at present has not changed much in decades. The research in science builds on previous knowledge and there should be an element of caution in using both very new science literature as well as very old. Advancement in science appears to be developing exponentially and keeping up with current knowledge can be a challenge. It is wise to ensure that your literature review reflects the most current information available in your chosen topic but also refers to literature older than 10 years. This will help to reinforce your understanding and also indicate to your tutor that you have considered all relevant times of writing. It is important to note that the most up-to-date studies and information can be found in journals as they are published regularly. Text books may have older publishing dates. Websites may not include the date of publication so it can be difficult to judge how up to date the information is.

These aspects should always be considered when carrying out a literature review:

▶ Always use more than one source of information – there are many sources available on the same subject material including tutors, text books, internet, newspapers, magazines, journals and television.

▶ Carefully choose your research titles – when searching for information on acid/base titrations, for example, be specific. You may get a lot of information about titrations in general which will take a lot of time to read and may not all be relevant.

▶ Select material from an authoritative source – professional experience and qualifications of the authors(s) is important and when more than one source agrees in content, you can develop confidence in sources you are using.

▶ Avoid plagiarism of other people's work – sometimes you may need to include a direct quotation of a researched note to help with your report. Keep these to a minimum, do not change the quotation at all and include a suitable reference to indicate where the article came from.

Knowing what material is useful to your field of study and what material to discard is difficult if your knowledge of the topic is not detailed. However, a large proportion of your literature review will involve you reading the background of your topic and helping you to develop your knowledge.

It is helpful to draw up a list of inclusion and exclusion criteria before starting your literature review and to adjust the list as you progress through the investigative project.

The following provides some guidance as to what you could ask yourself before you decide on inclusion or exclusion of the information you have reviewed:

▶ Do the sources chosen agree on the science involved?
▶ Are you able to identify a common methodology for your investigation?
▶ Is the information sufficiently detailed?
▶ Does a source provide information not given by any other source?
▶ Is the source publication well known or recognisable?
▶ Does the source material make use of information provided by well qualified writers?

> **Research**

Choose one topic from physics, chemistry and biology in Table 6.1 and carry out a brief research into the subject topics. Identify which source(s) of information:
- provide the clearest and easiest to understand information
- give an indication of alternative investigations which may be linked
- give reference to other sources
- is the oldest source of information.

Nature of study

Science is a theory-driven practical subject. If you have plans to follow a career in any of the scientific disciplines or advance your education, you will need to know how the knowledge you have acquired in the lecture room has been developed over time. You need to experience the science involved first

hand, by experimentation and investigation. Figure 6.1 shows potential physical areas of study.

Sources of information

> **Link**

Unit 7 – Contemporary Issues in Science (sources of information, e.g. journals, articles, news reports).

The research information used for your planning stage in the investigation should be correctly referenced in your notes. If you used a number of sources, then your eventual working plan will contain information which will have been judged to be common to more than one source, a skill which takes time as your knowledge develops.

It is useful to identify sources of reliable information quickly and to tabulate or record them in an appropriate manner. You may also wish to include any sources which you have discarded, providing reasons why. The example in Table 6.2 may help.

▶ **Table 6.2** How to tabulate sources of reliable information in an appropriate manner

Information source	Brief details of information	Useful	Not useful
Data & data handling for AS level biology, Bill Indge	Suggested use of glass container of water placed between lamp and cambomba (aquatic plant: from internet search) to help eliminate the effects of increased temp. on rate of photosynthesis	✓	

Fieldwork
This involves any work carried out beyond the confines of your educational building. It includes organised or independent trips to gather evidence, physical or observed, which will form the basis of your project. Example – determination of water flow rates at various points in a river profile.

Laboratory-based
Most forms of science investigation performed in a school, further education college or university will make use of a laboratory. It provides a supervised, well-resourced, well-organised platform for meaningful and relatively safe investigative work. Example – investigating various conditions for optimum plant growth.

Workshop
Studies using science workshops can be used to explore how the population relate to science and how learning is achieved.
Workshops are able to provide areas of study linked to practical activity and research. Example – identifying the positive aspect of group discussion prior to scientific problem solving.

Sports facility
The science of sport is a popular and increasingly important area of study. The availability of a sports complex and variety of equipment provides opportunities to investigate a range of aspects related to the sciences. Example – observing the effect of exercise on the body.

▶ **Figure 6.1** Physical areas of study

Extraction of information

When carrying out **extraction** of information, sources include:

▸ libraries and resource centres

▸ government organisations

▸ science organisations

▸ charities.

> **Key term**
>
> **Extraction** – a means to obtain information from different sources.

The method of taking information from each of these example sources will vary in accordance with their individual procedure which should be made available to the customer.

Resource centres and libraries will no doubt have a thorough system in place to log each available item in their collection. A visit to your school or college resource centre is a useful starting point if you haven't already done so.

Here you will notice that the resources are set out within subjects and further ordered into alphabetical sequence. This is essential to ensure that students are able to find relevant information quickly and to become independent in their research.

Items available to support your literature review will include:

▸ academic and leisure magazines

▸ network computers

▸ non-fiction resources such as journal papers or textbooks

▸ DVDs in science and general topics

▸ audio-visual sections for use when practising presentations.

Your research often depends on gaining access to either people or data from an organisation. As a result, you may need the cooperation of the organisation before you can extract the information needed. This may mean asking their permission or approval before you can continue with your research and use their information in the literature review. Such organisations can include: animal research, Universities, Animal and Plant Health Agency.

Most government and charity organisations allow access to their findings for research purposes, provided the information source is noted in the work and that any findings are not changed or misrepresented.

Source listing protocol

Your research sources must be listed at the back of your final report in a recognized manner. Most countries use the **Harvard referencing system**, or variations of it, which depend on the type of article, portion used and number of authors.

> **Key term**
>
> **Harvard referencing system** – a style of referencing system used to mention sources of information which have been looked at to inform your work.

▸ Suggested format for webpage referencing (journal article example): Author surname, forename or first initial (where applicable), year published, title, website URL *Example:* Smith, J, 2016, Journal of Everything, [online], P.140 Available at: http://www.pearson.edu.au/guides-services/research.html [accessed 10 Jan.2016]

▸ Suggested format for textbook referencing: Author surname, first initial, year published, title, edition (if not the first), city where published, publisher, page(s) *Example:* Smith, J, 2016, Textbook of Everything, 2nd Edition, London, Pearson Publ., pages 6–10.

▸ Suggested format for in-text referencing (citations): Author surname, first initial, year published, edition, city of publication, publisher, page. *Example:* If we agree that Black is an absence of colour (Smith, J, 2016, Textbook of Everything, 2nd edition, London, Pearson Publ., page 6) then we can assume …

All reference material used must be listed at the end of your report in alphabetical order. You should note that if author surnames begin with the same letter, then the first to be listed in your references will be dependent on the second letter in their names and so on.

Example: You have used reference material from two texts, one written by author J. Smith and one by author B. Smythe. The first to be listed in your references will be J. Smith, because the third letter of the surname is 'i' which comes before 'y' in the alphabet.

Once completed, your literature review must be written in a format which resembles that of a science investigation (see Figure 6.2):

▸ introduction

▸ main body

▸ conclusion.

Introduction:

1 Nature of the topic
2 Why you chose the literature
3 What the topic includes

Main body:

1 Mainstream views and alternative ideas
2 Background to the hypothesis
3 Definitions of key terms
4 Research studies performed, if any exist
5 New discoveries, if available
6 Questions asked
7 Methods used and to be followed by you
8 General conclusions

Conclusion:

1 Summary of general conclusions
2 Have you been able to focus your hypothesis clearly following review?
3 Points where you agree/disagree with the literature

▶ **Figure 6.2** Structure of review

Discussion

Have a general discussion in your group on the benefits of carrying out a laboratory-based investigation as opposed to a fieldwork investigation. You could focus on equipment, other resources, personnel and time.

Investigative project proposal

Thorough completion of the literature review will ensure that you will be in a position to provide a project proposal for investigation. This section of your work should include an outline of what you intend to investigate, the reasons for your choice, information in support of your topic choice, what you hope to achieve and problems you expect to encounter.

Rationale

Having chosen your area of study, based on your literature review, you must now write down the reasons why you have chosen this project and what you intend to gain from its completion.

This section of your project proposal is an opportunity for you to explain why you are performing this investigation and perhaps describe how the investigation results may be of further use or what implications your study may have on your present understanding of the topic.

▶ Identify the background information concerning your chosen topic that you have found by research.

▶ Identify how this information relates to your study topic – is any of it relevant?

▶ Comment on the science principles involved.

▶ Use appropriate and a sufficient number of sources of information to help you plan your investigation.

▶ Link your current knowledge of the subject in question to your investigation.

Remember

- Keep your rationale concise and short.
- Use clear language.
- Identify the main point of the investigation.
- Use and reference appropriate sources.
- Outline the scientific principles involved.
- Explain why you have chosen this topic.

Background

The information you need to include in the background section should identify and describe what the subject of your proposal is and its scientific history. You should refer to the literature review information and include the context of the work you are about to undertake. The background should present the scientific information which currently exists, outlining your knowledge development and explaining what is known about the science of the subject.

A well written background section will help other readers and your tutor determine how well you understand the subject in your project proposal. Demonstrating a clear understanding will promote further confidence in the quality of your overall investigative project. Summarise all you know about the subject, highlighting literature using **citations** and write it as though you are telling an interesting and engaging story.

Key term

Citation – a quotation or reference from a paper, article, book or specific author.

Background of the study

Cockroaches are common pests in the tropics. They have been known to cause allergic reactions to most people and chew holes in clothes. According to Bato Balani for Science and Technology, Vol.14, No.2, the real danger of cockroach lies in their ability to transmit sometimes lethal diseases and organisms such as Staphylococcus spp., Streptococcus spp., Hepatitis viruses and Coliform bacteria. They have been known to contaminate food, at the same time infect it with the bacteria they carry. The bacteria they spread in food can cause food poisoning. People have used various instruments to control the cockroach problem in homes. The most popular is the commonly used insecticide sprays. Most of these can destroy cockroaches but they can also do serious damage to humans as well. According to the experts of the website toxalbumin carcin. Along with other ingredients like sap, onions and weeping willow leaves, are Tubang Bakod seeds feasible to be used as cockroach killer? www.bayer.co.th, the active ingredients in these sprays, like tethramethrin and petroleum distillates can cause severe chest pains and cough attacks when inhaled. The second most popular instrument is the cockroach coils. These coils can kill roaches yet the active ingredients in these coils like allethrin, pynamin forte, prothrin and pyrethrin, can cause harm to humans when inhaled. It also has an ozone-depleting ingredient. The third most popular is the flypaper. The concept of the flypaper is simple. The roaches just stick into it. But when they are stuck the roaches die and carcass can spread more bacteria. In addition to the side effects of these materials, the costs of these insecticides are high. All these set aside, the question on everyone's mind is: 'What can be an effective and natural insecticide?' Tubang Bakod (Jatropha carcas) is a common plant in the Philippines. According to the website www.davesgarden.com, its seeds contain a certain toxic substance known as Toxalbumin Carcin, along with other ingredients like sap, onions and weeping willow leaves, are Tubang Bakod seeds feasible to be used as cockroach killer?

Source: https://www.scribd.com/doc/60715657/Background-of-the-Study-Sample

---

PAUSE POINT   Read the above passage carefully and re-write it so that it will be more easily accessible to your fellow students.

> **Hint**   Bullet point or number the key points in the passage but retain the referencing and scientific wording.

> **Extend**   Outline the main methods proposed to control the problems which cockroaches introduce. Identify and comment on the main aspects associated with natural and man-made methods at controlling transmission of diseases by cockroaches.

> **Remember**
> - Position the 'background' in the report abstract section and make it brief.
> - Outline what information is already known about the subject.
> - Outline what your investigation intends to find out or test.

Hypothesis

A hypothesis is an assumption based on your knowledge, understanding of the topic and observations. When carrying out an investigation, you can hypothesise about the outcome of the investigation and you may change this when other observations are made. Your hypothesis will then lead you to a prediction or predictions which you will be able to test.

To produce a meaningful hypothesis, you will need to identify the scientific question.

Example: *'Do plants need fertiliser to grow bigger?'*

Your hypothesis will be developed from research you carry out and knowledge which you already have on the subject.

Example: *'Plants need a variety of nutrients to grow well. Adding fertiliser containing these nutrients helps them to grow bigger.'*

Your prediction(s) will involve you applying the information you have researched to a situation which you will be able to investigate and test.

Example: *'Using two sets of tomato plants, if I add fertiliser containing nutrients to one set of tomato plants and do not fertilise the other set, there should be a significant increase in height and leaf size of the fertilised set of plants.'*

Null hypothesis

Your hypothesis is an explanation of what you have observed and can be supported or refuted by the evidence in your investigation. This is generally demonstrated by confirming a relationship between two variables. A null hypothesis would state that there is no relationship. By disproving the null hypothesis, you provide support for your hypothesis, i.e. there is a relationship between the variables. There are a range of statistical tests that can be used to determine the probability that your null hypothesis is incorrect and the degree to which there is a significant relationship between the two variables. These include chi-squared and t-test.

Example

It is accepted that concentration is related to the quality of sleep that you get. Your hypothesis to be tested could be 'A better quality of sleep increases your concentration levels.' Your null hypothesis could be 'Concentration levels are unrelated to quality of sleep'.

By testing and disproving the null hypothesis, you can conclude that there is a relationship between levels of concentration and quality of sleep.

Hypothesis checklist:

▶ Is it based on information from reliable sources about the topic in question?

▶ Can the hypothesis be tested in an experiment?

▶ Can you identify at least one prediction from your hypothesis?

▶ Does this prediction have both an independent variable and a dependent variable?

Aims and objectives

The purpose of the activity undertaken, and identification of what the activity is attempting to achieve, is called the **aim** of the investigation. The aim should be kept brief and to the point.

Steps to be taken to achieve the aim of the investigation are referred to as the **objectives** and are details of the specific tasks intended to help achieve the goals outlined in the aim. In many scientific reports, they may also be referred to as 'outcomes'.

Key terms

Aim – overall general statement of the purpose or intentions of the study.

Objectives – stages to be completed to successfully achieve the aim.

Example aim

A thermometric titration could involve titrating sodium hydroxide (NaOH) solution with hydrochloric acid (HCl). The change in temperature would then be recorded for every addition of a quantity of the acid and the highest temperature recorded would indicate the 'end-point' of the titration. This can then be used to find the concentration of the acid.

Aim *To measure and record the maximum temperature during the reaction of sodium hydroxide with hydrochloric acid solutions. The volumes of solutions which have reacted at the maximum temperature signify the titration 'end-point'.*

Example objectives

When microorganisms are initially introduced into a growth medium, they make sufficient use of the nutrients supplied by synthesising the available enzymes. This does not happen immediately but rather slowly at first, then quite rapidly. This is called the growth phase. By counting the bacterial population at suitable time intervals, a graph of population size (log scale) against time can be developed, from which the bacterial colony growth rate can be found.

Objectives *1. To develop data on bacterial growth over a suitable time period. 2. To identify the exponential growth of bacteria in a growth medium. 3. To be able to calculate bacteria growth rates from measurement.*

Case study

Tom Michaelson is a project advisor responsible for the development of text books and other published resources for science related to schools and colleges at KS3, KS4 and KS5. His team are currently deciding what information should be included in the latest edition of their textbook, *Science at Advanced Level*.

One member of the team suggests that, in order to determine what should be included, it is first necessary to understand what the aims and objectives of the book actually are.

Check your understanding

Using this text book as your example.

1 Outline the aim of the book in no more than four lines.

2 Identify the main objectives of the book.

The following aims and objectives passage has been taken from a research report into 'Evaluating the success of a public engagement project for the conservation of the Ural Saiga population in Kazakhstan':

> This thesis **aims** to evaluate the success of the EE [Environmental Education] campaign which provided local communities with information on the ecology and conservation status of the Uralsk Saiga 3 population. No previous Saiga education has been carried out in the region, so the campaign offered a unique opportunity to establish a baseline and to assess any changes in attitudes, knowledge and behavioural intent. The study provides insight into not only how attitudes effect conservation, but also how 'external' conservation measures and processes are judged by local people.
>
> Research **objectives** 1. Evaluate if/how levels of knowledge, attitudes and behavioural intentions toward Saiga changed, during the study period. 2. To assess any differences between socio-demographic groups regarding their experience of the campaign, in addition to knowledge of and interactions with Saiga. 3. To understand local people's perceptions of threats to Saiga and their conservation requirements and their own potential future role in Saiga conservation 4. To make recommendations for future awareness campaigns and Saiga conservation within the target villages.
>
> *Source:* http://www.iccs.org.uk/wp-content/thesis/consci/2011/Samuel.pdf

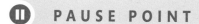

PAUSE POINT   Read the passage and simplify what the study was all about.

> **Hint**   You will firstly need to determine by research what 'Saiga' refers to – unless you already know.
>
> **Extend**   Identify any similarities and differences between this passage and the examples of aims and objectives shown in the section above.

Limitations

Even the most experienced and established scientific researchers will encounter aspects to a practical investigation which will need to be clearly identified and planned for well in advance of carrying out the activity.

The limitations of *your* study will generally depend on a number of different factors. Three of the most important considerations are outlined in Figure 6.3.

> **Implications for resources:**
> Most laboratories have an allocation of finances which can limit the number of different resources they may have. A booking system is standard practice and ensures that all are able to carry out their investigations eventually. Knowing exactly what is needed, by producing a resource list and presenting this to the technician early, can offset possible delays in completing the practical work.

> **Use of facilities:**
> You will need to identify the kind of laboratory and the types of instruments necessary, since some are designed with either chemistry or general science as a focus. Again, the use of specific equipment and laboratories is limited by factors such as timetabling or priority investigations. Early planning will help to reduce the possible need to re-think the course of your investigation.

> **Time constraints:**
> The time available to learners for carrying out scientific investigations has been reduced over the years, leaving highly constrained investigations to be preferred by most educational establishments. This approach does not allow more genuine, speculative science to develop. Planning your time is vital if the work is to have purpose and meaning. Negotiate with your tutor and technical staff, outlining the project proposal and the timetable of events so that you can successfully complete the work.

▶ **Figure 6.3** Limiting factors on study

Produce a simplified single A4 page literature review on the topic of variation of plant growth as a result of different light sources for possible further investigation.

You will need to include:

1 A full listing of the sources you use for research, their dates and relevance to the topic such as articles, websites, journals etc.

2 Complete and accurate referencing from the source information using the Harvard referencing system.

3 Identification of the form of study that is most suitable for the topic, outlining any previous studies which have informed your decision and the direction of your study.

Plan
- What is the task? What am I being asked to do?
- How confident do I feel in my own abilities to complete this task? Are there any areas I think I may struggle with?

Do
- I know what it is I'm doing and what I want to achieve.
- I can identify when I've gone wrong and adjust my thinking/approach to get myself back on course.

Review
- I can explain the results obtained from the task.
- I can apply the activity to other situations.

B Produce a plan for an investigative project based on the proposal

Producing your investigative project proposal is the first step in the process of providing a clear direction in your topic of study. Your project plan must now be considered in terms of the timing of each part of the project, outlining the method and relevant use of resources. In addition to this, you must now include important aspects of health and safety, identifying the hazards involved and the level of risk.

Schedule

Timeline for project

Any master chef will tell you that the key to producing a top-class meal as opposed to a culinary disaster is the timing. Although, in your studies of science at Level 3, you may not be expected to cook any food, producing and adhering strictly to a timeline for your project is vital if you are to complete the activity with sufficient time allocated for specific tasks within the investigation.

Ensure that you have given due consideration to the key aspects of time shown in Figure 6.4.

If all the information and times of start, completion and milestones are noted, the possibility of problems arising – such as running out of time, the need to carry out repeat experiments, being unable to use the laboratory because of a lack of prior booking etc. – will be reduced a great deal.

Start date:
After discussion with your tutor and technicians, set a suitable date for starting the practical part of the project. This will involve assembly of the equipment, collecting data or results, collating the evidence, analysing and presenting the data. Note the times when you are undertaking tutorials and laboratory hours.

Milestones:
Important sections of your practical work which need to be recognised. This could include; initial assembly and checking operation of equipment, identifying H&S aspects, performing the main functions and observing, recording results, analysing data, presenting data.

Completion date:
Dependent on the centre timetable, identify a suitable date to draw the physical aspects of the activity to a close. Remember that you will now need sufficient time to be able to finalise your report with a full evaluation of your findings.

▶ **Figure 6.4** Important time factors

Link

See *Unit 2: Practical Scientific Procedures and Techniques* (Fig. 2.27, page 134, keeping a detailed laboratory notebook).

Plan

Producing a working plan or method for an investigation involves the use of all your notes, literature review, tutor guidance and preliminary testing (where appropriate). If the working plan is thorough, the investigation should produce no real surprises but useful results. You should try to include detailed descriptions of the following points in your working plan.

▸ A hypothesis (if possible) – suggestions for explaining what may be happening.

▸ Theory – literature review notes on the topic of your project proposal.

▸ Apparatus to be used – include labelled diagrams, include the number and sizes of pieces of apparatus.

▸ Notes of preliminary tests (if appropriate) – if you have time planned into your project to carry out a preliminary test, include all the notes to guide the main investigation. This could include trying different methods or refining a method so that it will give meaningful results.

▸ Step by step instructions – details of how the investigation will be performed, generally in numbered or bullet point format. It may well be worth attempting to provide a method of a relatively basic procedure as a practice run, such as 'making a cup of tea', to give you some insight into the need to identify every specific point if others are to follow your method. The method should be in third person past tense and should include enough detail so that someone else can replicate the experiment exactly as it was performed, without having to ask any questions about what was done.

▸ A prediction – what you think will happen, guided by your research notes.

▸ Variables – which will be kept constant and how they will be kept constant; which will change.

Your investigation planning stage will take time to develop. Your completed plan needs to be monitored by your tutor and/or technician regularly before the investigation can be given the go-ahead. What may look feasible to you may not be appropriate to experienced members of staff.

Resources

Scientific investigation resources, in the context of the plan, refers to all those commodities available to the investigator within the confines of the laboratory. This includes:

▸ equipment and instrumentation

▸ materials

▸ participants.

Providing an equipment or apparatus list may not sufficiently address the use to which they are to be put or the reasons for their use in a certain activity and so you need to ensure that this is included in the resource section of your plan. It is important to outline the nature of the experiments to be undertaken, detailing the standard procedures applicable and whether they would need to be modified in any way to suit the requirements of the investigation.

Consider the example in Figure 6.5.

▸ Equipment and instrumentation – the tube would need to have markings set at an appropriate distance to ensure that sufficient time is allowed for measurements to be taken as the marble is dropped into the liquid and flows through the liquid to the bottom. A preliminary set of tests will need to be performed. How is the marble retrieved after falling, for example? Instruments used to measure the diameter of the marble will need to be precise, for example a micrometer screw gauge, but practice is required to ensure that this instrument is used correctly. Timing devices will need to be responsive to touch and in good working order. Should a light gate type be used instead?

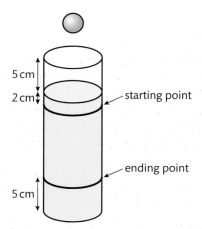

▸ **Figure 6.5** Viscosity (or velocity) tube method of determining the viscosity (resistance to flow) of liquids using a marble. The speed of the marble through the liquid is related to the viscosity of the liquid.

- Materials – the type of spherical object used may play a part in the results obtained. Using spheres of different mass will give different readings, but does this affect the aim of the investigation? The density of the liquid medium will necessarily need to be altered to demonstrate that the spheres fall at different speeds through different liquids and illustrate valuable points concerning liquid flow. Should the liquid be kept at constant temperature or varied?

- Participants – how can one person drop the ball and begin the timer at the precise time necessary? Using other people to help in measurement is important in many forms of investigation to obtain useful results. In this activity, the additional participant would need to be used to either drop the marble or start and stop the timer at the precise times. This additional participant will need to be firmly briefed beforehand and given preliminary tests to ensure that there is good understanding between all parties.

Contingency planning

'The best laid schemes o' mice an' men / Gang aft a-gley.'

This line, taken from a Robert Burns poem and adapted by writer John Steinbeck in his novel *Of Mice and Men*, is often quoted (albeit in present-day English) to indicate that whatever plans have been put in place and whatever the intentions, not all are started or carried out in the manner intended.

Contingency planning (often referred to as 'Plan B') is now built into all large projects by multinational companies and governments to ensure that all possible manner of events are identified to manage the possible risks. Not everyone makes a **contingency plan** because they may:

- be overconfident about the completion of their work at the first attempt
- not have considered the possibility of problems occurring.

> **Key term**
>
> **Contingency plan** – a plan or action designed to be introduced in response to circumstances which may or may not actually happen.

> **Case study**
>
> The crew of Apollo 13 (unluckily on 13 April 1970) suffered a large explosion in one of their oxygen tanks while on their way to the Moon. They survived by transferring themselves into the Lunar Landing Module for much of the trip back to Earth, a contingency plan worked out one year earlier and almost dismissed as impractical by NASA at the time. Nevertheless, NASA developed written procedures for the possibility of this occurring.
>
> ### Check your understanding
>
> 1 With a partner, discuss any time in the recent past when you have needed to refer to your contingency plan. This may not be an obvious plan B, but could be something as simple as a necessary change of direction when walking to school or college or decision to change a playing position following the absence of a member of the hockey team.

Your contingency planning and any remedial action you may have to take will need to be written into your initial science plan and probably informed by carrying out some preliminary testing. The following list provides some guidance.

- Keep your contingency plan simple and use clear language which is easy to follow by others.
- Identify possible triggers of your contingency plan and the actions to take. Who will be responsible at each stage?

- Quickly identify whether you need extra resources. If your initial method needs amendment, do it and continue.
- Change your schedule of events and timings if you need to. You should try to identify any operational areas which you regard as inefficient.
- Manage the risks, looking for ways to reduce risk so that full contingencies need not be used if possible.
- Repeat readings sufficiently to help eliminate the need for a contingency.

Health and safety and ethical considerations

Hazards

The final stage (and probably the most important) in your preparation for carrying out the physical investigation of your proposed project is the development of a detailed and well-considered risk assessment based on correct identification of hazards which may be encountered during the activity.

A hazard is something which has the potential to cause harm and it is very important that you identify all the possible hazards in order to minimise or even eliminate the possible occurrence of harmful conditions. Of course, listing all possible hazards is virtually impossible since almost every piece of equipment or action could give rise to an accident or danger, either predictable or purely an unforeseen event.

Table 6.3 lists the more common types of hazard that can occur in the school or college laboratory.

▶ **Table 6.3** Potential hazards in a laboratory

Hazard	Details
Flammable substances	Take note of any materials which can catch fire.
Chemicals	Skin and eye burns can occur with many acids and alkalis. Organic chemicals such as phenols are corrosive.
Excessive heat	Thermal burns occurring from a Bunsen Burner flame or heat produced by chemical reactions.
Cuts	A very common accident, generally a result of removing tight stoppers on glass tubing or careless use of beakers.
Chemical absorption	Some chemicals when touched can be absorbed by the skin, cause severe dermatitis or seriously affect the eyes.
Chemical inhalation	Solvents, for example, are extremely toxic if inhaled and other compounds can affect the eyes, nose, throat and lungs.
Explosion	Unvented systems for chemical reactions can be explosive, e.g. distillation.
Chemical ingestion	Entering the mouth by hand contact, pipette use, contaminated food and drink.
Trips and falls	Occur frequently where bags, materials and other items are not removed from the site.

Link

Unit 3: Investigation Skills, page 146.

Unit 4: Laboratory Techniques and their Application, pages 228–235.

Biological hazards are also commonly encountered in some specialist laboratories (see Figure 6.6). This may involve the use of bacteria and viruses for testing and different animal blood types, including human blood. In the school or college science environment, contact may be made with organs, blood, tissues and other fluids from animal dissection. If not dealt with carefully, diseases could be easily spread.

▶ **Figure 6.6** Standard hazard symbols

Technical requirements

Equipment which should always be used in scientific investigations to protect from health and safety risks is called **Personal Protective Equipment (PPE)**. This includes: safety goggles, laboratory coats, face shields, gloves and helmets. Most laboratory environments are equipped with a fume cupboard or laminar flow cabinet, designed to prevent contamination of samples such as delicate electronic semiconductors or biological samples.

Investigations in all three science disciplines carry hazards and potential risks but chemistry has more than most. When carrying out a chemical investigation, use every available resource (**COSHH**, **Hazcards** from CLEAPSS, 'Topics in Safety' by ASE) to help you prepare your risk assessment and to understand fully what potential risks are associated with the substances that you are about to use. COSHH Regulations, 2002, are responsible for carrying out research and publishing data providing information such as the amount of chemical to use and length of exposure time for industry workers, before their health may suffer. Employers have a legal duty to report all incidents, however slight, and to record these incidents in a suitable register. If an incident needs investigating, the Health and Safety Executive (HSE) will require access to the register.

The Health and Safety at Work Act provides the law which covers all aspects and areas in the workplace. Organisations with more than five employees must develop their own Health and Safety Policy, which must be endorsed by the HSE, who give guidance for each section of the policy, including: writing the policy, Risk Assessment, facilities, training of staff and poster displays.

Key terms

Personal Protective Equipment (PPE) – equipment designed to protect the wearer by limiting the risk of injury or infection.

COSHH – Control of Substances Hazardous to Health Regulations in place for education and industry to limit the exposure to workers of chemical effects.

Hazcards – a set of documents from CLEAPSS (Consortium of Local Education Authorities for the Provision of Science Services) giving information on storage, disposal and potential risks of chemical and biological substances.

Research

Under supervision, identify one chemical in the school or college laboratories which has one or more of the hazard symbols on its label. Refer to 'Hazcards' to outline the main dangers of the chemical when being used and how it should be handled.

Risk assessment

Before you perform any practical work, complete a risk assessment thoroughly and have it checked by your tutor. Your place of learning has a responsibility to ensure that this is carried out following the Health and Safety at Work Act of 1974. Remember, a risk assessment is your way of minimising the potential risks which can occur during all your activities and to identify how you would deal with any risk occurring, not just when spills and breakages happen. This would include lab coats, fume cupboards, tongs etc . . .

Ensure that your risk assessment is completed thoroughly and in clear terms. It must include the type of hazard, the level of risk, and how the risk of hazard can be prevented or minimised. Before any investigation:

1 Identify the equipment and substances that you intend to use.

2 Research the hazards and the potential risks which can occur.

3 Outline the measures needed to deal with spills and breakages.

4 Have your risk assessment checked properly by a member of the science staff.

Remember

Before any investigation, you should:
- ask your teacher / tutor for the standard risk assessment paperwork used in your college or school
- identify the equipment and substances
- research the hazards and potential risks and include the type of hazard, the level of risk, prevention methods and ways of minimising the hazard
- identify measures to deal with spills and breakages
- have your Risk Assessment checked.

Link

Unit 4: Laboratory Techniques and their Application, pages 230–1 (a useful procedure for completion of a risk assessment when using sulfuric acid after guidance from the relevant Hazcard pages).

Ethical considerations

It is accepted that the range of ethical considerations involved with laboratory science investigations in a school sixth form or college may be limited. It is unlikely, for example, that there will be serious ethical implications when carrying out many physics investigations, but there may be ethical considerations in specific biological activities where small animals are tested or when using other people within the study.

Since ethics is concerned with the essential rights and wrongs or moral principles of a person's behaviour, the issue becomes important in many forms of science investigation. Some examples are:

▶ using live animals, e.g. daphnia in caffeine solution

▶ effect on an ecosystem where live samples are taken

▶ using people for timed experiments after stimulants are used to determine change in reflexes

▶ using personal comments or details in survey information

▶ adjusting data to suit the investigation outcome.
   If you have an idea of the expected outcome of an investigation, it is very easy to mistakingly drive your experiment to provide the data which matches your expectation while not accepting the 'real' data. This introduces 'Bias'.

If you intend to use other people as part of your investigative project, for example, in reflex testing when caffeine has been introduced, or exercise to test heart rate changes, you should obtain informed consent. This indicates that the subjects have been given a good explanation of the activity and have willingly agreed to take part. When asking for consent from an individual, you must inform them of their rights, the purpose of the investigation, what procedures will be carried out, what the data will be used for, the possible risks involved and the potential benefits of the study.

Confidentiality of information must be ensured in the above investigative examples, ensuring that important and personal information on health issues or other personal aspects are not divulged to persons without the express permission of the individual concerned in the investigation.

Assessment practice 6.2   B.P3   B.M3   B.D2

Chemical titration is an activity which involves many pieces of equipment and chemical solutions. It needs to be well planned because of the potential hazards which are inherent in the procedure.

1 Produce a comprehensive risk assessment for a chemical titration of an unknown concentration solution of hydrochloric acid with a 0.5 mol dm^{-3} solution of sodium carbonate.

2 Use a suitable document for the purpose.

3 Identify the types of hazard involved and the level of risk.

4 Identify preventative measures and how hazards are minimised.

5 Use the appropriate 'Hazcard' information.

Plan
- What is the task? What am I being asked to do?
- How confident do I feel in my own abilities to complete this task? Are there any areas I think I may struggle with?

Do
- I know what it is I'm doing and what I want to achieve.
- I can identify when I've gone wrong and adjust my thinking/approach to get myself back on course.

Review
- I can explain the results obtained from the task.
- I can apply the activity to other situations.

C Undertake the project, collecting, analysing and presenting the results

All the background research, literature review and planning has now been completed and the practical element of the investigative project is about to begin. This section deals with the essential aspects related to the physical collection of results from setting up your apparatus to your final presentation of the data collected. The setting up stage of the scientific apparatus is of fundamental importance to the investigation and you must ensure that the health and safety information gathered has been understood and is in place. Your skills in using the equipment and observation of important experimental events will need to be perfected so that your results are both **valid** and reliable.

Key term

Valid – the degree to which the method and results obtained reflect the real results.

Experimental procedures and techniques

Assembly

Having produced your literature review, research and planning for the investigative project you have proposed, it is now time for the practical work. Your laboratory and equipment have been booked according to the timescales you have identified and you may need to confirm that your detailed list of apparatus is both available and relevant. Ensure that the tutor and technician are happy with your choice of equipment and that there are no other aspects to be explored before you begin setting up for the activity.

To do this, it is wise to set out your equipment, especially glassware, before you on the same bench where possible and to tick them off on your checklist. It is very easy to 'lose' items in the array of others available and so placing them back after using them, in the very same spot that you originally placed them, will always help to keep the investigation in order.

You should also provide specific information when placing your orders for equipment to the technician. This will include concentrations of acids and alkalis, number and volumes of beakers, etc. Doing this will help to minimise confusion and avoid possible time delays.

Finally, have your bench and equipment/apparatus display checked by your tutor and the technician so that you may begin your investigative project in practical terms.

▶ Students and tutor carrying out scientific investigation observing safe practices.

Adhering to health and safety, rules and regulations

Before carrying out any practical work, it is always wise to revisit your risk assessment and make yourself fully aware of the hazards involved and the potential risks.

The health and safety of you and others in the laboratory should be considered at every major stage in your activity.

Your planning stage will have helped you to understand clearly the hazards and possible risks involved with the activity that you are about to undertake. Revise them and if necessary, review them. Perform your investigation using the utmost caution and with respect to the people around you – including yourself.

Apply your knowledge of materials, equipment, chemicals, biological and physical hazards to ensure that you do not take unnecessary risks.

This checklist may help as a reminder:

▶ Refresh your knowledge of the Hazcard information.

▶ Check your Personal Protective Equipment for your own personal safety.

▶ Are your chemicals easily to hand but also safely positioned?

▶ Is your flame source in clear line of sight and not interfering with other equipment?

▶ Inspect your apparatus, especially glassware, before use – are they clean or do they have small cracks?

▶ Keep a large printed note on the bench in case you have to leave the investigation.

▶ Are your safety goggles available?

▶ Are you fully aware of the safety procedures in case you need to use: eye wash station, fire alarm, fire extinguisher and first aid kit?

You can then notify your tutor and technician and begin your investigation.

Handling skills

It is very important that you know the correct method to transfer materials from one place to another and what amounts are safe to transfer. If in any doubt, you must seek further advice. Liquids and chemical solids present some hazards and difficulties during transfer and should be moved with caution, but also with a high degree of firmness in your grip. Ensure that the distance to be transferred is limited, possibly even to a slight rotation of your body or even just your arms in many cases. Keep focussed on the transfer you are making and make sure that your apparatus does not impede your movements. You should ensure all of the equipment is at hand and laid out on the desk in the correct order before transferring any materials. It is well worth considering a "practice run" of transferring without including the actual material to be transferred.

Some equipment and materials may be very delicate and should be treated with great care, whilst some may be more resilient and require firm handling. Be confident in your handling technique, demonstrating safe and appropriate handling of all glassware by holding with both hands for larger pieces of equipment.

Precision instruments will need further care when handling, to avoid possible damage to delicate springs in a top-pan balance or sensors and probes, for example. Unnecessary damage during the investigation stage is both costly in financial terms and time. You don't want to postpone your investigative project after putting in so much work during the planning phase.

When transferring biological material, in particular bacteria and other microorganisms, the procedure to be followed is necessarily detailed and must be followed exactly. This will ensure that no possible contamination of your culture occurs. This process of transfer is termed 'aseptic techniques' and will involve such procedures as sterilising inoculating loops used to transfer microorganisms (by passing the metal loop through a Bunsen burner flame), for example.

Use of equipment

Notes and diagrams from your planning stage are now ready to be put into practice with the setting up of apparatus. Your tutor and/or technician will need to observe your use of equipment for the assessment of your practical skills. You should ensure:

▶ that you have complete awareness of health and safety issues

▶ that you can demonstrate competence in your assembly of equipment

▶ that you have the ability to manipulate the equipment to obtain results

▶ that you have skills in observation and record keeping practices

▶ that you adhere to **good laboratory practice** (GLP)

▶ that you can take measurements with **accuracy** and **precision**.

Key terms

Good laboratory practice – a system of regulation which ensures that tests carried out in non-clinical laboratories are well planned, reliable and have hazards suitably assessed to reduce risks to the public and environment.

Accuracy – the closeness of the readings to the actual value.

Precision – the degree of uncertainty of a measurement linked to the size of the measured unit.

The way in which you use the equipment available needs to be well practised and suitable for the laboratory environment. Mistakes at this stage will have serious effects on your results and render the investigation invalid because the conclusions drawn will be meaningless.

Repeat readings to obtain useful **means** of the results and conduct preliminary tests if the investigation is suitable. Use the correct 'base' and 'derived' units in all your measurements with the correct symbols. Keep a clean and well-organised record of notes and diagrams in a laboratory note book.

Key term

Mean – the average of all the numbers within a set of results. It is obtained by totalling the results and then dividing the total by the number of results.

▶ **Table 6.4** How to take measurements and use instruments and sensors accurately

Taking measurements	Instruments and sensors
Pour liquids into test tubes and measuring cylinders slowly and at eye level.	Check or recalibrate any sensors or instruments which have not been used often.
Use wash bottles to ensure all solid residue is used.	Carry out test readings to get familiar with values expected.
Hold thermometers at the top and rotate gently in the liquid.	Don't underestimate the importance of analogue instruments or rely totally on digital forms.
Repeat readings using digital and analogue meters to establish those results to be discarded.	Double check electrical meters. There can be considerable differences in voltage and current readings using a variety of instruments.

▶ **Table 6.4** Continued

Taking measurements	Instruments and sensors
Gently shake test tubes when heating liquids until desired temperature is reached.	Ensure that probes are clean.
Check top-pan balances carefully before measuring masses.	Record the readings of more than one power supply to identify any which may be providing inaccurate results.
Read liquid volume measurements at the bottom of the meniscus.	Take repeat readings to confirm instrument or sensor results.
Perfect your use of pipettes.	Handle all sensors according to the manufacturer's strict guidelines.

Observation skills

Good observational skills in the laboratory are essential if you are to successfully identify and interpret a range of reactions, phenomena or other details to draw meaningful conclusions from your studies. During your studies you may observe fluids, tissue sections, cellular specimens, colour changes, consistency changes, meter readings and other physical changes. In some instances it is useful to have a reference chart to compare things to when making observations. For example, if you are observing a colour change you could have a colour chart to help you to make accurate judgements.

Many of these observations will be easily seen but others may only be observed through a microscope, and you need to give yourself the time needed to observe such science related aspects so that your results can be more firmly based on sound evidence.

Becoming a good observer requires training. The more scientific observations you make, the better at observing you will become. When asked to draw what they see on a slide of onion cells, the differences shown by two students can be quite significant.

Discussion

In pairs, give yourselves 60 seconds and count the number of 'red' words in this colourful sentence.

'*First **of all**, **I** would **redefine** the concept of the green synthesis, because **I don't** want **to** reduce this to **an obvious** statement and **swear** that black **is** white*.'

If your observational skills are good, you should arrive at 10.

Adherence to relevant legislation

Good Laboratory Practice (GLP)

The principles of GLP were first developed in the USA in 1978 and later adopted by the European Community. Sections of the regulations are updated when necessary.

In the wider context, GLP ensures that tests carried out in non-clinical working laboratories are reliable and well regulated, assessing the hazards and risks to the public and the environment.

Industries using these principles include:

▶ industrial chemicals
▶ food and food additives
▶ pharmaceuticals
▶ cosmetic chemicals
▶ agrochemicals (used in agriculture, for example, pesticides or fertilisers)
▶ veterinary medicine.

In the context of your school or college, GLP relates to the general, well-established principles of carrying out a practical investigation using safe procedures and suitable scientific techniques.

The general guidelines are set out below. You may already be fully aware of many of those listed.

▶ Produce a detailed risk assessment.
▶ Tie back long hair securely.
▶ Wear goggles always and a laboratory coat where relevant.
▶ Wash hands before and after the investigation procedures.
▶ Wear waterproof dressings on existing wounds.
▶ Do not take food or drink into the laboratory.
▶ Regularly monitor electrical equipment.
▶ Use a Bunsen burner safety flame when not in constant use.
▶ Handle glassware with care and caution.
▶ Prevent liquid spills on electrical appliances.
▶ Identify areas where firefighting equipment is kept.
▶ Identify eye washing facilities.
▶ Identify the gas and electricity emergency cut-offs.

Good Manufacturing Practice (GMP)

Good manufacturing practice (GMP) is 'that part of quality assurance which ensures that medicinal products are consistently produced and controlled to the quality standards appropriate to their intended use and as required by the marketing authorisation (MA) or product specification. GMP is concerned with both production and quality control' (source: www.mhra.gov.uk, a government agency set up to regulate quality and safety in the manufacture of medicines and medical devices).

> **Key term**
>
> **Good manufacturing practice** – a system of quality assurance to ensure that medicines and medical products are manufactured to the highest quality standards.

The agency is committed to ensuring:

▶ benefits to the public justify the risks taken

▶ availability of information

▶ quicker access to medical products and treatments

▶ thorough investigation of defective medicines and appliances.

Good Clinical Practice (GCP)

Good clinical practice (GCP) is 'a standard for the design, conduct, performance, monitoring, auditing, recording, analyses, and reporting of clinical trials that provides assurance that the data and reported results are credible and accurate, and that the rights, integrity, and confidentiality of trial subjects are protected' (source: www.ct-toolkit.ac.uk, Department of Health and Medical Research Council, 2004).

> **Key term**
>
> **Good clinical practice** – a set of quality standards for studies involving human beings.

In general, GCP is a quality standard for studies involving humans. It became effective from 1997 in the European Union, Japan and USA, setting down the principles of sound ethics, conduct and record keeping. It helps to ensure that clinical trials are reliable, confidential and safe.

Collect, collate and analyse data

Recording results

The way in which you record data obtained during a scientific investigation is important and is a fundamental part of your method. If your results tables do not show clarity and organisation, you can easily lose sight of the general patterns which may be available in the data.

Collecting the data must be completed with a high degree of 'Integrity', which means that you must record the results exactly as they are found and not as you expect or would

like to see. Record your data with 'Precision', to the correct number of decimal places based on the equipment used.

In science, there is generally more than one outcome or result from an experiment and there may be many repeats necessary, depending on the type of investigation undertaken. Tabulated recording of results is the most suitable in most cases and even where one outcome is to be recorded, the principle of repeat experiments to check the validity of your results determines the need for you to produce a table. Of course, in many cases you will need to devise your own tables in accordance with the investigation you are performing.

Here are some examples of tables for different experimental results.

Length (m)	Current (A)	Voltage (V)	Resistance (Ω)

Load (or tension) (N)	Extension (mm)

Water-soluble compound	Acidic, alkaline or neutral solution?
Small-chain aliphatic acids	Strongly acid, pH < 7
Small-chain alcohols	Neutral pH = water pH
Small-chain aldehydes	Neutral or slightly acid due to oxidation
Small-chain ketones	Neutral pH = water pH
Small-chain amines	Weakly alkaline, pH > 7

Maintenance of laboratory notebook

Completing a laboratory notebook is considered a lost art, but nevertheless a very important document which shows the work carried out during investigations. Many well-known scientists have had their work interpreted and analysed later from notebook recordings which they completed during practical investigations (the photograph on the next page shows an extract from Charles Darwin's notebook).

▶ Charles Darwin's notebook used during his voyage on the 'Beagle'.

To use a laboratory notebook successfully, follow these guidance points.

- ▶ Protect your lab book with a cover.
- ▶ Use a dark coloured ink, not a pencil.
- ▶ Don't rip pages out or use correction fluid for mistakes. Cross out mistakes with a single straight ruled line through the error.
- ▶ Make good notes on safety issues.
- ▶ Put the title, task outline and date in your book.
- ▶ Start each task on a clean page.
- ▶ Label drawings well.
- ▶ Produce drawings of specialised equipment, not essential items such as small pipettes or beakers.
- ▶ Use bullet points for notable points and comments.
- ▶ Clearly label tables and graphs.
- ▶ Show relevant calculations where useful.
- ▶ Highlight general conclusions and link them to the hypothesis.

Organisation of data

In many scientific investigations, a large amount of data in the form of numerical figures can be generated and so must be organised in such a way that you can analyse the data.

The following example shows the steps needed to organise data for analysis.

'Consider a class set of 20 agar plates showing bacterial colony counts in identical conditions after 24 hours.'

1, 2, 4, 3, 6, 7, 6, 8, 3, 9, 6, 7, 7, 6, 5, 4, 5, 6, 5, 8

Using the first set of figures for the 20 agar plates of bacterial colonies after 24 hours, the mean is 5.3.

Table 6.5 gives the **frequencies** for the data.

▶ **Table 6.5** Frequency table

Score	Tally	Frequency
1	I	1
2	I	1
3	II	2
4	II	2
5	III	3
6	IIII I	5
7	III	3
8	II	2
9	I	1
		20

The figure that occurs most often is 6, so the **mode** is 6.

> **Key terms**
>
> **Frequency** – how often a particular value occurs within a set of values.
>
> **Mode** – the data value that occurs the most often in a set of values.

When the data values are much larger, it is more useful to group the figures into ranges of values which are called class intervals.

In this example, the same agar plates were kept in suitable conditions for a further 24 hours. The colonies grew in number and were counted again:

30, 46, 45, 43, 53, 42, 51, 55, 61, 44, 50,
52, 35, 37, 54, 62, 68, 58, 56, 46

The range is from 30 to 68 (a full range of 38). A suitable class interval width is 5 for this value, so you need 38/5 = 7.6 (rounded up to 8) class intervals.

▶ **Table 6.6** Class intervals

Class interval	Tally	Frequency
30–34	I	1
35–39	II	2
40–44	III	3
45–49	III	3
50–54	IIII I	5
55–59	III	3
60–64	II	2
65–69	I	1
		20

From Table 6.6, we can see that the modal class interval is 50–54.

This tells us that most bacterial colonies grow to 50–54 in these conditions over a 48-hour period.

Methods and uses of data processing and analysis

Figure 6.7 sets out a standard deviation on a normal distribution curve of agar plates/bacterial colonies.

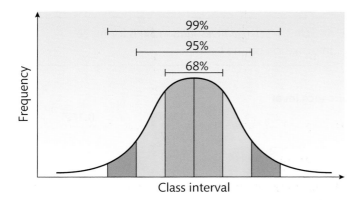

▶ **Figure 6.7** Standard deviation graph of agar plates/bacterial colonies. Generally, about 68 per cent of data lies within 1 standard deviation, 95 per cent within 2 standard deviations and 99 per cent within 3 standard deviations.

> **Link**
>
> Go to *Unit 3: Science Investigation Skills* for more on methods of data processing and analysis.

> **Theory into practice**
>
> Calculating standard deviation (using the formula in the worked example).
> - Calculate the mean.
> - Subtract the mean from each of your datum values to get the standard deviation.
> - Square these numbers and add them all together.
> - Divide this figure by one less than your sample number.
> - The standard deviation is the square root of this value.

Worked Example

You are working in a horticultural laboratory which is testing a mixture of nutrients on the growth of tomato plants. Having measured 10 of the plants, you must now determine their standard deviation.

Results (cm): 10.2, 10.4, 10.3, 10.5, 10.4, 10.8, 10.6, 10.9, 10.6, 10.6

Step 1: Calculate the mean ($\bar{x}$)

$\bar{x}$ = (10.2 + 10.4 + 10.3 + 10.5 + 10.4 + 10.8 + 10.6 + 10.9 + 10.6 + 10.6)/10 = **10.53**

Step 2: Produce a table that shows all the figures at a glance.

A	B	C
Height	$x - \bar{x}$	$(x - \bar{x})^2$
10.2	−0.33	0.1089
10.4	−0.13	0.0169
10.3	−0.23	0.0529
10.5	−0.03	0.0009
10.4	−0.13	0.0169
10.8	0.27	0.0729
10.6	0.07	0.0049
10.9	0.37	0.1369
10.6	0.07	0.0049
10.6	0.07	0.0049

Adding column 'C', $\Sigma(x - \bar{x}) = $ **0.421**

Standard deviation: $s = \sqrt{\dfrac{(x - \bar{x})^2}{n - 1}} = \sqrt{\dfrac{0{\cdot}421}{9}} = 0.22$

The standard deviation for the height of the tomato plants is 0.2 cm and is given to the same number of decimal places as the original set of data.

> **Theory into practice**
>
> As part of a statistical analysis of the flight distance of honey bees on one day from the main colony, a tiny transmitter was carefully attached to a selection of 20 bees and the following set of data has been found.
>
> Maximum distance of bee from the main colony (km):
>
> 1.2, 1.4, 1.6, 2.4, 3.5, 5.7, 6.3, 5.4, 3.2, 4.1, 4.2, 5.2, 3.7, 1.4, 1.5, 6.2, 4.2, 1.8, 2.3, 3.8
>
> Calculate the standard deviation.

Student t-test

This statistical method is used to compare the means of two samples and is effectively an indication of how separate two sets of data are. The final figure is then checked in t-tables (see Table 6.7) to determine the percentage probability in terms of how significant the differences in means are. There are two types:

▶ unmatched pairs (two separate groups are used in the study, e.g. two selected groups of people given a new pharmaceutical for testing blood pressure)

▶ matched pairs (e.g. one group of people tested for their reflexes before being given a new stimulant and then after being given the new stimulant).

Link

See *Unit 3: Science Investigation Skills* for more information on the t-test.

▶ **Table 6.7** The t-test table

Degrees of freedom	Significance level					
	20%	**10%**	**5%**	**2%**	**1%**	**0.1%**
1	3.078	6.314	12.706	31.821	63.657	636.619
2	1.886	2.920	4.303	6.965	9.925	31.598
3	1.638	2.353	3.182	4.541	5.841	12.941
4	1.533	2.132	2.776	3.747	4.604	8.610
5	1.476	2.015	2.571	3.365	4.032	6.859
6	1.440	1.943	2.447	3.143	3.707	5.959
7	1.415	1.895	2.365	2.998	3.499	5.405
8	1.397	1.860	2.306	2.896	3.355	5.041
9	1.383	1.833	2.262	2.821	3.250	4.781
10	1.372	1.812	2.228	2.764	3.169	4.587
11	1.363	1.796	2.201	2.718	3.106	4.437
12	1.356	1.782	2.179	2.681	3.055	4.318
13	1.350	1.771	2.160	2.650	3.012	4.221
14	1.345	1.761	2.145	2.624	2.977	4.140
15	1.341	1.753	2.131	2.602	2.947	4.073
16	1.337	1.746	2.120	2.583	2.921	4.015
17	1.333	1.740	2.110	2.567	2.898	3.965
18	1.330	1.734	2.101	2.552	2.878	3.922
19	1.328	1.729	2.093	2.539	2.861	3.883
20	1.325	1.725	2.086	2.528	2.845	3.850
21	1.323	1.721	2.080	2.518	2.831	3.819
22	1.321	1.717	2.074	2.508	2.819	3.792
23	1.319	1.714	2.069	2.500	2.807	3.767
24	1.318	1.711	2.064	2.492	2.797	3.745
25	1.316	1.708	2.060	2.485	2.787	3.725
26	1.315	1.706	2.056	2.479	2.779	3.707
27	1.314	1.703	2.052	2.473	2.771	3.690
28	1.313	1.701	2.048	2.467	2.763	3.674
29	1.311	1.699	2.045	2.462	2.75q	3.659
30	1.310	1.697	2.042	2.457	2.750	3.646
40	1.303	1.684	2.021	2.423	2.704	3. 551
60	1.296	1.671	2.000	2.390	2.660	3.460
120	1.289	1.658	1.980	2.358	2.617	3.373
∞	1.282	1.645	1.960	2.326	2.576	3.291

The following worked example calculation is for unmatched pairs.

Worked Example

From microscopic analysis, the heart rates of two sets of water fleas (daphnia) were recorded in cool river water, group A and group B. Caffeine solution of 0.01 per cent concentration was also added to the water for group A. There were 10 fleas in each sample group. (n = number of sample, i.e. 10.)

Heart rate A	Heart rate B
113	68
111	56
136	62
121	78
108	82
109	64
117	66
122	78
132	77
116	81

Key term

Degrees of freedom – number of variables that are used to make a calculation.

Means [$\bar{x} = \sum x/n$]:   A = 118.5   B = 71.2

The sum of the squares of each value in each table [$\sum x^2$] :   A = 141 225   B = 51 438

The squares of the totals $(\sum x)^2$:   A = 1 404 225   B = 506 944

$(\sum x)^2/n$]:   A = 140 422.5   B = 50 694.4

$\sum d^2$ using [$\sum x^2 - (\sum x)^2/n$]:   A = 802.5   B = 743.6

Standard deviation (σ^2) using [$\sum d^2/n - 1$]:   A = 89.2   B = 82.6

Variance of difference between means (σd^2) using [$\sigma 1/n1 + \sigma 2/n2$]:

A = 8.92   B = 8.26   Answer = 17.18

$\sum d$ using [$\sqrt{\sigma d^2}$] : $\sqrt{17.18}$ = 4.14

t = **11.43**

Looking at t-tables with sample number of 2, the value obtained is much higher than even the 99.9 per cent probability **(0.1 per cent significance)** which shows as **3.92** in the t-tables.

The heart rates of daphnia in group A are much higher than those of group B, which can be linked in this case to caffeine.

Ⅱ   PAUSE POINT

The following sets of data were recorded by a student who tested the impact of light on photosynthesis for a water plant 'cambomba'. He used 10 plant stems in total, five were subject to light from a 100 W lamp and five subject to light from a 40 W lamp. He counted the bubbles produced for each plant. Determine if there is any significant difference between the samples.

Set A – 37, 42, 39, 50, 48

Set B – 26, 29, 31, 23, 30

Hint   Use t-test for unmatched pairs and also quote your 'null hypothesis' for the experiment.

Extend   Suggest some possible next steps in the investigation to confirm your findings.

Chi-square test (X^2)

This statistical test is used to test or support a particular scientific hypothesis or to identify a relationship between two quantities. Chi-squared is used to compare observed data to expected data. It allows you to see if there is any significant difference or whether the difference is due to chance alone.

> **Link**
>
> The full procedure and explanation of how to perform the chi-square test is provided on pages 159–161 of *Unit 3: Science Investigation Skills*, Student Book 1.

Using the tables for confidence levels at 5 per cent and 1 per cent shown in Table 6.8, you can determine the 'degrees of freedom' for all columns of data from an investigation.

▶ **Table 6.8** Degrees of freedom

		p	
		0.05	**0.01**
Degrees of freedom	1	3.841	6.635
	2	5.991	9.210
	3	7.815	11.345
	4	9.488	13.277
	5	11.070	15.086
	6	12.592	16.812
	7	14.067	18.475
	8	15.507	20.090
	9	16.919	21.666
	10	18.307	23.209
	11	19.675	24.725
	12	21.026	26.217
	13	22.362	27.688
	14	23.685	29.141
	15	24.996	30.578

⏸ PAUSE POINT

In a conservation woodland analysis, species of four birds native to the UK woodland were observed visiting a carefully constructed feeding station and photographed for identification. Given their similar dietary requirements, sizes and other characteristics, it was expected that there would be approximately equal numbers of each bird type observed. The results over a weekly period were recorded:

chaffinch – 37; house sparrow – 52; great tit – 44; robin – 14

The null hypothesis states that there is no difference between the observed and expected frequency of birds. From the tables, at 3 degrees of freedom ($n - 1$), a confidence level of 5 per cent is 7.82.

Use a chi-square test to determine if the null hypothesis should be accepted or rejected.

Hint Produce a suitable table for $(O - E)^2/E$ for each bird type. Find a value for X^2. Is this value larger or smaller than the critical value for a 5 per cent confidence level? Should you reject or accept the null hypothesis?

Extend If another survey were carried out in other, similar woodlands, would you expect to find equal numbers of the same types of birds?

Correct units

We come into contact with units in all aspects of life. A supermarket with an offer on potatoes, for example, always provides the units. This is important because there is a considerable difference between selling potatoes at £1.00 per kilogram (kg) and £1.00 per pound (lb).

In all your measurements and calculations for science, you should begin to use the correct and appropriate units associated with the numerical figure as soon as possible.

It is a common mistake to leave units out when performing calculations or making brief notes but this can cause confusion when you are writing your final reports from lab notes and also when work is being assessed.

Using the appropriate sub-unit of measurement is equally important. The distance between cities is obviously measured in kilometres (km), not centimetres (cm), even though they are both units of distance measurement. Table 6.9 lists some units of measurement.

▶ **Table 6.9** Common units of measurement

Standard International (SI) base units	Derived units	Commonly used SI derived units	Common prefixes
• Second (s) • Kilogram (kg) • Metre (m) • Ampere (A) • Mole (mol) • Kelvin (K) • Candela (cd)	• Force (N) • Acceleration (m/s²) • Volt (V) • Speed (m/s) • Energy (J) • Charge (C) • Pascal (Pa)	• Volume – litres (dm^3), millilitres (ml) • Area – m^2, cm^2, mm^2 • Electrical resistance – Ohms • Density – kg/m^3 • Frequency – Hz	• Nano – (10^{-9}) • Micro – (10^{-6}) • Milli – (10^{-3}) • Kilo – (10^3) • Mega – (10^6)

Assessment of accuracy, reliability and precision

Your investigation findings should now be assessed in terms of the accuracy of results and precision of readings or measurements. A section in your final report will be devoted to your explanation of dealing with accuracy and precision and so the way in which you ensure that both these aspects are in place should be clearly shown in your laboratory note book.

> **Link**
>
> Guidance on understanding the difference between accuracy and precision has been given on page 148 of *Unit 3: Science Investigation Skills* Student Book 1.

Figure 6.8 gives a useful recap of the main points using separate measurements of a verified 52.0 g mass on a top pan balance.

The figures shown in (A) have a range from 51.5 g to 52.2 g, a total range of 0.7 g. All 10 readings are very close to the actual value of 52.0 g, whilst the reading furthest from the actual value is 0.7 g. These reading are very accurate as they are close to the true value.

(A) Accuracy	(B) Precision
51.6	50.3
51.5	50.5
51.7	50.4
52.1	50.1
52.1	50.5
51.8	50.3
51.6	50.2
52.2	50.4
51.8	50.5
51.5	50.3

▶ **Figure 6.8** Accuracy versus precision of measurements

Figures shown in (B) have a much smaller range of values, from 50.1 g to 50.5 g – a total range of 0.4 g. This range of readings is much lower than those in (A) but is much further from the actual measured value of 52.0 g. These readings are precise as they are close together.

The degree to which the measurements can be depended upon is an important factor when you are trying to determine the accuracy and precision of your results. For data to be reliable, there must be a small variation within the values, even though there is always some variation in any set of measurements.

Ⅱ PAUSE POINT

The two data sets were obtained by two students from different titration experiments. This involved neutralising an unknown concentration of hydrochloric acid (HCl) and a 1.0 mol dm^{-3} concentration of 25.0 cm^3 sodium hydroxide (NaOH). The actual concentration of the acid was tested by a technician and known to be 1.1 mol dm^{-3}. Which are:

a) most accurate   b) most precise?

Data Set 1: Volume of titre (cm^3) – 24.7, 25.0, 24.8

Data Set 2: Volume of titre (cm^3) – 23.6, 24.1, 23.8

Hint   Calculate the individual values of concentration, average them and present them to three significant figures.

Extend   Find the average titre value which would provide a concentration of 1.1 to two significant figures.

Resistor components for electronic circuits:

Resistor value – 1000 Ohms
Tolerance – 10%
Range – 900 to 1100 Ohms

Measurement of thin constantan wire:
Diameter measured value – 0.38 mm
Probable error – 0.02 mm
Range – 0.36 to 0.40 mm
Quoted measurements on the label – 0.38 mm ± 0.02 mm

▶ **Figure 6.9** Examples of probable error

Validation of method and results

The scientific method is meant to be a logical set of steps that need to be followed carefully so that you can draw conclusions about an area of study. It is a valuable means to ensure that you are able to organise your thoughts and scientific procedures.

The scientific method you have used should have been well designed and well planned. If this is the case, experimental errors or bias can be greatly reduced or eliminated and your overall confidence in the activity and its results can be high. To establish whether the method used is fit for purpose, you must ask yourself, 'Did the method allow me to fully investigate the hypothesis?' and 'Could I have made any improvements?'

If someone else can reproduce your method and results almost exactly at a later date, then your investigation is **repeatable**. If the data is similar after many repeats, then you can be confident that the results are reliable. This is the basis of scientific work. You can then build on your

theories until they are firmly established. To establish whether a science investigation is valid, you need to ask, 'Will someone else be able to repeat my investigation to get the same or similar results?'

> **Key term**
>
> **Repeatable** – the consistency of a set of results.

When measurements are taken, you need to take into account the possible error value in the reading, particularly when dealing with smaller scales of measurement.

In most cases, the probable error in measurement is quoted so that the value obtained can fall within a range of values and still be acceptable in science (see Figure 6.9).

Sources and magnitudes of errors in readings taken

It is common practice to estimate the probable error using the precision of the scale from which your readings are to be taken. Here are some examples which show the measurements and the probable error.

▶ Pipette scaled in ml (±0.5ml)

▶ Analogue DC ammeter scaled in 10mA (±5mA)

▶ A typical Celsius/Fahrenheit thermometer scaled in 2 °F (±1 °F) and 1 °C (±0.5 °C)

Top-pan balances and other pieces of technical precision instrumentation are thoroughly checked for operational capability in manufacture and subjected to calibration before dispatched to customers. Most will need to be re-calibrated during their lifetime. A top-pan balance that has a 0.1 g error for a considerable length of time will transfer this error to all readings taken with it. This becomes a systematic error and may be difficult to detect. If the operation of a top-pan balance is suspect, the balance needs to be re-calibrated by comparing a set of mass readings against a known standard weight. Adjustments can then be made to the balance.

If doubts are raised during scientific investigation, check the measurements with an identical or similar form of equipment or analyse your measuring technique thoroughly. Remember that repeat readings cannot eliminate systematic errors.

Errors that may occur as a result of reading analogue meters or imprecise scales can be reduced if the following important points are noted.

▶ Ensure that you fully understand the principles involved in your investigative project.

▶ View glass measuring devices, meters and gauges at 90°.

▶ Make repeat readings where appropriate.

▶ Accept that digital meters are no more reliable than analogue versions.

▶ Check the calibration of equipment where appropriate to avoid systematic errors.

▶ If in doubt, check readings with a second meter to avoid further systematic errors.

Data presentation

Expert professional evidence and an array of academic studies have concluded that graphs, tables and numerical displays play a vital part in enhancing the overall quality of the scientific report. It is generally accepted that visual displays of data will provide valuable information to the reader in a much shorter time scale, provided the displays are kept brief, contain the essential information intended and are produced with clarity.

Range of data presentation

Wherever possible, you should use a number of different ways of displaying the data obtained from a scientific investigation to help visualise the information provided and to 'bring out' the relationships or comparisons which are present. Options include:

▶ graphs
▶ charts
▶ sketches
▶ tables
▶ photo/video.

Choosing the correct and most appropriate graph for the data obtained is important both in the context of the ease with which the information can be understood by the reader and also the need to display the relevant information with a sufficient degree of accuracy.

The most common method to display data is using graphs (see Figures 6.10–6.13). The type of graph used is based on the kind of data that is to be displayed, as shown in Table 6.10.

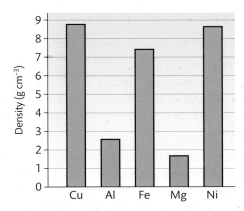

▶ **Figure 6.10** Bar chart showing densities of metals (g/cm³)

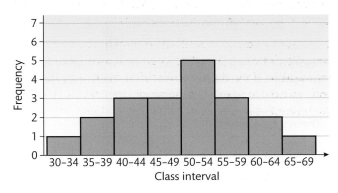

▶ **Figure 6.11** Histogram of bacteria growth in class intervals

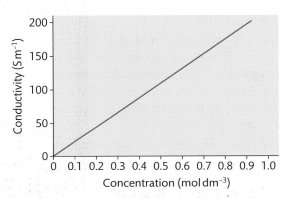

▶ **Figure 6.12** Line graph showing conductivity/concentration for HCl

▶ **Table 6.10** Types of data

Type of data	Description	Examples
Discrete data	• Also referred to as 'discontinuous' • Labels on axes are for measured values and whole numbers for count or other aspects • Pie chart or bar chart typically used	• Number of masses added • Number of layers of insulation used
Continuous data	• Horizontal axis scale from appropriate values • Usually a histogram or line graph • Histogram bars touch to show continuation of values	• Temperature • Mass • Length • Time • Volume
Categoric data	• Specific word labels such as 'Animals – Dogs, Cats, Horses . . .' Axes labelled the same as for discrete charts • Bar charts are typically used	• Names of metals • Names of plants • Names of compounds • Names of animals

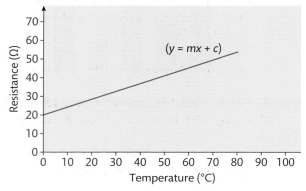

▶ **Figure 6.13** Line graph of $y = mx + c$ showing that the y-axis is intercepted at a point above zero. In this case it indicates that the wire coil has significant resistance at 0 °C.

Theory into practice

Producing appropriate graphs from sets of information and data is an important part of many employment positions in the science industry. Unless you have the ability to present the information in a format that can be easily visualised by others and that makes the data clear to understand, an employer will limit your tasks to more mundane activities.

In your role as a junior member of the technicians' team in a large city college, you have been asked to put together suitable graphs of the following two sets of data which are to be included in a PowerPoint presentation by the Head of Department during an open evening for the college.

The data sets are to illustrate some of the activities and research that are undertaken by the college's Science department:

a) resistivity testing of conductive wires

b) levels of CO_2 recorded between 1950 and 2000.

a)

Metal	Resistivity Ωm
Copper (CU)	1.7×10^{-8}
Constantan	4.9×10^{-7}
Aluminium (Al)	2.8×10^{-8}
Silver (Ag)	1.6×10^{-8}

b)

Year	CO_2 levels (ppm)
1950	310
1960	325
1970	340
1980	350
1990	365
2000	275

Graphs should be able to stand alone in your report. This means that the information about the investigation, such as the title and other important factors, should be placed with the results on the same document.

Including the following aspects will ensure that the focus of your investigation can be viewed immediately from your graph document.

▶ Heading – the title of the investigation and the variables plotted.

▶ Labelled axes – this should include the correct units and scales of measurement.

▶ y axis non-zero indicator – sometimes, the scale on the y-axis does not start from zero. A 'zigzag' is usually placed between the origin and the first scale point so that the proportions are viewed correctly.

▶ Error bars – these show the maximum variations in measurements on the x-axis, y-axis or both axes on the same graph. The length of the bar represents the error or uncertainty in a measured value.

Data format explained

We are all different in our perception of numerical data and how we identify key points in a set of data. For this reason, it is useful to ensure that you provide a variety of displays for presentation of the data and to outline the reason for your chosen presentation.

You may wish to incorporate data into the main body of the text, use the data to produce a graph or chart, or present the data in a suitable table. Your ultimate choice will depend on:

▶ the amount of data or numbers that you are attempting to present

▶ the complexity of the data (significant figures, for example)

▶ the ease with which your data can be visually interpreted

▶ whether you can describe the information in a few words in your table/graph titles and labels.

Assessment practice 6.3   C.P4 C.P5 C.M4 CD.D3

A competent technician needs to make the correct choices for instrumentation and apparatus, using them with a high degree of precision and care. In your studies for this qualification, you will probably have many practice attempts at science investigation for many different units. Your skills and knowledge will almost certainly improve.

Browse over some previous practical activities that you have completed:

1   Explain the function of the equipment and instruments that you have used.

2   Outline your accuracy and precision arrangements.

3   Comment on your evaluations for the activities. You should find that you are able to make many improvements on earlier versions.

Plan
- What is the task? What am I being asked to do?
- How confident do I feel in my own abilities to complete this task? Are there any areas I think I may struggle with?

Do
- I know what it is I'm doing and what I want to achieve.
- I can identify when I've gone wrong and adjust my thinking/approach to get myself back on course.

Review
- I can explain the results obtained from the task.
- I can apply the activity to other situations.

Review the investigative project using correct scientific principles

Your literature review and research have given you a suitable area of study and your planning stage has ensured that all the work undertaken (experimentation, results gathering, observational and data analysis) has been carried out to the best of your ability. Now you must produce a coherent and comprehensive report based on your work.

Your scientific report will encompass all that you have achieved during your investigative project and identify areas where you have particular strengths. It will also identify aspects in the investigation where you may have certain weaknesses. Your report should be well-written, have a logical structure and be appropriately referenced. Results must be evaluated and suitable conclusions drawn. You must consider areas that could be improved if the investigation were to be repeated sometime in the future, and link your findings clearly to the hypothesis.

Scientific report for the vocational investigative project

Correct scientific principles

The following list sets out a standard structure used in scientific investigative reporting. There may be some slight variations in use, dependent on differences in some scientific establishments and the investigation type.

▶ **Title:** The name of the investigation.

▶ **Abstract:** A summary of the whole investigation, to be completed last and should include the summary of the results as well as what was done to produce them.

▶ **Introduction:** The whole purpose of the investigation with some background. The hypothesis should be explained here.

- ▶ **Method:** A step-by-step procedure on how to complete the investigation. It should include the sizes and numbers of the pieces of equipment.
- ▶ **Results:** Identifying comments and tables with figures/observations.
- ▶ **Accuracy:** An in-depth account of how you ensured accuracy in your results.
- ▶ **Discussion:** Detailed comments on your findings, graphs, problems that you encountered, **reliability**, validity.
- ▶ **Conclusions:** General statements of what your investigation was able to show with proposed areas for further study.
- ▶ **References:** A correct list of research sources that directly contributed to your investigation.
- ▶ **Bibliography:** A full list of all research sources that may or may not have been used in your investigation.

Key terms

Reliability – how well a set of experiments, tests or measurements is able to be repeated with similar results.

Bibliography – generally regarded as a list of sources that may or may not have been found useful in providing information for an activity.

Collate your rough notes and experimental diagrams from your laboratory notebook and decide which are most useful to include in your final report. Set out your report in the structure shown above, using the sub-headings shown. It is advisable to begin a new page for each section of the report, although not always necessary. Consider the format of your report from the list below:

- ▶ handwritten bound report
- ▶ word-processed report on suitable electronic storage device or printed bound report
- ▶ PowerPoint presentation, if accepted by your tutor in support of your main report.

Video and photographic evidence will need to be referenced within the report and made easily available.

Terminology

Using the correct words to describe an item or procedure in science is vital to make sure that the whole scientific community is able to understand what you are communicating and so that others will have confidence in the science that you are trying to convey.

Many words used in science are particular to science and will not transfer easily to everyday language, so be careful in their use. Instruments, processes and all items of apparatus in science will have specific words, names or titles which you need to understand and use correctly when in conversation or when completing reports.

Research

Learn and use words for the following:
- glassware
- instruments
- sensors, probes and meters
- scales
- measurements in scientific procedures
- scientific processes
- chemicals
- specific biology/physics/chemistry terms.

Past tense

Since your report has been fully completed after the work has been done, your sentences need to reflect this. All sentences should be in the past tense, as shown in Figure 6.14.

Third person

Reports in science are always written in the **third person** so that there is no use of 'I', 'we', 'you', 'our group', etc. If you do not already use this style of writing, it may take some practice before you automatically use it in your scientific reports. Some examples of statements made in first person and their third person corrections are shown in Figure 6.15.

Key term

Third person – a means of writing an account of an activity without referring to anyone who performed the activity.

"The apparatus was set up as shown in the diagram…"

"R_f values were calculated using the formula –
$$\frac{\text{distance travelled by compound}}{\text{distance travelled by solvent}}$$

"Results were recorded in the table shown."

▶ **Figure 6.14** Examples of the past tense

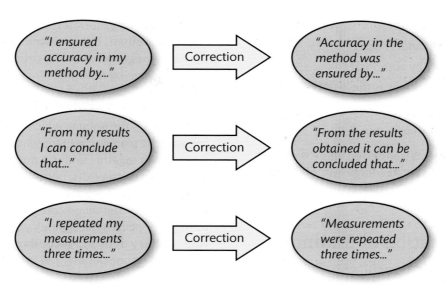

▶ **Figure 6.15** Correcting from first person to third person

---

Ⅱ **PAUSE POINT**

Convert the following scientific passage to third person.

'Our percentage yield of 73 per cent was not as we had expected from our research and preliminary tests. We can assume that some of the loss of yield was caused by the number of transfers we made during our investigation.'

Hint Identify all the words that relate to the investigators.

Extend Rewrite the information in third person, adding changes to the sentences where appropriate and making the passage more fluid and understandable, but not altering the essential meaning.

Remember

To write a good report, you should:
- check your spelling
- check your grammar
- explain abbreviations and acronyms
- use past tense
- use third person.

References and bibliography

When carrying out research which may help you in developing your investigative project, the sources you use must be clearly listed in your completed report to acknowledge that the information is not your own and to provide recognition of the work carried out by the original author(s). The references that you produce can then be located by the reader of your report so that they are able to confirm the details that you have included.

The Harvard referencing system is the most widely used method of identifying the source of information that has contributed to a report. The information taken is cited from sources, which are usually shown at the end of a report or possibly within the actual text and are called in-text citations.

Examples of both forms of referencing are shown below together with the generalised definition of the bibliography. Both references and bibliography must appear in the **appendix** of your report – the supplementary section at the end.

Key term

Appendix – a list of subsidiary material at the end of a report.

Harvard referencing should be in the following format:

▶ **Publication**
A. Fothergill, V. Berlowitz, 2011, Frozen Planet, London, Ebury Publ. BBC Books, Page 205

▶ **Online**
J. Clarke, 2007 (modified 2016), [online], Page 1 available at: Paper Chromatography, *http://www.chemguide. co.uk/analysis/chromatography/paper.html. (accessed 20/09/2016)*

A bibliography should:

▶ list all sources browsed and/or used

▶ be listed alphabetically by author(s)

▶ be written in the same format as the 'references'

▶ appear after the 'references' section in the report.

An example of in-text referencing is as follows:

> Whenever energy is transferred, there is usually some form of unwanted energy. *'We often refer to these unwanted forms as "wasted" energy.'* (Pearson, 2002)

Scientific evaluation of findings

Evaluation of statistical results

You will need to scrutinise your data processing and analysis of results obtained during the practical aspect of your scientific investigation so that you can assess the overall quality and value of the data in relation to the full investigation. This is your evaluation. At this point you will need to identify whether there are any links or correlation between the variables you have tested and so you will need to look closely at what your statistical analysis tells you about the data.

> **Remember**
>
> The null hypothesis can be regarded as the opposite of what you are asking in the investigative project. A *null hypothesis* would state that there is no relationship. By disproving the null hypothesis, you provide support for your hypothesis, i.e. there is a relationship between the variables.

The t-test: this will help you to establish if there is a correlation between the sets of data you have collected. If the means of the two sets of data are statistically the same, then there is no great difference between them. An example of this would be to establish a correlation between gender and height in a non-paired population study. Having calculated the relevant significance level from the sample, if the t-test indicates a value equal to or less than this, we can reject the null hypothesis.

The chi-square test: this will allow you to determine if there is a difference between the values you expect from a quantity and the number you observe. In an example of the general population to determine if there is a correlation between gender and colour blindness, the null hypothesis would state that there is no difference between the occurrence of colour blindness in males and females.

If the chi-square value is less than the critical value, then the null hypothesis can be accepted. If the value of the chi-square test is greater than the critical value, then the null hypothesis can be rejected.

Conclusions drawn

This section is the point at which you can bring your investigative planning, experimental technique and data records together and you must now take your statistical analysis, errors, uncertainty and other data into consideration when drawing your conclusions. Assuming that you have maintained correct and appropriate scientific principles throughout, you will need to review your **primary data** and **secondary data** with a view to developing conclusions and include this into your evaluation of data.

> **Key terms**
>
> **Primary data** – information that you have collected directly during your investigation.
>
> **Secondary data** – information collected by someone else that you have used in your investigation.

It is easy to ignore the importance of the final conclusions and evaluation of your investigation after conducting such an extensive, time consuming and practically demanding study. However, this section represents the culmination of your efforts and, if it were reported on in a scientific article, would form the focus of the first statement.

Some focus questions to produce your conclusions and aid in your evaluation are as follows.

▶ **Is there a link or relationship between the variables?**
No correlation – no obvious pattern between the data.
Weak correlation – points of data are not too close to the line of best fit.
Strong correlation – most points of data are close to line of best fit.
Positive correlation – as the value of x increases, so does the value of y (a positive slope).
Negative correlation – as the value of x increases, the value of y decreases (a negative slope).

▶ **Do the results support the hypothesis?**
If your method was followed closely and correctly, you may not necessarily have results that support the hypothesis.

▶ **Have you identified errors in your investigation?**
Identify anomalous readings and errors and make a valid comment on how you have dealt with them.

▶ **Were there any experimental problems?**
Make valid comments on the difficulties of measurement or suitability of the equipment if you feel they were significant.

▶ **Can you outline improvements?**
Perfect investigations do not exist. There will always be something that can be improved. Make a comment which tries to link your work to the context.

Limitations of investigation

This section of your report should be your chance to identify the strengths and weaknesses in your investigative project. In science activities, the need to measure particular objects or events with great precision and/or accuracy can sometimes be impossible with the type and the limits of the equipment available.

In measuring bacterial colonies under a high power microscope, for example, we can only measure the two-dimensional area of the bacteria, even though the volume of the bacteria may be necessary for us to report on. In this example, the microscope itself introduces certain limitations in the investigation.

Investigation limitations are generally identifiable in three areas:

▶ investigation design

▶ instrument limits

▶ practical considerations not identified.

To deal with investigation limitations, you should:

▶ Identify the limitations: Ask whether your findings show quality in your work and whether you were able to answer your hypothesis. If 'yes' then your limitations (practical skills, instruments, measurements etc) may not be too important. Try to reduce this point to those areas that you feel would have had the most impact on your work.

▶ Explain the nature of the limitations: In your sampling method, for example, you can explain the limitations in numerical terms. Were there enough samples? Did you repeat the investigation enough times? Was there other, more suitable apparatus available? Were there alternative samples in a qualitative study? Can you justify your choices?

▶ Suggest how to reduce limitations: This section is a chance to describe what can be done to overcome the limitations that you have identified. As a result, this work will be preparation for your decisions on recommendations for further research.

Assessment and relevance of information sources

The literature review that you completed for learning aim A has been used to help you develop an understanding and a working plan for the subject area for which you have recently completed an investigative project. At this point in your final report, you must now look back at the information gathered from your list of references and decide whether the information taken was valuable for the purpose of the investigation.

▶ Were your sources reliable and do they reflect current knowledge?

▶ Was your information taken from a reputable publication such as a scientific journal or independent source?

▶ Is there a clear indication that your information was written by informed experts in the field?

▶ Were any of your other sources, such as newspaper or non-technical magazines, also useful?

Your report will now contain a substantial word count and a lot of relevant information that will form the culmination of your accrued knowledge and experience.

> **Reflect**
>
> Read your report thoroughly and identify, in bullet points, all those areas in your work that can also be identified in your research notes from the literature review.
>
> Are the number or points quite extensive and do you have confidence in the information provided from your initial literature review? Identify all sources that provided relevance to your investigation findings.

Evaluation of proof of hypothesis

As a scientist, you should not discuss your hypothesis in terms of being 'right' or 'wrong', but look at whether or not your data and evidence support your hypothesis. A single investigation is never usually enough to provide a categorical answer. You should not rely on this, because the data may be misleading unless other independent investigations are carried out and are in agreement with your initial findings.

If the evidence you have collected does not support your hypothesis, make this clear in your report and look closely at your experimental data to ensure that you have reached the correct decision. Remember that in science your evaluation of the hypothesis will be judged by the quality of the overall investigation and not simply the final decisions that you have made.

'Form a hypothesis and try to prove it wrong, not right! Then perhaps formulate a new hypothesis based on your results.'

In summary, ask yourself what the 'strengths are in your hypothesis?'.

▸ Does it encompass exactly what you set out to test?

▸ Does it fit with the assignment brief?

▸ Can the hypothesis be interpreted in a different way by someone else?

▸ Does it match with your initial assumption of the experimental outcomes?

▸ Was your hypothesis suitable or does the evidence suggest that another hypothesis could have been used?

Remember to include in your abstract:

▸ your hypothesis statement

▸ whether the evidence supports it

▸ how significant your findings are to the hypothesis.

Recommendations for further research

Finally, consider what you would do to improve or change aspects of your investigative project if you were able to repeat it. You should not necessarily aim this section of your report at yourself, but could address the points to other learners who may be able to continue your work. Set your ideas out into a clear list of formal recommendations that are firmly supported by your study. Remember that further research in this context does not mean 'more of the same' but rather alternative areas that could be explored that are linked to the investigative project which you have carried out.

Assessment practice 6.4

D.P6 **D.M5** **CD.D3**

At this stage of the process, your full investigative project is now coming to an end. This is now the opportunity to compile all your work into a clearly organised set of headings to provide a means from which your work can be assessed.

Use this as a checklist for the final completion of your report. Make sure that all points are included:

· correct structure and format
· correct terminology
· written in third person, past tense
· correct referencing and bibliography
· evaluation of statistical analysis
· conclusions
· assessment of information sources
· evaluation of your hypothesis
· recommendations for further research.

Plan
· What is the task? What am I being asked to do?
· How confident do I feel in my own abilities to complete this task? Are there any areas I think I may struggle with?

Do
· I know what it is I'm doing and what I want to achieve.
· I can identify when I've gone wrong and adjust my thinking/approach to get myself back on course.

Review
· I can explain the results obtained from the task.
· I can apply the activity to other situations.

Skill development within project work

Your final report is now complete and it's time to ask yourself *'What have I learned and what skills have I used and developed over the course of completing my project?'*

For many, the identification of their personal skills does not come readily. It is more usual that the answers are provided by an individual following direct questioning from another person, such as your tutor at school or college or supervisor in a working context. Becoming confident with critical self-appraisal and questioning of abilities is an important part of the development of all of us, because in order to develop our abilities further and to make progress in any forthcoming endeavours we need to be realistic in our acceptance of personal abilities and understand our limitations. In short – know your strengths and weaknesses!

Time management and organisation

Think back over the course of your investigative project to the very start of your literature review. Estimate how much time you have spent on the project in total, covering both formal and informal input into the activity. Was it more or less than you had anticipated?

For each of the stages outlined below, identify if it took a longer or shorter time than expected and note any aspects during each stage which could have accounted for the difference (e.g. you needed to change your hypothesis following consultation with your tutor or working partner):

▶ literature review

▶ investigative project proposal

▶ schedule

▶ plan

▶ health and safety and ethical considerations

▶ experimental procedures and techniques

▶ collect, collate and analyse data

▶ data presentation

▶ scientific report for the vocational investigative project

▶ scientific evaluation of findings.

▶ **Figure 6.16** What skills have you learnt and what skills have you used over the course of your project?

Discussion

How much time difference have you been able to account for? Which stages were the most difficult to determine in terms of the length of time needed? Where could you have saved time and where was time wasted?

Adhering to and following appropriate standards and protocols

At this stage in the completion of this unit, you should have a very good idea about the standards in scientific investigation and the need to adhere to the general procedural method so that you were able to complete your investigative project successfully and, hopefully, without incident.

Irrespective of the subject of your investigation, the protocols followed would have been developed over a very long period of time and have been fine-tuned by generations of scientists to ensure that all eventualities have been taken into account. This scientific practice takes time to develop. We are not born with the natural ability to foresee everything!

The following questions provide some guidance for you to consider in order to complete your self-appraisal concerned with adherence to science standards and protocols during your investigative project work:

▶ Was my equipment choice appropriate?

▶ Was my equipment handling good at all times?

▶ Did I consult relevant documentation, such as procedures and Health and Safety, during the procedure?

▶ Did I follow my plan closely or adapt correctly when necessary?

▶ Was my attitude mature and professional at all times?

▶ Did I encounter problems and deal with them appropriately using contingency planning?

▶ Was my laboratory notebook used correctly?

Taking responsibility for completing tasks/procedures

Whether working individually, in pairs or even within a group, all of us must take responsibility for the individual tasks that we are performing. This does not simply imply that we are to carry out the activity, make a few notes and 'wash our hands'.

Committing to a science investigation is an important undertaking and must always be given the utmost consideration and attention during the planning and experimental stages. It does not matter what your particular responsibilities were in the overall activity, what matters is the manner in which you completed it.

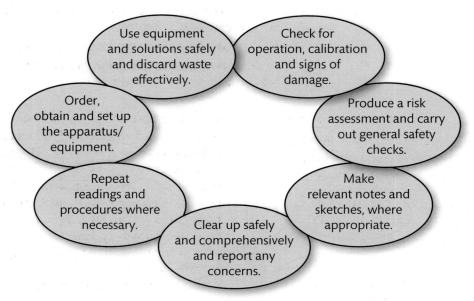

Use equipment and solutions safely and discard waste effectively.

Check for operation, calibration and signs of damage.

Order, obtain and set up the apparatus/ equipment.

Produce a risk assessment and carry out general safety checks.

Repeat readings and procedures where necessary.

Make relevant notes and sketches, where appropriate.

Clear up safely and comprehensively and report any concerns.

▶ **Figure 6.17** Questions to ask of yourself

If you have worked alone on the investigative project, then you alone must bear the full responsibility of the tasks that you performed. Ask yourself 'Did I . . .?':

Making judgements within defined parameters

Working in the discipline of science involves a greater degree of personal responsibility than in other subject disciplines which do not involve the need to be extra vigilant with equipment. You will generally be expected to undertake work with a higher level of complexity, linking physical and chemical apparatus and equipment to the subject information to determine a satisfactory outcome under strict scientific conditions.

Working independently or within small groups means that you make more decisions when carrying out your work. The judgements you make will have important influences on the eventual outcome of your investigation. You must be clear about the scope of your investigation at the start so that you can work within the measurable limits (defined parameters) of what you were finding out.

It is important that you were able to identify what the defined parameters were, how they were to be measured and what, or how, you needed to observe in order to measure them.

For the purpose of assessing your skills development, it is useful to list these parameters, record the judgements you made and comment on your performance. Are there any judgements or decisions you made with a low level of confidence? Are there any you made with a high level of confidence?

Application of safe working practice

With the very best of intentions and the most careful planning considerations, it is still true to say that not all safe working practices are followed ALL of the time.

The breadth and detail of safe working practices you have applied will depend on the type of investigation you have completed. With hindsight, you should now be in a very good position to appraise critically your safe working practice based on your pre-investigative planning. Correct and honest identification of those safe practices you fully observed and those which you 'let slip' from time to time, will help you to establish a firm knowledge base for future investigative work and may well protect you from a possible dangerous incident.

Your completion of a risk assessment, identification of hazards, use of suitable PPE and adherence to COSHH regulations may have fulfilled the requirements set down in health and safety legislation, but your thoughtful consideration of safe working practices which you did not perform consistently can determine whether you have seriously learned from this activity.

⏸ PAUSE POINT   Look back over your health and safety considerations in your project plan. Identify those which you adhered to and those which you did not.

Hint   Were there times when you removed your goggles and delayed putting them back on? Did you carry out some practical work alone for any given time? Did you carry delicate glassware appropriately?

Extend   How could you ensure that all those working in a laboratory environment adhered strictly to high standards of safe working practice? Would you change or add to 'Notices'? Could you introduce sanctions, perhaps?

Give and receive constructive feedback

The manner in which we are able to provide feedback to others and to receive it ourselves will eventually determine the extent of our own future improvement and the degree of personal development in both practices and understanding. Remember, 'feedback' is not simply advice. It is information given freely in an attempt to identify what went well and what went not so well, so that improvements can be made.

When offered correctly, and from an objective point of view, the feedback can be very constructive, leading to sound improvements in practices and understanding. It is this form of feedback which is supplied by the trained professionals in your school or college and should be accepted positively. In a similar manner, feedback which you provide to your investigation partner should also be objective and developmental, explaining the reasons behind your comments and giving guidance to the way forward in an activity. When you provide feedback to your tutor or technician about your investigation, you should also adhere to objective statements in a manner befitting the professional environment you are working in.

Case study

Tamelia is a second year sixth form student in a very successful urban college of further education. She has been performing many science investigations in her time at the college and has produced some outstanding reports based on biological, chemical and physical scientific investigations. Her most recent investigation involved a study into the effects of pollution from acid rain on the germination of cress seeds for which she needed to produce standard solutions and titrations to determine the concentrations used.

During and following her practical activity, she was given feedback from a supply tutor about her investigative procedure:

"How on earth did you break that beaker!? You haven't used enough seeds in each dish and your acid concentrations weren't accurate. Don't ever scoop the broken glass into the paper bin and why weren't you wearing goggles at all times? Where's your risk assessment? Shanice was supposed to be with you but you did it all – why? Your notes are very poor. Hasn't Mr. Evans shown you how to carry out science practicals before? I want it done again, only right, this time!"

Check your understanding

1   Re-write the feedback comments to reflect a more professional tone.

2   Tamelia was at fault in some aspects of the investigation. Identify them.

Identify, organise and use resources effectively to complete tasks

So, what were the resources at your disposal and did you use them in a suitable manner to complete the investigation? (See Figure 6.18.)

Utilising channels of communication

By what means do we communicate in science, before, during and after completion of an investigative project?

Consider the effectiveness of your overall communication to others in your work. How well did you explain what was needed and how well was your communication received? Identify areas where there may have been misinterpretation with colleagues or staff and times when communication was simply lacking between you and those around you. What would you have communicated differently? Did you use all available communication channels (see Figure 6.19)? How could your procedural communication be improved? Did you need to re-phrase anything to anyone?

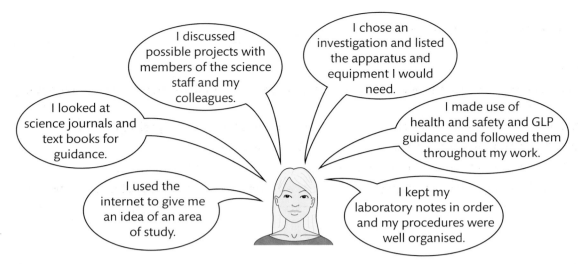

Communication channels
Verbal (questions, equipment orders etc)
Written in pen (lab. notes, equipment orders etc)
Word processed/e-mail/texting (project plan, equipment orders, guidance from staff etc.)
Physical gesturing (simple thumbs-up when level of solution has been reached, perhaps)

▶ **Figure 6.19** What communication channels did you use?

Being resourceful and using initiative

Being resourceful is an attribute which some people have naturally and others may acquire over a long period of time, given the guidance and tuition. It is the difference between completing a project to a satisfactory standard and completing one to a high standard, using an abundance of materials, resources and innovative ways to incorporate them into your work.

When working on a task, whether individually or within a group, it is usually apparent who has the greatest level of resourcefulness and the initiative to put their plans into action.

Review your entire investigative project and ask yourself the following questions. It will help you to learn about the general failings in the way that you performed the work, but also help to provide valuable information about **YOU** and the way in which you will be able to improve your performance for the tasks awaiting you in future years.

▶ Did I realise immediately who the expert in the subject was and use their abilities from the start? (This is important in providing you with the fundamentals of the investigation and also objective and additional information which may steer you in a slightly different direction.)

▶ Were there other ways I could have achieved the result and did I pursue them? (This may depend largely on the extent and depth of your initial literature research which can be underestimated in its importance to providing comprehensive information on the subject material.)

▶ Was I content with confirming what the outcome was going to be or did I try to find something different? (Most scientific investigations in school or college are based on tried and tested methods and so will usually produce the established results – but there may be some which produce results that do not always conform.)

▶ If you worked in a group or partnership, did you adopt a culture of listening and encouraging all possible thoughts and initiatives? (Many ideas we share or keep to ourselves may not be particularly useful or correct, but sometimes even the most unusual ideas can be beneficial to a given problem and should always be considered.)

Remember

When something is important to us, we are all capable of achieving a solution to our problem.

Resourcefulness is a product of necessity and our creativity and persistence.

Further reading and resources

Websites

www.sop-standard-operating-procedure.com website which provides information on standard operating procedures in a laboratory

www.rsc.org/learn-chemistry resources to assist tutors and learners in chemistry related topics

www.hse.gov.uk website of the regulation authority for health and safety

www.npl.co.uk National Physics Laboratory website for ionising radiation and ultrasound

www.physicsclassroom.com a virtual physics classroom with a variety of resources at different levels

THINK ▶FUTURE

Richard Young Science technician

I work in the science section of a very busy urban college on the outskirts of one of the largest cities in England and have been working as a science technician for about 7 years. I am currently studying for my Higher National Diploma whilst working in the college on a full-time basis.

My role involves dealing with the needs of all laboratories in the science department, but I specialise in the chemistry section. The daily routine can be very changeable because of having to respond quickly to the demands of the individual science disciplines. Much of my time is taken up with the Biology and Chemistry sections of the department, but there are also opportunities to cross over to the purpose-built wing that houses the specialist Physics equipment. As an experienced science technician, I am now called upon to produce annual notes for the department. These are added to the departmental report sent to the Head of Centre.

A typical day in the science department begins with a general check on the maintenance of the laboratory equipment, particularly if it was used late in the previous day. Some maintenance forms part of the daily and weekly specific routines: equipment calibrations, glassware, electrical checks and checking of biological equipment such as microscopes. It is my responsibility to ensure that all junior technical staff are supervised and are able to perform their duties in the science sections to which they are allocated. I also provide induction and continued training for technical staff as required by the senior departmental manager. Much of my time, however, is spent tending to the needs of the science lecturers and making sure that they have all the equipment and apparatus necessary to perform the investigations and demonstrations required for teaching and student project purposes. I find that I am regularly called upon to test a new method of performing a particular experiment or an alternative experiment sometimes devised by the lecturers themselves. This is one of the more interesting aspects to my role, since it also tests my understanding of the subject material and demands more engagement on my part. My work is overseen by the Head of Department and Senior Manager, who check on my continued professional progress and understanding of procedures.

At present, I am increasing my qualification status by studying software applications in technical sciences at the college part-time in the evenings and hope to enrol onto a full degree programme if this is successful. I am also developing my skills with people, communicating more effectively in writing and face to face conversations.

Focusing on Skills

Think carefully about the role of a science technician in your school or college.

1 What scientific knowledge will you need and what additional skills are required to perform your tasks successfully? Can you identify what skills you currently have?

2 What sort of tasks will you be carrying out on a daily basis and which ones will be carried out less often?

3 Who will you be working with and how will you support them in their roles?

4 What can you do to improve your knowledge and skills in the science laboratory? Are there likely to be good opportunities for further education or experiences?

Getting ready for assessment

Corey is working towards his BTEC National Diploma in Applied Science. His first assignment was based on producing a literature review that would be used to develop his investigative project proposal. The title of the assignment was 'What am I interested in?' The assignment expects learners to research areas of science with a view to finding a topic of interest that can be investigated independently. A full range of resources is to be used and recorded in the appropriate referencing manner. Once the topic for investigation has been decided, Corey will be expected to write a rationale for his area of study in a report that will be added to during his investigative project proposal. The report must include the following: identification of sources of information used, what type of study is proposed, e.g. field work, laboratory work etc., why he chose the area of study and the background to the topic, a hypothesis based on his research and a set of aims or objectives.

Corey shares his experience below.

How I got started

First I made sure that my laboratory notebook was available and that I had a clear understanding of what I needed to do. I discussed the type of research with my tutor and science technician so that I had sound advice and began a general search on the internet and in reference books on areas of science which I find interesting. I had already decided that I would like to use my centre's new sports facilities, so that the limitations of my project were reduced, and was then able to focus my investigative project into physiological aspects of testing. This was important because I then had a clear direction and was able to research the subject in greater depth. Having taken relevant notes from the research sites and listing them as possible reference sources, I began to fill in a table to list them.

How I brought it all together

Writing the rationale for my proposal was quite straight forward since I had a particular interest in sport and keeping fit. I began to find further background information on my proposed topic so that I could focus my investigation into an area that could be tested. At this point, I needed to identify exactly what I would be testing and if it was measureable. My tutor reminded me that, if other learners were used to provide the data for my project, then I would need their permission. I came up with three possibilities for my proposal.

After considerable thought and further research, I decided upon the effect of exercise on a person's reaction times because of the data that would be generated. Producing a hypothesis such as 'exercise increases a person's reaction times' gave me a clear direction for my study and resulted in a simple null hypothesis: 'there is no difference in a person's reaction times observed after exercise'.

My aims and objectives followed from my proposal and I needed to identify the issues regarding use of people in the sample, the length of time that would be needed for the testing and the availability of the physical equipment especially since the fitness room is a very popular area for most learners and staff.

What I learned from the experience

I became very involved with the project proposal and the information that I was beginning to find from research sources. I was surprised to discover that I had a very keen interest in the science behind the subject area. I was nervous when asking a group of other learners to take part because I was a little unsure as to what I needed to tell them. My tutor provided additional and important guidance on this.

Finding relevant information was easy and there were many similar web pages that gave information on my project proposal with some providing an actual method. I learned how to use the Harvard referencing system, although this needed a lot of practice at first. The background information available was staggering. Much of it was a little beyond my understanding, but I was able to narrow down the information to the basics I needed to help me focus on the direction of the project. By proposing to use the centre's fitness equipment, I was able to decide which pieces of equipment my 'subjects' would be using and also book these in well in advance.

Think about it

▶ Have you decided how many sources of information you intend to use and where they are from?

▶ What age of sources are you going to use and which type are you going to exclude?

▶ What is the nature of your study?

▶ Have you clearly referenced sources from a wide variety?

▶ Can you demonstrate a good understanding of your project proposal and the subject involved?

Contemporary Issues in Science 7

Getting to know your unit

Our knowledge and understanding of all the different areas of science is developing fast, helped by advances in computer technology. As scientific development and innovation increases, our need to discuss and address the issues involved becomes ever more important and complicated. Study of this unit will give you an insight into why science reporting and investigation must be firmly grounded in scientific evidence. You will gain an understanding of the ethical, social, economic and environmental impacts of some of these developments, including their potential benefits, disadvantages and risks. You will discover the influence of large organisations such as governments, private companies and universities on the direction of scientific technologies. You will consider how scientific evidence is presented and how scientific information is interpreted, analysed and evaluated. You will develop an appreciation of the way in which science is reported through different media and you will develop an ability to analyse and question the way in which scientific evidence is presented.

How you will be assessed

This unit will be assessed through a written task set by Pearson.

The assessment is designed to test:

▶ Your knowledge and understanding of contemporary scientific issues

▶ Whether you can apply your knowledge and understanding of contemporary scientific issues to real-life scientific scenarios

▶ Whether you can make valid judgements based on interpretation, analysis and evaluation of different sources of scientific information

▶ Whether you can apply and synthesise scientific ideas from several different sources and adapt to other real-life scenarios.

The questions in your task book are likely to include the following areas:

▶ Discussing the impacts/implications of the scientific issues in the articles

▶ Identifying the different individuals/organisations mentioned in the articles and commenting on how they might influence the scientific issues

▶ Discussing (with evidence) whether an article has made valid judgements

▶ Suggesting potential areas for further research and development relating to the scientific issue in the articles

▶ Writing your own article about some aspect of the science for a specific target audience

Part of this unit gives you an insight into contemporary scientific issues. The rest helps you decide how to recognise good science from bad, and how to judge if the science you are presented with is valid or not. As you work through it, keep in mind the skills you need to develop to help you succeed in your assessment. You will find lots of activities to help you. Completing these activities doesn't give you any credit towards your unit, but you will have carried out useful research or preparation which will be relevant when it comes to completing your assessment.

Getting started

The concepts and ideas involved in scientific research are wide ranging and also complex in terms of the science which helps to explain them. As the development in 'cutting-edge' science continues, the number of applications to which it is put increases. This, in theory at least, also helps people all over the world. Working with a partner, produce a questionnaire to be used on learners and staff at your school or college to help you appreciate the general level of scientific understanding in a small sample of the population. Your questionnaire should be a maximum of 20 questions designed to show if people understand a contemporary science issue, e.g. the difference between non-renewable and renewable energy sources or between embryonic and adult stem cells, or the value of space exploration.

A Contemporary scientific issues

This section outlines a number of the fundamental scientific issues which are linked to areas of scientific development. Study of this section will focus on: energy sources, medical treatments, pharmaceuticals and chemicals, food technology and nanotechnology. In each topic you will be provided with an outline of the contemporary scientific issues involved and the impact of some of the technological developments associated with the science.

Understand the scientific issues in terms of ethical, social, economic and environmental impact

Science is moving forward fast. There are developments in areas ranging from the nature of the universe to new ways of treating both inherited and infectious diseases. Much contemporary science – especially the new developments which result from our expanding scientific knowledge – has an impact on human life and well-being. Whenever you look at a contemporary scientific issue you need to consider what ethical, social, economic and environmental impact it may have – if any. What do these terms mean?

▶ **Economic impacts:** Scientific discoveries and technological developments often affect the financial situation of individuals, companies or countries in unexpected ways. For example, a genetically modified crop which has a much higher yield can have a positive economic benefit on farmers and indeed a whole country. A new medical treatment may make a lot of money for the company which produces it but make a health service or individuals who want it much poorer.

Often the economic impacts of a scientific issue are complicated and work at very different levels.

▶ **Environmental impacts:** Both global and local environments are fragile. A new scientific development may harm the environment – for example, when CFCs were developed as refrigerants, no-one could have imagined they would cause a hole in the ozone layer. However, scientific developments can also protect the environment – for example solar lamps used in Africa both reduce harmful gases being released into the atmosphere and improve human health.

▶ **Ethical impacts:** When people try to decide the ethical impact of new science it will depend on many things. For example, many people think that treatments which help infertile couples have children are a very good thing, but some people think it is unethical either because of their religious beliefs or because they think the human population is too big already.

▶ **Social impacts:** Every time a scientific or technological development changes how people live, it has a social impact. For example, antibiotics have had a big impact on life expectancy and family size in many countries by preventing deaths from bacterial diseases. GM crops containing high levels of nutrients or vaccines could have a similar impact in areas such as sub-Saharan Africa. Driverless cars could greatly reduce road deaths and enable old people to remain independent longer.

Energy sources

Energy transfers are used to power our cars, boats and planes and for lighting, heating and cooking in our homes and industries. Most of the energy we use comes from fossil fuels such as coal, oil and natural gas. Even the electricity we use in our homes has often been produced by burning fossil fuels. Our use of fossil fuels raises two huge issues:

1 Fossil fuels are finite – they cannot be replaced and sooner or later we will use them all up.
2 When fossil fuels are burned, the chemical reaction between the fuel elements and oxygen in the air follows a very similar pattern in all those fuels which are based on hydrocarbons. The process for methane (natural gas) is shown below:

$$CH_4(g) + 2O_2(g) \rightarrow CO_2(g) + 2H_2O(g) + Energy$$
Methane Oxygen Carbon   Water   Heat
                 dioxide

Carbon dioxide is a **greenhouse gas**. This means it acts like the glass in a greenhouse to prevent infrared radiation escaping from the Earth's atmosphere, keeping the surface warm enough for life as we know it. The carbon dioxide in the atmosphere allows light from the Sun to reach the surface of the Earth – but the wavelength is changed as it passes through the atmosphere to the surface and bounces back. So some of the infrared light cannot escape. But as the levels of carbon dioxide in the atmosphere increase, more of the infrared radiation is trapped and the temperature at the surface of the Earth is increasing. This is known as global warming. Most scientists are convinced that this is the result of us burning fossil fuels (see Figure 7.1). Global warming appears to be causing climate change and the melting of the polar icecaps.

One of the biggest scientific issues of our time is to find alternative energy sources to fossil fuels – but those alternatives may also cause some problems.

Renewable and non-renewable fuels

Scientists all over the world are working hard to develop alternative, renewable energy sources to be used for electricity generation or as fuels for our many different types of vehicles. Most of our systems for generating electricity rely on rotating a coil of copper wire (armature) at very high speed inside a curved magnet (see Figure 7.2). Currently most electricity generation depends on using energy from burning fossil fuels or nuclear energy to rotate the coil.

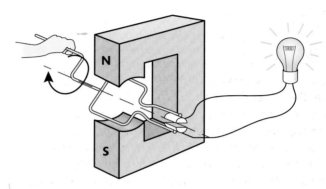

▶ **Figure 7.2** Simple electrical generator principles – rotating a coil of wire inside a magnetic field produces electricity.

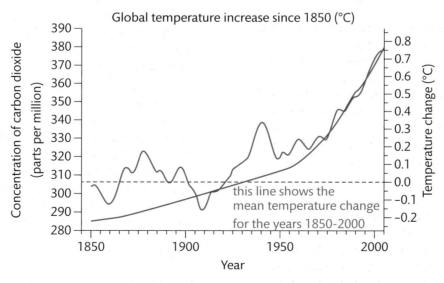

▶ **Figure 7.1** Graph (the blue line) showing the average global change in temperatures from 1850 to 2000. The red line indicates the variations in recorded CO_2 concentrations in the atmosphere over time but demonstrates an upward trend. What is the apparent correlation between the CO_2 levels and temperature rise? What more would you want to know before you could think about a causal link?

The source of energy we use to generate the electricity will determine the availability of the electricity produced and the subsequent costs and effects on the environment, because non-renewable sources will eventually run out.

Table 7.1 shows a summary of the types of renewable and non-renewable methods currently at our disposal and the aspects which need to be considered for each energy source.

▶ **Table 7.1** Renewable and non-renewable methods

Renewable	Ethical points	Social aspects	Economic impact	Environmental impact
Wind	Possible health effects (little evidence) and environmental effects, but important alternative to greenhouse gas production.	Home grown industry feeds into the economy, little effect on farming. money generated internally.	Jobs created during construction and maintenance, free fuel, low cost energy.	Damage to local environment during construction, noisy and unsightly, damage to birds in migration/flight, but clean energy.
Hydroelectricity	Uses a clean fuel source – water. Locally produced electricity so no reliance on imported energy. No greenhouse gases produced during operation.	Minor job creation in rural areas, creation of usable lakes for recreation and fishing, tourism.	High initial costs but cheap operating and maintenance costs.	Large-scale flooding of landscape, changes to wildlife habitat, damage during construction of dam.
Geo-thermal	Great reduction in greenhouse gas emissions, possibility of geological damage from extraction.	Improved home heating capacity in suitable areas, cheap energy for all, land disputes, and short-term disruption.	Cheap to produce after initial outlay, job creation on large plants.	Drilling and siting of large industry materials, unsightly, non-polluting, renewable.
Solar	Clean, renewable and helps to reduce dependence on fossil fuels.	Solar panels use up local land and create few jobs in maintenance.	Expensive to produce and intermittent energy supply, low costs afterwards.	Some disruption and displacement of wildlife, unsightly, minimal pollution.
Biofuels	Deforestation and biodiversity reduction, contributes to lower greenhouse gas emissions, some human rights violations in crop-growing regions.	May help to revive towns with crop growing, land used for growing biofuels instead of crops affects food supply.	Unsustainable currently as costly, job creation for poorer countries.	Large areas taken to produce crops, no long term ecosystem, lower levels of greenhouse gases given off.
Non-renewable				
Coal/oil/natural gas/shale gas	Damage environment when extracted, release greenhouse gases when burned, underpin economies of many countries.	Provide many jobs in extraction/production, provide source materials for many industries/jobs. Cause health problems which cost money.	World economies initially flourished but this may be a slowing trend. Fossil fuels still provide economic benefits.	Production of greenhouse gases, sulfur dioxide producing acid rain, particulate materials and other gases. Causing global warming, climate change and massive environmental impact.
Nuclear	Scientists accept as safe but unrealistic public perception of risk. Problem of disposal of nuclear waste but potential for safe and long-term energy.	Perception of serious risk to life, public confusion and lack of communication between industry and public. Potential for clean long-term energy supply.	High production of electricity, job creation, expensive means of energy production until established (well-established in some countries).	Damage from initial construction. Small risk from disposal of waste, leaks or accident. No greenhouse gas production.

From www.dailymail.co.uk 22 April, 2008

'The shocking picture that shows how a wind farm has disfigured one of Britain's loveliest landscapes'

(By R. Camber and D. Derbyshire)

During a long and bloody history, it has withstood more than a dozen sieges and held firm against the army of Bonnie Prince Charlie. But now Stirling Castle has been surrounded by a new and very modern army – of towering wind turbines.

This extraordinary picture (*right*) shows a sprawling wind farm dramatically overshadowing the famous city where Mary Queen of Scots was crowned in 1543.

Standing a staggering 328 ft high, the 36 looming turbines dominate the skyline of the Braes O'Doune and have angered many local residents, who claim they have blighted one of Scotland's classic vistas.

Check your understanding

1 How would wind energy benefit the surrounding area?

2 Did the local authority or Scottish parliament have any alternatives to wind power?

3 Are there any other sizes of wind turbines?

4 Research why this site was chosen.

5 Based on this article, what viewpoint do you think the newspaper takes on alternative energy production?

▶ Wind farm situated behind Stirling Castle on the River Forth, Scotland. What do YOU think about the turbines on the hill?

What other viewpoints on alternative energy production can you find?

Fuel and transport

The fuel types used to provide energy for our cars, ships, trains, planes and other powered vehicles is usually referred to as 'transport fuel'. Transportation of people and materials accounts for up to 20 per cent of the world's energy consumption and it is estimated that up to 80 per cent of this fuel is derived from petroleum, itself produced from refining crude oil. The main fossil fuels used for transportation are diesel, petroleum (petrol), marine fuel, liquified petroleum gas (LPG) and aircraft fuel.

Non-fossil fuels used for transportation include biofuels and synthetic fuels. Biofuels are fuels made from biomass. Ethanol is made from fermentation of sugars rich in carbohydrates and biodiesel is made from a mixture of alcohol, animal fat, vegetable oils and recycled grease and has a lower energy content. Synthetic fuels are made from direct conversion of coal, biomass and gases to liquid fuels.

▶ Transporting goods around the world is both essential to world economics and damaging to the environment. This tanker uses the highest polluting waste oil – left over from the oil refining process – to power its massive engine which uses about 380 tonnes of oil per day at sea. It has been estimated that only 15 ships of this size release the same amount of sulfur dioxide gas (which dissolves in water droplets in the atmosphere and produces acid rain) in one year into the atmosphere as all the world's cars. Suggest two alternative ways of transporting these goods, with their pros and cons.

Inject CO₂ into porous rocks

Trapping the CO_2 deep underground as oil and gas are trapped naturally. This method is both relatively cheap and environmentally friendly.

Inject CO₂ into the oceans

Carbon dioxide is soluble in water and can be stored in the oceans which already naturally take in the gas. This is relatively inexpensive but the long term effects on the ecosystems in the sea are not clear.

Mineral carbonation

In nature, CO_2 reacts chemically with minerals in the Earth's rocks. This happens very slowly so the overall effect on emissions will be quite low but costs could be very high in energy terms.

▶ **Figure 7.3** What can we do with the CO_2 once we have captured it?

Carbon capture

'**Carbon capture**' or 'Carbon dioxide Capture and Storage' (CCS) describes processes used to remove the carbon from the CO_2 emissions from fossil fuel power plants and store it in underground storage areas – just as the Earth has been doing naturally for millions of years. Figure 7.3 shows what can be done with the carbon dioxide once it has been captured in this way.

> **Key term**
>
> **Carbon capture** – the process of capturing and storing CO_2 from fossil fuel power stations, for example, preventing it from entering the atmosphere.

In the natural world carbon is stored in the biomass of plants and marine algae, in coral reefs, in solution in the sea, in fossils and in carbonate rocks.

Artificial carbon capture technology is in its infancy but continued research is being shared between universities and research centres around the world, including the UK, the USA and the Middle East.

The amount of CO_2 captured by current systems installed at fossil fuel power plants in the USA can be as high as 90 per cent, if the CO_2 is quite pure. Unfortunately at the moment the additional cost involved in CCS means that the electricity produced where CCS is used will be more expensive. The cost can rise by as much as 85 per cent and will be passed on to consumers. We need to decide how much we are willing to pay to guarantee the future of our planet.

In summary, CCS is a feasible way of reducing our effect on the Earth's atmosphere, but there remain significant problems in cost and technology which need to be overcome. It is generally agreed that CO_2 levels in the atmosphere can be greatly reduced using a combination of technology in CCS and world agreement on limiting CO_2 emissions.

Medical treatments

Medical treatments have progressed a very long way since we amputated limbs without anaesthetics or applied mouldy bread to infected cuts. Over 3 billion pounds is spent in the UK alone each year on medical research of all types (UKCRC Health Research Analysis 2014). Medical and surgical treatments are changing fast. However, new treatments can be very expensive, and there can be ethical dilemmas about their source and availability. Should society pay for new treatments or are they the responsibility of the individual patients? Our search for better and more effective treatments can have wide-ranging impacts as well as health benefits. Some examples follow.

Proton beam therapy

Protons are positively charged particles in the nucleus of atoms. They can be accelerated to very high speeds using a particle accelerator called a 'synchrotron' and then focused into a very small area. This allows them to be used in the treatment of cancerous disease, particularly in the underdeveloped brains of young children. The proton beam damages molecules in the cancerous cells but there is less damage to healthy tissue surrounding the tumour.

> **Key term**
>
> **Proton** – positively charged particle in the nucleus of all atoms.

The advantage of using a beam of protons to kill cancer cells is that the protons have no more energy once the target is hit. This means there is less damage to the surrounding healthy tissue than in most treatment methods. Proton beam therapy is a suitable method in the treatment for cancer which may lie close to important

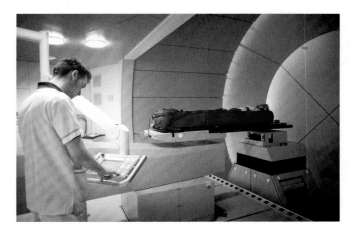

▶ A proton beam therapy room and equipment. At present this treatment is unavailable in the UK

organs or nerves, such as the optic nerve attached to the eye, because less normal tissue damage may occur with proton beam therapy.

Proton beam therapy is a form of highly sophisticated and expensive treatment that is not easily available in the UK at present. Patients are assessed by a medical panel to find out if they can be funded by the NHS to have treatment abroad. Since 2009 the NHS has funded the treatment for 370 patients (the majority children) but each case has a unique set of criteria, and so not all patients who would benefit from proton beam therapy will have treatment funded. The cost of a course of treatment for one individual to the NHS can be as much as £100,000.

The UK now has two proton beam facilities under construction, and both should be in use by 2019. The government have provided £250 million for the development of both sites. Private companies are also developing proton beam centres, but the costs may not be affordable for NHS patients.

Case study

The following article appeared in the *MailOnline* – Published: 13:08, 30 January 2016 (Updated: 10:07, 31 January 2016

So Ashya's parents were RIGHT: Proton beam cancer therapy that forced family to go on the run to Spain because they couldn't get it on the NHS is as good as chemotherapy – and has fewer side effects.

- Ashya King's parents triggered manhunt after removing him from hospital.
- They fled Southampton Hospital to seek proton beam therapy abroad.
- Now study says proton beam treatment has fewer side-effects than radio.
- It also has 'similar survival outcomes' to photon-based radiotherapy.

They were arrested after taking their son abroad for brain tumour treatment, as the NHS initially refused to offer proton beam treatment.

The case of Brett and Naghmeh King, who took their son Ashya from Southampton hospital and travelled

to the Czech Republic for proton therapy, sparked an international manhunt and made the headlines in the summer of 2014.

Eventually, the boy was treated at a Prague hospital and declared cancer-free in March 2015.

Now, a study published in the Lancet Oncology journal states that proton beam therapy causes fewer side-effects in child cancer patients than conventional radiotherapy.

Researchers said common 'toxic' effects of photon-based radiotherapy on the heart, lungs and stomach were not seen in those treated with proton treatment.

Check your understanding

1 Do you think that Ashya's parents were right to look for treatment in the Czech Republic?

2 What do you think are the underlying factors when the NHS is deciding whether or not to grant financial help for this form of treatment?

3 Look into the case yourself. Has any law been broken by Ashya's parents?

 **PAUSE POINT**   Identify the main advantages and problems associated with proton beam therapy.

Hint   Outline the medical benefits based on the science used and the issues faced by people from the UK, for example in terms of costs and availability.

Extend   Find out the percentage of the annual NHS budget which would be used up if 100 UK citizens were given a course of proton beam treatment abroad.

Prosthetics

A prosthesis is an artificial replacement, such as a tooth, bone, or other body part. These parts may be removable such as prosthetic legs, dentures or a false breast following a mastectomy.

▶ In this modern, fast-paced and more accepting age, the site of prosthetic limbs is becoming increasingly woven into the fabric of society and generally accepted as a part of life.

Reflect

Do you know anyone who has a prosthetic limb and what would you think if you needed one yourself? Do you think you would feel like the same person?

Evidence of the first known use of prosthetic limbs or devices dates back about 3000 years in Egypt. Their development has continued through the use of basic materials such as wood and metals to the present day plastic and carbon fibre composite materials. The use of prosthetics is not really surprising since our basic biological design of balance and dexterity relies on the overall stability of the body. To many, it also brings a certain degree of emotional comfort, but to others they are simply a practical application.

At present, there is a very wide range of prosthetic limbs available for almost all external parts of the body and many internal parts. As technology makes continued progress, the availability and range of replacement parts will increase significantly and could include the development of artificial skin to make the prosthetic limb look and feel more real, for example.

Research

Find out the most important benefits of using prosthetic limbs for those people involved. You may wish to interview someone you know or research information using a variety of sources. Your focus should be on the economic benefits and emotional well-being of people who use prosthetic limbs.

Stem cells

Stem cells are unspecialised cells which can differentiate to form specific types of adult cells in the body.

Key term

Stem cells – undifferentiated cells which can develop into many different cell types.

There are three main types of stem cell.

▶ Embryonic stem cells – found in an embryo at the early stages of development after fertilisation. These cells are pluripotent – they have the potential to develop into almost every type of cell in the human body.

▶ Adult stem cells – found in many different body tissues in relatively low numbers. They are sometimes referred to as 'Somatic stem cells' and are already partly specialised for the tissues in which they are found. They are multipotent – they can only form a limited number of new adult cell types.

▶ Induced **pluripotent** stem cells (iPSC) – these are normal adult cells which are genetically modified and so reprogrammed to form pluripotent stem cells. They are currently used in research, drug testing and some trial treatments.

Key term

Pluripotent – referring to a stem cell which has the ability to become almost any cell type in the body.

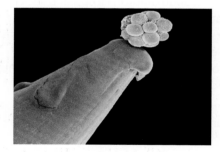

▶ A human blastocyst situated on the tip of a pin. Does this collection of cells have 'person' status?

There are benefits and associated problems for stem cell research and treatment.

▶ Benefits: Regeneration of cells and tissues for teeth, bones, heart, liver, eyes, pancreas, brain, trachea, blood and immune system. Blood and bone marrow transplants have been successful for 50 years.

▶ Associated problems: Embryonic stem cells are controversial because of ethical arguments concerning the possible development of an embryo. Immune rejection can occur where the stem cells are not matched to the patient. Adult tissue stem cells are few in number, limited in type, and their effectiveness is not fully supported.

▶ Further dilemmas: There is currently limited research into the impact on the health of the recipient of stem cell treatment. The cost of the technology is very high and people wishing to have stem cell treatment are being exploited by private companies when the full research is in its infancy. Many believe that the medical profession is "messing with nature" or "playing God" by altering genetic material.

The main cause of concern for some people is the source of the cells. They come from very early human embryos, often donated from infertility treatments. Some people have ethical objections to this.

Stem cell research and treatment forces us to make a moral choice between helping people with their medical conditions and suffering and the preservation and value of human life. The issue lies in the destruction of the early formed embryo when the embryonic stem cells are extracted.

Theory into practice

Scientists working at the cutting edge of development in embryo research and its uses are faced with many arguments from different organisations about their work. A key point which has surfaced and has been voiced by 'Campaign Life Coalition' is that the embryo used in research is *scientifically* a human being.

The group argues that using an embryo to help save lives is no different to deliberately ending the life of a terminally ill person to use their organs to help save the life of another patient. They also argue that stem cell research has produced no cures for illnesses and that it doesn't really work.

As a junior technician working in the biological laboratories of a large research company, you have been asked to present a scientific case 'for' embryonic stem cell research to a group of your peers for part of your biannual appraisal. You will need to establish when human 'life' is considered to be legally accepted to begin.

Genetic engineering

Scientists can change the genetic material of organisms ranging from bacteria to humans. The techniques they use are becoming more and more sophisticated and refined. Using **genetic engineering** we hope to be able to repair damaged genes in humans to treat genetic diseases, and add new, improved features to bacteria, animals and plants. Some people have strong objections to this new technology, but scientists and many people around the world increasingly see genetic engineering as the solution to both genetic diseases and the problems of feeding an ever-growing world population.

Here are some of the current successes of genetic engineering:

▶ Genetically modified bacteria can make human medicines, e.g. human insulin, human growth hormone, vaccines, etc.

▶ Introducing pesticide resistance into plants to reduce the need to spray chemicals which are expensive and can damage the environment

▶ Modifying plants to contain high levels of nutrients, e.g. golden rice with high levels of beta carotene needed to make vitamin A

▶ Modifying plants to survive in different conditions to cope with global warming, e.g. flood resistant, drought resistant

▶ Quick-growing trees which are disease resistant and provide a rapid source of wood

▶ Pollution-absorbing plants, e.g. grasses which absorb explosive residues, poplars which clean polluted water

▶ Salmon and other fish which grow faster than usual to reduce production costs

▶ Genetic modification of human cells to treat genetic diseases, e.g. current research trials on sickle cell anaemia, cystic fibrosis

Cloning

Clones are identical copies of the original. In terms of biological science, a clone is a group of cells or organisms which are genetically identical, having the same DNA.

In nature, identical twins are clones, and in 1996 scientists produced the first clone of a large mammal – Dolly the Sheep. Dolly was 'born' at the Roslin Institute in Edinburgh and died after just 7 years, younger than average, with a variety of problems. Since then, there has been discussion concerning the possibility of human cloning which has led to intense debate. Read the lists in Table 7.2 and decide which you agree with.

Cloning organisms is often used in combination with **genetic modification** – GM bacteria, plants and animals are all cloned.

▶ cloning endangered species may help them to survive, but so far there are few success stories

▶ animal and plant clones play a very important part in scientific and medical research. They provide genetically identical models for different types of treatment

- clones of genetically modified animals or bacteria are used to produce human proteins badly needed as medical treatments
- some people are prepared to pay a lot of money to have much-loved family pets cloned.

▶ **Table 7.2** Some of the reasons for and against cloning humans

Reasons for cloning humans	Reasons against cloning humans
Removing genetic diseases.	A 'sub-class' of persons may be created.
Solving infertility for couples.	Many miscarriages and deformities could occur before success.
Produce a good supply of organs for donation.	Ongoing ill health and complications in the clones.
To bring back the dead.	Development of 'designer-babies' with features chosen by parents.
Develop eternal life by cloning in old age.	Embryos sold on the black market.
Producing a steady supply of stem cells.	

Key terms

Genetic modification – the process of altering the DNA in a plant or animal using laboratory techniques.

Genetic engineering – a process which allows the transfer of desired genes from one organism to another (not necessarily of the same species) artificially, to enhance the characteristics of the organism.

Pharmaceuticals

Phamaceuticals relates to the manufacture and production of substances used as medicines. There are many products available.

The pharmaceutical industry combines the branches of health and chemical science. It dates back thousands of years, since many of the properties of pharmaceuticals are available in herbs and plants, which grow naturally and would certainly have been discovered by our earliest ancestors. The word 'pharmaceuticals' derives from the Greek 'pharmakon' which means 'medicine, drug or poison' and was first used around 500 years ago. It costs over 1 billion pounds to develop a single new medicine.

▶ **Figure 7.4** An international sign for medicine. This is a common sign showing a bowl of Hygeia and a serpent. Hygeia was the Greek goddess of health – hence 'Hygiene' which became the basis of modern health care.

Since the discovery of 'penicillin' in 1928 by Sir Alexander Fleming, and the subsequent production as a drug by Australian Howard Florey and German scientist, Ernst Chain, scientists have produced a vast array of medicines which act as 'antibacterials' or 'antibiotics'.

The type of antibiotics currently used will depend on the bacterial effects and medical condition of the patient. *Bactericidal* antibiotics, for example, kill bacteria directly, while *bacteriostatic* antibiotics stop the bacteria from growing. *Narrow-spectrum* antibiotics are effective against a narrow range of bacteria and *broad-spectrum* antibiotics are effective against a broad range of bacteria.

It is in the interest of pharmaceutical companies to promote their new products in order to justify – and finance – their research and further development in new products. It is also ethical for scientists working in these pharmaceutical companies to continue to develop medicines and to find new medicines to reduce suffering of the population from current and future illnesses and ailments resulting from bacterial infection.

Unfortunately bacteria are becoming increasingly resistant to antibiotics. There are many reasons for this, including over-use of antibiotics, people stopping using antibiotics before the end of a course, poor hygiene in hospitals and care-homes, and the use of antibiotics in animal feeds in some parts of the world.

Research

Produce a list of known or suspected cases of bacteria which have been shown to become resistant to antibacterials. Methicillin-resistant *Staphylococcus aureus* (MRSA) and the recent case of antibiotic colistin-resistant *E. coli* could provide some information which may help you understand the problem.

The future is uncertain. Much will depend on the organization and full agreement of processes between governments, physicians, the pharmaceuticals industry and the front-line health care professionals if there is to be a significant slow-down and eventual stop in the development of **antimicrobial** resistant bacteria.

> **Key term**
>
> **Antimicrobial** – a substance that can kill microorganisms or inhibit their growth.

Any new antibiotic will be saved for use in emergencies when bacteria become resistant to everything else. That means it will be hard for pharmaceutical companies to recoup their investment. Perhaps society needs to fund the work to discover new antimicrobials?

Performance enhancing drugs

Performance enhancing drugs (PEDs) are chemical substances that may provide athletes with an advantage over others by enhancing or improving their capacity to train or remain more alert, or reducing the pain felt following injury or tiredness. The most common are stimulants and hormones and the practice is referred to as 'doping'. Figure 7.5 shows some PEDs and their effects.

> **Will Russia compete at the Olympics?**
>
> Russia's athletes face a nervous wait to see if a blanket ban will be imposed with the IOC

announcing it will retest all of its athletes who competed at the 2014 Winter Olympics in Sochi.

That's because an independent report published by Canadian law professor Richard McLaren found urine samples of Russian competitors were manipulated across the 'vast majority' of summer and winter Olympic sports from 2011 through to August 2015.

Russia came top of the medal table at Sochi — winning 33 medals, 13 of them gold.

But McLaren concluded Russia's 'Ministry of Sport directed, controlled and oversaw the manipulation of athlete's analytical results or sample swapping, with the active participation and assistance of the FSB, CSP, and both Moscow and Sochi Laboratories.'

Source: http://edition.cnn.com/2016/07/19/sport/russian-doping-ioc-rio-2016-olympics

This comment is taken from the CNN web page before the banning of 118 Russian track and field athletes and all of the Russian Paralympians at the 2016 Olympic games in Rio de Janeiro, Brazil. This example was different to other examples of malpractice or 'doping' in that the athletes were given performance enhancing drugs by their coaching staff and responsibility was at the highest levels.

Doping scandals include:

▸ 1988 Seoul Olympics 100 m – Ben Johnson
▸ 2012 Cycling. Seven times winner of Tour de France – Lance Armstrong

---

Anabolic steroids
These help athletes to build muscle by training harder, but can lead to the athlete becoming more aggressive in their approach and kidney failure.

Analgesics (narcotic)
These are very addictive and can cause further injury because they reduce the level of pain felt by the athlete.

Stimulants
These are also addictive and help to reduce the feelings of fatigue making the athlete more alert. Continued use can cause heart failure.

Diuretics
These make athletes pass urine more frequently, reducing their weight and masking other drugs. They also cause dehydration and significantly affect the functioning of the kidneys.

Peptides and hormones
EPO can help to increase blood cell count and raise energy levels, but may increase the risk of stroke or heart problems. Human growth hormone can help to increase muscle but may increase the risk of many other health issues.

Other means
Beta blockers are used for heart conditions and reduce the heart rate, reducing movement in the hands during archery and shooting. Blood injected back into the body after being removed earlier allows more oxygen to be carried.

▸ **Figure 7.5** Types of PED and the effects they have on the body

- 1994 World Cup USA – Diego Maradona
- 2008 Bejing Olympics – Nesta Carter won a gold medal for Jamaica in the 4 x 100 m relay. He was stripped of his medal following a re-test on his blood and urine sample in 2017. All the relay team, including Usain Bolt, lost their medals as a result.

Discussion

Find out about the doping scandals for the individuals listed above. Discuss in a group the key differences between these and the case of the Russian athletes in the Rio Olympics in 2016.

The ethical point concerning 'doping' is clear – drugs taken by some athletes will artificially enhance their performance over the efforts of other athletes who do not use them. In everyday life, the addictive nature of many PEDs can transform the day to day lives of many ordinary people, making them train in the gym harder and more regularly. This aspect can have serious effects on relationships and families if not identified or limited.

Money in sport is plentiful. With such high earnings available and the incentive to win large sums of money, the temptation to use PEDs in all sports becomes greater. The effect is to add pressure on athletes and sports professionals to ignore the legal implications in order to increase their advantage over rivals.

Research into the effects of PEDs, such as anabolic steroids, released into the environment is currently being carried out. It was always believed that such chemicals which find their way into waterways after flushing, for example, would degrade over time and become a low ecological risk. This understanding is now no longer fully accepted and further research is continuing.

An estimated 1.1 million sea birds and marine creatures die every year due to plastic ingestion or entanglement.	There are currently 5 large swirling masses of plastic accumulation in the world's oceans, the size of large countries. They are called 'Gyres'.
The Great Pacific Garbage Patch is a Gyre the size of Spain found off the west coast of N. America.	Approximately 6.5 million tonnes of plastic finds its way into the oceans each year, most of it washed into or blown from the land.

▶ **Figure: 7.6** Plastics in the oceans

Chemicals

Everything is made of chemicals - but this term is often used to refer only to chemicals manufactured and used by humans on an industrial scale.

Plastics

Polypropylene and polyethylene (polythene) are the most common form of plastic used by most of us. Plastics are made by combining small organic molecules known as monomers into long-chain molecules termed polymers.

Plastics, while being very convenient for carrying items, storage, and a whole host of other uses, take many hundreds of years to degrade (non-biodegradable). It is estimated that 90 per cent of the plastics in our possession are only used once and then discarded. Figure 7.6 shows some facts about plastics.

▶ Plastic rubbish has been found on some of the most remote beaches in the world.

The problem of plastic entering our food is also a real and measurable issue that is highlighted in this excerpt of a report which featured information from the Chair of the Commons Environmental Audit Committee:

> The Thames is becoming polluted with microplastics that are finding their way into oysters and other foods, a new study has found . . . Some 35,000 particles were identified in the samples . . . 'Pieces of plastic will be carried downstream and will be eaten by oysters in the Thames estuary'. . . . 'If you eat six oysters, you've probably consumed 50 particles of microplastics . . .'

Source: London Evening Standard, 26 August, 2016

The environmental impact of plastic waste will be identified in the effects on both animal and plant species. The longevity of the material in the environment and slow breakdown of plastic molecules means that they will remain a problem for many hundreds of years. Animals are already susceptible to eating small plastic pieces in soil and at sea, and are suffering from choking or trapping of the plastic in their intestines and mouths. This in turn affects the feeding habits of other creatures and entire ecosystems will be affected. 'Down-cycling' (or using the material for another purpose) seems the only option for the immediate future.

From an economic view point, whilst plastics are inexpensive and a useful product in many ways, the costs of dealing with the problems of waste will be counted in both financial terms and the impact on the environment. The human race has an ethical argument to address: do we continue in this way, leaving the problems to future generations to solve, or do we change our whole attitude and stop the use of plastics in modern society?

One potential solution is to recycle more plastic. In Japan, the world leader, they recycle almost 80% of all the plastic they use.

Acidification of oceans

It is now accepted that the rise in CO_2 levels in the atmosphere is going to significantly change climatic conditions over the course of the foreseeable future. The effect of the increase in CO_2 on the pH level of the oceans is now also becoming a reality.

Historically, the oceans of the world have played an important part in absorbing CO_2. Carbon dioxide is essential for the production of calcium carbonate for the shells of sea creatures and the chemistry of the waters. It has taken millions of years for sea creatures to evolve to suit the conditions in our oceans, conditions which involve very narrow ranges in the levels of CO_2 and pH.

pH levels recorded, pre-Industrial Revolution, were reliably given as 8.2, which is slightly basic. There was no change for the next 150 years. Levels recorded in the last few decades have shown a small but significant drop in pH of 0.1 (see Figure 7.7). While this may not seem a large figure, you need to understand it in the context of the evolution of sea creatures and their sensitivity to the conditions in which they have developed over millions of years.

The possible effects of lowered pH on the oceans are as follows:

▶ Shells – these will become thinner as a result of **acidification**, making the creatures which rely on them for protection more vulnerable to predators.

▶ Coral reefs – already dangerously susceptible to environmental changes, these reefs and their dependent organisms would be damaged further by acidic sea water.

▶ Larger creatures – deep sea dwellers may need to expend extra amounts of energy to compensate for a change in pH. They have developed with a finely tuned acid/base balance and may struggle in the new environment, becoming smaller as a species or not surviving the changes.

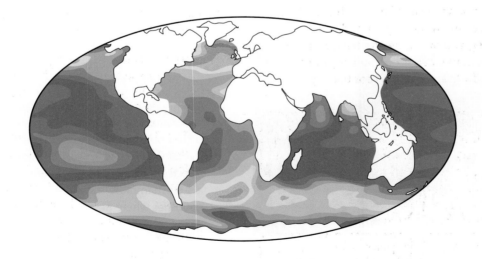

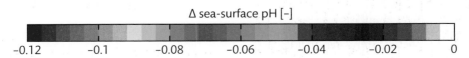

▶ **Figure 7.7** Projected pH sea-surface changes since the 1700s. Note the significant areas of green indicating drops of between 0.06 and 0.08.

▶ Mid-level creatures – migratory sea creatures which cross the boundaries from deep water to shallow water may be forced to change their patterns of movement. This will impact on the whole sea food chain. Lower oxygen levels in some areas may force these creatures to seek shallower waters and become more prone to predation. The effect on fish stocks will be devastating.

Key term

Acidification (oceans) – the lowering of pH levels by the uptake of CO_2.

❚❚ PAUSE POINT

Find and draw an example of an ocean food web which includes all the important levels of organisms interdependent on each other. Develop the food web by introducing information on to your diagram which will illustrate the changes that will take place at each level as pH levels fall. Include the ethical point of allowing acidification of the oceans by the uptake of CO_2 and question the possibility of addressing the acidity by other means.

Hint Begin by identifying creatures most susceptible at the bottom of the food web and working your way up to larger species.

Extend What could happen to the feeding habits of some creatures which may help them to survive changes in pH? Provide examples.

Insecticides

Insecticides are a form of pesticide specifically used to target insects. The increase in human consumption of food has led to the development of large scale agricultural activity. The need to protect crops from pests in order to maximise yields has resulted in farmers using a range of insecticides.

Broadly, insecticides are toxic chemical substances which kill or control insect populations. They may be either synthetic, inorganic or natural organic chemicals and are sprayed onto the plant surface to affect the pests.

▶ A farmer spraying crops with an insecticide. Without this, the productivity of the farm will fall significantly.

The type of insecticide used will depend on the method by which the insect will be affected. Some insecticides are absorbed through the outside of the insect, some are inhaled into the breathing system and some need to be eaten by the insect to have their effect.

The type of synthetic compound used is determined by the insect species to be targeted.

The downsides of using insecticides are as follows.

▶ Environmental damage – Food chains can become very complex and involve a wide variety of animal species interrelated by their need for food. Contaminated insects will be eaten by larger species, and so on. The insecticide enters the food chain, becoming more concentrated with the number of animals that it passes through. This can cause serious harm to the carnivores higher up the food chains and may even affect humans.

Continued treatment of crops causes run-off into the soil and waterways. Banning of some insecticides, such as dichlorodiphenyltrichloroethane (DDT), was introduced in many countries in the 1970s following severe effects on bird of prey populations. This substance affects mammals in a variety of ways, but is still used in countries where pest control is vital to crop growth.

▶ Resistance to insecticides – some insect populations can no longer be effectively controlled using insecticides as they have developed a resistance to the chemicals. Since many insecticides have similar functions and actions, insect species that develop a resistance to one type may also develop a resistance to other types.

The action of insecticides on 'friendly' insects will also increase the population of crop-damaging species by reducing the numbers of useful insects which would normally prey on them. Therefore, more work has been carried out on 'integrated control', which uses additional methods, such as adding predators or parasites to pest species, to limit crop damage.

Nanotechnology

This is the study and application of very small objects, typically atoms and molecules, which can then be used in all forms of science. The term 'nanotechnology' was first used by Professor Norio Taniguchi in the late 1960s and the concept first discussed by physicist Richard Feynman some ten years earlier.

In scale, the technology deals with particles with dimensions smaller than 10^{-9} m (1 billionth of a meter).

Manipulating such small particles of the chemical elements gives significant advantages, including lighter weight materials, higher strengths, chemical reactivity changes and effect on the way that the visible wavelength of light is seen. Using the elements in some applications is nothing new – stained glass windows in large medieval churches had gold and silver added to the glass mix. The different sizes and shapes of the tiny particles transmitted different colours. Of course, the artists of the day had no idea as to how this was being produced.

Space exploration

Although the space industry is always at the cutting edge of human advancement in terms of engineering, technology and research, there are limitations on the means of transporting people and machines into orbit around the Earth and further into space. The proposed plans to develop a 'manned' mission to our nearest planetary neighbour, Mars, in the early 2030s, is providing many branches of science with some of the most difficult problems to be overcome. For unmanned spacecraft, the problems remain, in terms of size limitations and distances of travel. Research into the use of nanotechnology in space is hoping to overcome some of these problems. Figure 7.8 gives a summary of the current developments in space exploration using nanotechnology.

To appreciate the difficulties of launching an object into space, you need to have a good understanding of the force necessary to overcome the gravitational forces acting against the rockets. As a rough estimate, it takes approximately £13,000 in costs to lift 1 kg of material into orbit around the Earth.

Lightweight materials:
Stronger and lighter weight materials used in the construction of machines, clothing and spacecraft, aerogels which increase strength of structural components and adhesive properties, improved insulation and even 'self-healing' spacecraft puncture repairs from micro-meteorites.

Nanosensors:
At the molecular level, being able to detect changes in spacecraft body work as a result of stresses or atmospheric changes in gases for life support, would be much more effective and speed up the detection of related problems.

Nanorobotics:
The development in better materials will enhance the building and uses of smaller robots, such as improved computers, better sensing and advanced power sources. These can be used to monitor astronaut health, both onboard and remotely, for example, and costs would be greatly reduced.

Future development in practice:
- Lighter and stronger materials
- Better radiation protection
- Stronger adhesive properties
- Advanced sensing equipment
- Health management improvements
- Computer advancement
- New materials
- Smaller spacecraft

▶ **Figure 7.8** Current developments in space exploration using nanotechnology

Worked Example

The Apollo 11 payload (command module, landing module and equipment) had a mass of 140,000 kg. At today's prices, how much would it cost to get the payload into Low Earth Orbit?

Using: £13,000 per kg, so 140,000 kg = 13,000 × 140,000 = £1.82 billion

This does not include the extra cost to propel the payload to the moon and back.

Diesel fuel

This form of engine fuel is used extensively in road transport and marine vehicles. When people come into contact with diesel fuel or fumes through ingesting, inhaling or contact with the skin, the effects range from mild nausea to severe vomiting and burning sensations. When burned in air, diesel exhaust contains particulate material and gases, such as nitrogen oxides, both of which can lead to coughing and reduced lung capacity. Of the air pollutants normally present in towns and cities, diesel exhaust particles account for a large proportion.

The addition of **nanoparticles** is now being researched for possible large-scale introduction. This would increase the efficiency of the combustion of diesel fuel and reduce the emission of particulates, carbon monoxide, carbon dioxide and nitrogen oxides. Research is also being carried out to try to determine the effects of including nano-particles in diesel fuel and its possible health related issues. At present, the effects are not certain but point towards low levels of risk at ambient temperatures and even possible benefits since there is an improvement in fuel efficiency, resulting in lower exhaust particle emissions. In terms of economics, the costs could be reduced for fuel consumption, and medical costs reduced for treatment of conditions associated with pollution in traffic congested areas.

Key term

Nanoparticles – particles approximately 10^{-9} m in size which have a large surface area for their volume.

Cosmetics

The cosmetics industry is huge. Table 7.3 gives an indication of the amount of money from actual sales of cosmetic products for just five of the top cosmetic companies.

▶ **Table 7.3** Sales for the world's top five global cosmetic companies (last annual figures 2015/16)

Company	Sales (£ billion)
Procter and Gamble	20.0
L'Oréal	19.5
Unilever	12.0
Avon	5.75
Beiersdorf	5.65

The use of nanotechnology in the cosmetics industry is now a very important part of the overall production of the skin and hair care products which are available. The regulations governing the safety and manufacturing of products made using nanochemicals are the same as for standard cosmetics.

There are currently two main uses of nanochemicals in cosmetics:

▶ Ultra-violet (UV) filters: These products are used to improve the skin's ability to reduce UV levels and so prevent damage. The most common nanochemicals are zinc oxide and titanium dioxide. They are called nanopigments.

▶ Delivery agents: These are chemicals which carry the active ingredients (such as vitamin A) deep into the skin. New polymer nanocapsules are better at carrying the active ingredients deep inside the skin. These are called nanoemulsions.

Current research and testing appears to conclude that nanotechnology use in the cosmetics industry produces no harmful effects to people or the environment. The cases of skin cancer would significantly increase if it weren't for the use of skin care UV protection products. It is also believed that nano UV products are probably more effective at UV protection. There is some concern, however, that the use of nanochemicals in cosmetics can result in DNA damage in cells, a claim which has not yet been fully tested. There is also very little known about the environmental impact caused when the product is washed away. The possible effects of nanochemicals on small organisms and to their ecosystems are not clear. At the social level, we should consider whether it is right to make short-term, aesthetic improvements to our skin and other body areas, simply to look good. There can be a short-lived gain in self-esteem but this may not help in our development of suitable social interactions and well-being.

Food technology

Food technology covers the science behind the growth, production and content of the food that we eat. It therefore covers all aspects of the way in which we are able to provide foods which are beneficial to our health, containing the ingredients necessary for good biological functioning, and ensuring that we can store food stocks for a significant time period. The latter is one important aspect which will ensure that food shortages are kept as infrequent as possible, given an ever increasing world population.

Food composition

We have become increasingly aware of the need to limit sugar, fat and salt content in our food, following many well-publicised campaigns to draw attention to excessive intake. But what has the food industry done to limit the amount of these components in our diets and why has there been a significant increase in cases of obesity?

Reflect

The world's population is an estimated 7 billion people. Approximately 1 billion of these are estimated to be obese or seriously overweight, more than 10 per cent of whom are children. What do you think are the main causes?

Obesity and certain other health related problems resulting from obesity have risen in number significantly over the last 25 years for a number of reasons. While some factors, such as environment and genetics, are not directly related to the diet and lifestyle of the individual, the general scientific community broadly agrees that unhealthy dietary levels of fats and sugars, coupled with a lack of physical exercise, are the main cause.

Current recommendations from the World Health Organization (WHO) on sugar intake for a 'typical' adult is no more than 30 g of added sugars per day. Added sugars refers to anything that is not found naturally in products. The greatest problem for consumers is learning to spot the added sugars and to check labels carefully. It is ethical for the food industry to include this information in easy-to-understand words and tables on food packaging to help consumers make informed choices concerning their sugar intake, but also to help reduce the financial burden on hospitals if unecessarily high sugar intake leads to illness. Many shoppers also wish to ensure that their food products are produced ethically, such as under good labour conditions and by environmentally friendly methods. As such, it is correct for governments to include as much information on labels as possible.

Salt content in foods has been linked to blood pressure, heart disease and stroke. The current UK recommended intake is between 5 and 6 g of salt per day. Keeping to this limit is not easy, especially since dietary habits have changed to incorporate 'ready meals' which are so quick and convenient for our busy life styles and which often contain high levels of salt.

Excess fat in your diet can also cause heart disease by raising cholesterol levels. Table 7.4 gives some interesting facts concerning our intake of fat.

▶ **Table 7.4** Advantages and disadvantages of dietary fat

Advantages	Disadvantages
Fats help our bodies absorb vitamins A, D and E.	Unused fats in the body are stored as body fat.
Fats are high in energy.	Excess energy from fats increases the risk of becoming overweight which in turn increases the risk of diseases such as type 2 diabetes.
Fat is a source of fatty acids such as omega-3.	Too much saturated fat produces higher cholesterol levels which restrict blood flow in the arteries.

Preservatives

Food is not always consumed immediately when it is bought. Smoking and salting have been used for hundreds of years in order to retain adequate supplies for families, and even entire villages, over long periods when food was not always easy to obtain.

In modern times, our increasing demand for more choice, better safety and convenience has brought about a need for our foods to be preserved using techniques aimed at preventing spoiling by any of the causes shown in Figure 7.9.

Current methods of food preservation include:

▶ use of packaging and vacuum sealing – removes the air within the package and reduces the growth of oxygen-dependent micro-bacteria and also general food spoiling

▶ canning – a process which heats the contents of the jar/can, killing micro-bacteria and vacuum sealing for long shelf life

▶ freezing – freezing foods quickly, so that micro-bacteria do not multiply and enzyme activity which can spoil the food is slowed down

▶ smoking – helps to prevent mould development on some meats and kills certain bacteria

Physical changes:
Food packaged in clear packs will be affected by light and heat. Quality, vitamin levels and oxidation levels can all be changed.

Biological changes:
Microorganisms such as yeast and mould will begin to grow on the food and secrete toxins which are poisonous.

Chemical changes:
Oxidation effects of chemicals reacting with oxygen in the air will cause changes in the food.

▶ **Figure 7.9** Changes which can happen in stored food

▶ **dehydration** (and **rehydration** to eat) – a process whereby moisture in the food is removed by drying, reducing its bulk and making storage easier and enabling the food to last longer, then adding water to the food to restore its bulk ready for consumption

▶ food additives (e.g. **antioxidants**) – preventing oxidation (reaction with oxygen) and slowing down food spoilage. These can include fats, gelling agents and stabilisers

▶ chemical preservatives (commonly seen as 'E' numbers on food labels) – include sweeteners, colouring agents, antioxidants and preservatives approved by the European Union (hence 'E' for use in foods.

Key terms

Dehydration – removal of water.

Rehydration – addition of water.

Antioxidant – an enzyme or vitamin added to food to reduce damage due to reactions with oxygen in the air.

The use of chemicals identified by their 'E' numbers has been the subject of much controversy, although most cases of concern have not seen a change to the practice

and there is a very low level of risk associated with their use in preservation of food products.

Genetically modified (GM) crops

Genetically modified (GM) commercially grown crops used to service the food and livestock industries have been produced for many years and, in global terms, this has been increasing steadily within approximately 30 countries. The practice is not carried out in the UK, but imports of GM food materials for animal feed are allowed.

The process involves identifying a particular desired characteristic of a plant and removing the gene in the DNA which is responsible for it.

A complex process then allows the gene to be introduced into the plant to be modified and for new plants to be created. The new plants now have additional characteristics and could develop useful substances which make them resistant to pests or weed killers (herbicides), for example.

Link

See *Unit 11: Genetics and Genetic Engineering* for more detail on genetic modification.

Benefits:
1 Increased crop yield to feed increasing population.
2 Resistance to herbicides.
3 Crops which can make extra vitamins.
4 Developing countries are able to grow sustainable crops.

Possible drawbacks:
1 The effects of GM crops on ecosystems and other crops are not yet understood.
2 High costs are involved but these are falling as cheaper and more efficient techniques are developed.
3 Developing countries become too dependent on GM crops.
4 Issues concerning dietary requirements for health or even religion are raised.
5 Development of pests which are not affected by insecticides.

▶ **Figure 7.10** Genetic modification of crops

Assessment practice 7.1

For each topic in the following list, identify one ethical, social, economic or environmental aspect or issue. Your document need not be more than two A4 sides and could be presented in the form of a journalistic study:

- carbon capture
- proton beam therapy
- resistance to antimicrobials
- plastic waste in oceans
- use of nanotechnology in cosmetics
- food composition.

Plan
- What is the task? What am I being asked to do?
- How confident do I feel in my own abilities to complete this task? Are there any areas I think I may struggle with?

Do
- I know what it is I'm doing and what I want to achieve.
- I can identify when I've gone wrong and adjust my thinking/approach to get myself back on course.

Review
- I can explain the results obtained from the task.
- I can apply the activity to other situations.

Understand the influence of different organisations/individuals on scientific issues

Science and scientific research do not exist in a vaccuum. People decide which research they want to do, and other people decide which research is going to get funded. Many different factors influence which scientific research actually takes place. In this section you are going to look at some of these influences.

Government and global organisations

Large global and governmental organisations have a lot of money behind them. They can affect the direction of scientific work in many different ways. Here are some examples.

United Nations

The United Nations (UN) was set up in October 1945, following the official end of the Second World War. It was developed to supersede the failing 'League of Nations' and to establish international cooperation and world peace so that events such as the Second World War would never be repeated.

The United Nations has its headquarters in New York, USA, with other main offices situated in Vienna, Geneva and Nairobi, and is headed by the Secretary General. It began with 51 member states, but now has 193. (The Republic of South Sudan was accepted as the latest member in 2011.) Funding is provided by member states using both fees based on wealth/influence and voluntary contributions. It funds research into international issues such as food security and climate change.

Primary 'organs' of the United Nations (UN) include:

▶ General Assembly (main assembly for discussions of member states)
▶ Security Council (assembly for discussion on world conflicts)
▶ International Court of Justice (department concerned with implementing justice in serious global cases)
▶ Secretariat (implementing research studies, provision of facilities and information for the UN itself).

System specialised agencies include:

▶ Food and Agriculture Organization (FAO)
▶ United Nations Educational, Scientific and Cultural Organisation (UNESCO)
▶ World Health Organization (WHO)
▶ World Bank Group.

World Health Organization

The aim of the WHO is to effectively coordinate and direct international work related to health and to ensure technical information is consistent and factual through continued research. At the heart of this aim is the objective to ensure that all people have access to sufficient health care – for physical, mental and social well-being – and not simply to provide measures to eradicate disease and physical impairment.

The WHO works directly with scientific research and development in an attempt to combat diseases such as influenza and HIV and noncommunicable diseases such as cancer and heart disease. They target science research into the health and well-being of children and mothers, ensuring the quality of air, food and water, and the availability of medicines and vaccines. WHO works with governments, agencies, foundations, non-governmental organisations, and those in the private and civil sectors, highlighting world health issues and developing strategies to deal with them. They employ thousands of health experts and scientific professionals from all over the world, raising awareness of problems and developing support for scientific research and eradication programmes.

Case study

Smallpox vaccination

In 1956, the WHO set out a serious commitment to eradicate the highly contagious 'smallpox' from the world. It was the intention to ensure that use of vaccinations would be stepped up in the fight against the disease and that no countries would be left out of the programme.

The disease is caused by the 'variola' virus and its first known appearance appears to have been as long ago as 3000 years BC. The mummified remains of Egyptian Pharaoh, Ramses V, dating back to 1157 BC, showed traces of the tell-tale marks of the virus. Once contracted, the disease itself had an estimated death rate of 20 per cent of those infected, but it had long-term effects on the body of all those who were infected. The symptoms were very distressing and difficult to treat.

Vaccination against smallpox was not a new idea. More than a thousand years ago, physicians in China had begun to use small amounts of the virus on individuals, taken from swabs of infected people, in order to introduce early exposure.

Since 1979, the world has been free of the disease, which was finally eradicated in its last stronghold of Eastern Africa. It had been responsible for the deaths of approximately 300 million people worldwide in the 20th century alone. The virus is now stored in two highly secure laboratories in America and Russia.

Check your understanding

1 Set out a list of procedures that you may follow when it is decided that you are going to eradicate a serious disease from the human population.

2 Why do you think that the physicians in China infected people with small amounts of the virus? What happens to the body's immune system when this happens?

3 Identify a contingency plan that could be brought into play if the smallpox virus were to 'escape' from the laboratories in America and Russia.

4 Research the methods used by the authorities when dealing with another, more recent disease – Ebola. Is there anything that you would do differently?

European Union (EU)

The European Union had a similar development to the United Nations, as it was created due to a wish to prevent the possibility of another major war in Europe following the aftermath of the Second World War. Both France and Germany were at the forefront of the need to establish a closer working and economic commitment and, together with four other mainland countries, formed the European Economic Community (EEC) in 1957 after signing the 'Treaty of Rome'. The UK joined with seven other nations in 1973.

The present-day European Union developed through an increase in membership applications from other European countries and the fall of the Berlin Wall when East Germany and West Germany were reunited. More recently, a national referendum in June 2016 resulted in the UK voting to leave the EU, which now has a membership of 28 nations, each having representatives or Members of the European Parliament (MEPs).

Just as our earnings are taxed by government to pay into a system whereby all of us at some time in our lives, and to

different extents, will make use of the benefits and facilities available, so too do the individual member countries pay into the EU. The exact amount varies annually, but is generally calculated on the economic strength of each nation. As an example, Germany paid more than 21 per cent of the 145 billion Euros EU budget in 2015.

The planned EU budget for the period up to 2020 is approximately 960 billion Euros. A large proportion of this will be used to fund many areas of science, both directly and indirectly. Many nations benefit from the current science research funding awarded by the European Research Council, especially Germany and the UK. There are excellent opportunities for researchers to take part in multi-state projects, allowing management experience and technological sharing, such as the Large Hadron Collider at CERN in Switzerland. In addition, clinical trials on medicines are well funded and all research performed is supported by a large network of specialists who share their expertise within a framework of shared science policies of the member states. Science areas where European Union financial assistance is provided include:

- university research
- water quality standards
- agricultural subsidies
- engineering
- GPS development systems
- emerging technologies
- nuclear/radiation research
- biotech and pharmaceuticals
- atomic research
- inter-European science collaboration
- space flight and aerospace development.

Discussion

Find out and summarise the main arguments put forward in the UK referendum debate on reasons to stay in and reasons to leave the EU. Include specific reasons linked to science.

Discuss which reasons were the most popular among voters and use website research to find out which points raised have been vindicated.

Environment Agency (EA)

This UK government agency is the largest environmental protection authority in Europe and so has significant input into the research, development and planning of UK government projects related to many aspects of scientific study. The amount of influence that the EA has depends greatly on the quality of its research programmes and the extent of its responsibilities (see Figure 7.11).

Climate change, for example, has been at the forefront of EA research for many years and is the driving force behind a number of planned developments relating to improving the UK flood defences and management of water run-off. By understanding the science, serious annual flooding and associated environmental damage can be limited.

▶ A familiar scene in the UK over the last number of years. This is a village in Somerset, 2014.

The Environment Agency has a budget of approximately £500 million to use on planning for, and coping with, the future effects of flooding in the UK. The organisation must be able to use all the available current scientific information on climate change and to accurately project possible scenarios of flooding over the course of the next 50 years. It is also using information from European and global statistical studies and relating the details to the effects on UK flooding over the last significant number of years.

Food Standards Agency (FSA)

The responsibilities of the FSA are outlined in Figure 7.11. This UK agency was established in 2001. It uses many independent committees and working groups to provide its scientific expertise and its advice to consumers relating to the production of crops, animal farming and food standards in general. The advice given by the agency

---

Environment Agency (EA)

Responsibilities:
- Climate change
- Water quality
- Fishing
- Land quality
- Water resources
- Air quality
- Marine environment
- Flooding and coastal risks

Food Standards Agency (FSA)

Responsibilities:
- Food safety
- Food hygiene
- UK meat standards regulation
- Food labelling policy
- Nutritional information labelling policy
- Food law code of practice

▶ **Figure 7.11** Important UK-based agencies

must be clear, accurate and based on the most recent evidence available.

Among its many successful campaigns to make the public aware, and to bring about changes by the food supply industry, is the influence that was imposed in relation to the salt content in food. A summary of the campaign is shown below:

▶ Phase 1 – 2004 campaign of public awareness begins using slogan *'Too much salt is bad for your heart'*.

▶ Phase 2 – 2005 change to campaign emphasis using scientific backing *'Eat no more than 6 g a day'*.

▶ Consumers advised to always check food labels.

▶ Phase 3 – 2007 reiterated previous slogans and added new one 'Is your food full of it?' referring to the percentage of salt already in the food bought.

▶ Evaluation of phase 1 – Older age groups (50 and over) recognised advertising used and there was evidence of moderating behaviour.

▶ Evaluation of phase 2 – Targeted women aged 35–60. Results showed significant increase in awareness, recognition of new TV advertising and checking of salt content labels.

▶ Evaluation of phase 3 – Continued rise in awareness of the campaign for all groups. Visits to the FSA website peaked at 66,000 in one day.

▶ An estimated 1.0 g of salt on average has been taken off the population's daily salt intake.

▶ Food and retail industries committed to reducing salt content in their products.

Research

Conduct a group research activity aimed at determining a sample population's understanding of the effect of excess salt on their health. Develop a suitable set of questions which would help you to assess the public's awareness of the issues and recommended daily intake of salt in their diet.

Non-government organisations, professional bodies and associations

This refers to organisations which are not part of government and are usually non-profit making; Figure 7.12 gives three examples.

Royal Society of Chemistry (RSC)
Professional not-for-profit association established in 1980. Committed to 'advancing the chemical sciences'. Headquarters in London. The **RSC** publishes:
- Journals
- Text 'Education in Chemistry'
- Online material
- Chemistry magazine - 'Chemistry world'
- Reference books
- Student books
- Chemistry history books.

The society awards medals and prizes annually to recognise professional and pioneering work in the field of chemistry.

National Physics Laboratory (NPL)
The UK's National Measurement Institute established in 1900. The world's leading centre for application of accurate measurement standards for science and technology. Situated in London.
- Administration by the Dept. of Business, Innovation and Skills.
- Partnership with Universities.
- Produces research, peer-reviewed papers.
- Post graduate research specialism.

The NPL works to advance business solutions in: technical research and development, instrumentation, manufacturing perfection and performance checking.

General Medical Council (GMC)
The GMC has main offices in the four capital cities of the UK plus Manchester. Aim 'To protect, promote and maintain the health and safety of the public'. It became a registered charity in 2001.
- Must maintain and update registrations of all medical practitioners.
- Sets standards of good medical practice.
- Provides medical education and training.
- Commissioning medical research.
- Two-way consultation activities.

The GMC needs to follow the UK Medical Act and European legislation. The Medical Act gives the GMC powers to act where there are concerns about the fitness to practice of registered doctors.

▶ **Figure 7.12** Non-government organisations, professional bodies and associations

They are set up by professional people, including chemists, physicists, doctors and other medical professionals, from all nations and may be financed privately, through businesses, foundations and, in many cases, from government funding initiatives from associated government departments.

Universities and research groups/teams

University research groups usually consist of a number of individuals within a specified department of the university where there is a serious focus of skills and expertise for a given area of science.

The research focus may be in an already established field or a new and emerging scientific aspect which may form part of the interests within the group. By linking up with other academics in a number of departments, they are able to establish a useful and coherent research network.

Most, if not all, universities publish an annual research review document that outlines developments in the current research areas for the year and opens up the possibility of additional research opportunities to future post-graduate students. The standing of the university and the calibre of the scientists it attracts affects the direction of the scientific research which takes place.

Assessing the impact of university research on new and established scientific fields is difficult because the implementation of the work and findings for a particular area of expertise may not be immediately realised or clear to identify. Some research work, however, may be introduced into the scientific field of study and possibly put into practice almost immediately.

The benefits of scientific research in terms of the ethical, social, economic and environmental context will be determined by the availability and suitability of the graduate and post-graduate programmes which universities can offer.

Examples of advancement resulting from university research and study include the following:

▶ Refinement of wind turbines to produce electricity – aerodynamic improvements, mechanical efficiencies, wind modelling, low and efficient maintenance, specialised blade designs, advanced materials/coatings.

▶ Albert Einstein's theory of gravitational waves of 1915 was realised by investigations funded by the National Science Foundation in September, 2015. The detection was of gravitational waves from the collision of two black holes some 1.3 billion years ago and came 100 years after they were predicted.

▶ Cambridge University study and report on implications of biodiversity as a result of ocean acidification.

Private and multinational organisations

A surprising percentage of the scientific and medical research carried out is done not by academic institutions but by private firms and multinational organisations. The research is in very specific fields which may range from drug development to nuclear energy. These organisations reinvest the money they make to move the science – and the technologies they develop from the science – forward as fast as they can. So, for example, our world is still relying on oil for many of the products and transport methods which we have come to take for granted since the first commercial oil well was developed in Pennsylvania, USA in 1859. By the outbreak of the Second World War, our dependence on oil was firmly established.

▶ The BP 'Deepwater Horizon' oil spill disaster in the Gulf of Mexico April 20, 2010. Approximately 800,000 m³ of oil was spilled in the sea.

Table 7.5 shows the approximate revenue of the five largest oil companies at the end of 2015.

▶ **Table 7.5** Approximate revenue of the world's five largest oil companies (end 2015)

Oil company and country	Country	Revenue ($ billion)
Sinopec	China	450
Royal Dutch Shell	UK	385
Exxon Mobile	USA	365
BP	UK	335
PetroChina	China	330

These oil companies reinvest a lot of this money to develop better technologies for extracting oil, methods of reducing the production of greenhouse gases when oil is burned, ways to deal with oil spills in oceans and alternative forms of cleaner energy for when the fossil fuels run out.

Case study

Climate change and oil company influence

Large oil companies have known about the science relating to climate change and the result of releasing greenhouse gases from fuel combustion since the 1970s. A well-known campaign group, 'Greenpeace', claims to have obtained information to show that funding was provided by an oil related company to an independent research scientist which may have resulted in arguments against the general scientific community who were in agreement that fossil fuels were to blame and that climate change would continue unless we changed our energy sources immediately.

Oil companies have been accused of spreading 'misinformation' (deliberate false information) about the causes of climate change for many years and have been very reluctant to accept any changes to their practices from government climate policies. It has been said that by keeping to this method of countering respected scientific research and obstructing policies in law, the acceptance of the problem is continually delayed in the minds of both politicians and the general public. As a result, doubts are raised on the scientific evidence and analysis of data obtained. The integrity of the science behind the findings is brought into question. The scientific community is now forced to defend its work and to investigate further in an attempt to add new or stronger evidence to support its claims, all of which takes a lot more time and money.

Check your understanding

1   Conduct a paired research activity. List all the items in *one* room of your school or college and identify what they are made of. Can you find a link for all of them with oil? This will include transportation and machine manufacture.

2   Identify any items which could be made without using oil in the process.

3   Write a short report titled 'Our Dependence on Oil'.

Pharmaceutical companies

For many years, large pharmaceutical companies have attempted to answer criticism of the costs of their drugs and medicines by pointing out that their research and development (R&D) costs are incredibly high and many products don't actually get to the 'market'.

Table 7.6 sets out the financial details of the world's largest pharmaceutical companies in 2014.

Table 7.6 illustrates the vast amounts of money which are generated from the sale of useful pharmaceutical products. Obviously, the industry employs very able scientific

▶ **Table 7.6** World's largest pharmaceutical firms

Company	Total revenue ($bn)	R&D spend ($bn)	Sales and marketing spend ($bn)	Profit ($bn)	Profit margin (%)
Johnson & Johnson (USA)	71.3	8.2	17.5	13.8	19
Novartis (Switzerland)	58.8	9.9	14.6	9.2	16
Pfizer (USA)	51.6	6.6	11.4	22.0	43
Hoffmann-La Roche (Swiss)	50.3	9.3	9.0	12.0	24
Sanofi (France)	44.4	6.3	9.1	8.5	11
Merck (USA)	44.0	7.5	9.5	4.4	10
GSK (UK)	41.4	5.3	9.9	8.5	21
AstraZeneca (UK)	25.7	4.3	7.3	2.6	10
Eli Lilly (USA)	23.1	5.5	5.7	4.7	20
AbbVie (USA)	18.8	2.9	4.3	4.1	22

Source: GlobalData / http://www.bbc.co.uk/news/business-28212223

PAUSE POINT

Produce a standard bar chart to show the profit of the 10 companies listed in Table 7.6 in $ billion dollars.

> **Hint**   Using the correct axes scales is important.

> **Extend**   Superimpose their profit margins onto the bar chart and ensure that there is an appropriate key.

professionals working in well-equipped laboratories to provide valuable research and development in order to produce the range of medical drugs and appliances which will enhance medical help to the general public and, of course, maintain the company's financial profit. If the company did not make a profit, it would fail and be unable to make the medicines so many of us need to keep us alive and well.

When a new product is developed, it is essential that it is assessed for use. This will involve chemical testing and medical trials, sometimes using the new product on thousands of patients. Problems can arise when the product is tested by professional individuals who are paid by the pharmaceutical company that has developed the drug. This introduces the risk of bias, even if it is not deliberate, so drugs continue to be monitored long after they have been passed for use by doctors.

Case study

GlaxoSmithKline

In 2012, the 'Washington Post' reported on a case which illustrates the problems facing trials of new pharmaceutical products following expensive development by large private pharmaceutical companies. GlaxoSmithKline (featured in Table 7.6 as GSK) had produced and tested a new product for the treatment of diabetes which they called Avandia. The company funded the trials and all authors of the scientific report, which fully endorsed the product, had received money from GSK.

Unfortunately, important points in the data of 4,000 patients tested in the trial were overlooked. It became apparent that the new drug could significantly increase the risk of a heart attack and it has been estimated that

more than 80,000 people have suffered heart attacks or deaths related to the drug since it was first put into circulation. It is no longer used.

Check your understanding

1. Produce a questionnaire of ten questions about your personal attributes or qualities and ask two of your friends and two professional people (tutors or other persons who know you well) to complete them.

2. Make a note of the answers and assess whether there may be an element of bias in the answers. Discuss this with your science group.

3. What can be done to reduce bias and perhaps change the funding source in the example above?

Fair trade organisations

Fair trade organisations are groups which have been set up to ensure that small producers of goods and commodities worldwide are able to make improvements to their lives and local communities while producing their goods. There are more than 350 fair trade organisations around the world which have a voice provided by the World Fair Trade Organization (WFTO) set up in 1989 and currently comprised of five world regions: Africa, Asia, Europe, Latin America and the Pacific Rim.

Essentially, the WFTO campaigns for the rights of exploited workers and promotes fair practices for producers in terms of their sales and benefits. This generally applies to agricultural and handicraft industries.

The WFTO is a non-profit making organisation and does not directly influence scientific development. However, the work carried out to provide support for small producers has a significant impact on the methods used in terms of environmental sustainability, in particular.

Production of coffee in Latin America, for example, is vital to the region's economy. As the crop became more important in the nineteenth century, the use of mechanisation, the need to develop larger areas of growth for coffee plants and the increase in open, unshaded areas of land meant that both plant and animal diversity was significantly reduced. Chemical pesticides and fertilisers became increasingly important and widespread in their use.

Fair trade certificates promote the continued and increased use of more environmentally beneficial methods to produce crops. This has a positive impact on the way in which farmers develop their practice and harvest their crops by:

▶ reducing chemical fertilisers and pesticides
▶ prevention of soil erosion
▶ forest protection and tree planting
▶ increased biodiversity.

Voluntary pressure groups

Sometime the funding for research comes not from governments or businesses but from voluntary organisations such as charities. In these cases, the research which is funded has to tackle an issue which is chosen by the organisation – which means that minority causes can miss out. It also means that maverick research which backs an unpopular or unsupported point of view can also sometimes find funding.

Charities

The definition of a charity is: 'A bequest, foundation, institution or suchlike that exists for the benefit of others'. The importance of charities to identify and campaign on important causes has increased significantly. One example is 'Prostate Cancer UK' which has highlighted the disease, particularly in men over 50. Funding is from donations, and campaigns such as 'Movember' (involving men being sponsored to grow facial hair in November) have raised more than £400 million for men's health causes.

Trusts

Trusts (and foundations) are groups that have money given to them by private individuals. They offer financial grants for many charitable causes (up to £2 billion annually) but do not provide services. Examples include the Big Lottery Fund, the Diana, Princess of Wales Memorial Fund, the Rank Foundation, Steven Gerrard Foundation, Wellcome Trust and the Wolfson Foundation.

Leaders of global businesses can become very rich. Sometimes, these individuals use their money to set up trusts to support research to help those less fortunate than themselves. People such as Bill and Melinda Gates (Microsoft) and Mark Zuckerberg and Priscilla Chan (Facebook) are philanthropists, both couples using their wealth to help cure and prevent serious disease around the world.

World Wildlife Fund (WWF)

The World Wildlife Fund was set up in 1961 to drive towards ensuring that humans live in harmony with wildlife. The organisation works with politicians, businesses and others to safeguard the natural development of wildlife. Publicised focus areas include giant pandas, tiger species, orang-utans, leopards, rhinos, polar bears, forests, oceans, carbon emissions and sustainability.

Greenpeace

Greenpeace developed in 1971 after a protest against nuclear testing in Alaska. They are very direct in their methods, which aim to stop large corporations from damaging Earth's natural environment. Publicised campaigns include protests against whaling, drilling in the Arctic, CO_2 emissions, toxic chemicals in fashion brands and deforestation.

Friends of the Earth

Friends of the Earth was set up in 1969 and became international in 1971 to establish clear campaigns and strategies to help everyone understand the need to protect the Earth. The group has identified key concerns in its long history, including climate change, countryside laws, recycling, green energy, use of fracking, protection of pollinators such as bees, and greenhouse gas emissions from biofuels.

B Interpretation, analysis and evaluation of scientific information

'Lies, damned lies and statistics.' This phrase was made popular by the US writer Mark Twain in the early part of the 20th century, who attributed it to the British Prime Minister, Benjamin Disraeli, although there is evidence for its use some time before this. It is commonly quoted when statistics are used to provide 'evidence' for quite weak arguments based on the results of an investigation or analysis. The phrase can also be used to cast doubt on statistics used by others to prove a point.

Information and accepted 'facts' rely both on the quality of the information or account of the evidence and the numerical data which has been compiled. How the information is then used depends greatly on its interpretation, which can vary significantly, even in scientific study.

Interpretation and analysis of scientific information

Qualitative evidence

Qualitative evidence in science refers to the particular 'quality' of the results obtained rather than expecting a definite numerical value. In many aspects of our lives, we are faced with qualitative judgements which are not always easy. In the sport of synchronised high board diving, for example, the points awarded for style and performance are given by a selection of panel judges who will have varying subjective views. The resulting points awarded are rarely contested, yet it is a difficult form of judgement to maintain consistently.

Key term

Qualitative evidence – information gathered about the qualities of a given investigation with no direct measurements.

When you are considering the science content of a report or article, the qualitative evidence presented is very important. This includes the references to other publications which confirm or support the findings being presented. The quality of the citations and references given is very important.

A citation is a specific source which is mentioned or quoted in the body of the text of a paper or article. Citations are very important because they give credit to other scientists or authors whose ideas have been used. They also provide a trail of references which have been consulted, show the level of background research which has been done, and help avoid any accusations of plagiarism.

The references are found at the end of a paper or article and list in full all of the sources which have been cited in the text. This enables you to look them up and read the same original sources as the author. Acknowledgements of individual scientists and institutions also help you to judge the quality of the research.

When searching various sources, you will need to assess whether the information provided is reliable. To do this, you should note the following:

▶ look for more than one reference source

▶ discuss the topic and suggested references with colleagues and tutors

▶ make a note of key facts which appear common in most of the sources used

▶ if a claim is made in one source which you have not previously encountered, research the information to gain a further understanding and to validate the information

▶ make a note of the development of the information over time; if it has been used for many years without change, it may be more reliable.

Key term

Reliability – a measure of how trustworthy the evidence or data may be in an investigation or research.

Remember that when you use information from another source to assess the reliability of information you have gathered, you must always acknowledge the work produced by other writers and researchers, recognising their intellectual property rights. Not to do so would constitute *plagiarism*.

Examples of appropriate referencing are as follows:

▶ Textbook:
Annets F, Hartley J, Hocking S, Llewellyn R, Meunier C, Parmar C and Peers A – *BTEC National Applied Science Student Book 1* (Pearson, 2016) ISBN 9781292134093

▶ Internet web page:
Wells, D. (2001) Harvard referencing, [online], Available: http://lisweb.curtain.edu.au/guides/handouts/hardvard.html [14 Aug 2001]. (Samantha Dhann, 2001).

▶ Journal article:
Smith, J (2015) Lack of iron meteorites on Earth, *Journal of the Unexplained*, vol. 2, pages 27–32.

Quantitative evidence

This aspect of scientific evidence relates to values which are measured, i.e. *quantities*. The determination of information from this form of evidence requires a good level of mathematical understanding and the ability to identify patterns which may not be clearly demonstrated when the data is presented in complex tables or statistical analysis. Analysis of scientific evidence also relies on your ability to use statistical analysis to ascertain how well evidence supports or refutes the hypothesis.

The usual visual method to present **quantitative evidence** is in tables and graphs and the numerical data is analysed by calculation and by statistical means.

Key term

Quantitative evidence – information gathered using measurement of quantities and producing numerical data.

Link

Unit 6: Investigative Project.
Unit 3: Scientific Investigation Skills, pages 149–172.

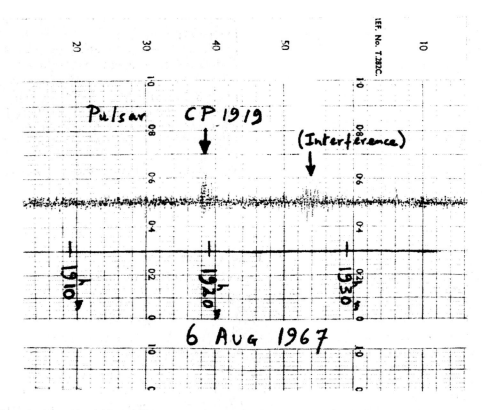

▶ Trace document from Anthony Hewish's Nobel Lecture in 1974 referring to a pulsar (rapidly rotating star made entirely of neutrons).

$$R_{\mu\nu} - \frac{1}{2}R\,g_{\mu\nu} + \Delta\,g_{\mu\nu} = \frac{8\pi G}{c^4}\,T_{\mu\nu}$$

▶ **Figure 7.13** Einstein's equation for general relativity is regarded by some as 'the most beautiful equation of all . . .' Not everyone is able to see this beauty and not everyone needs to. Calculations in science form part of the fundamental use of numerical data to tell us what is happening in our world and outside it. They do not necessarily need to be this complex to provide some important information.

Graphs and tables

Graphs provide an immediate visual representation of the data or other information which has formed part of the study in question. The exact type of graph to be most effective will be determined by the activity and the purpose of the study. Figures 7.14–7.18 give some examples of graphs.

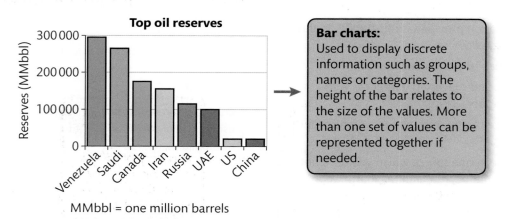

Bar charts:
Used to display discrete information such as groups, names or categories. The height of the bar relates to the size of the values. More than one set of values can be represented together if needed.

MMbbl = one million barrels

▶ **Figure 7.14** Bar chart

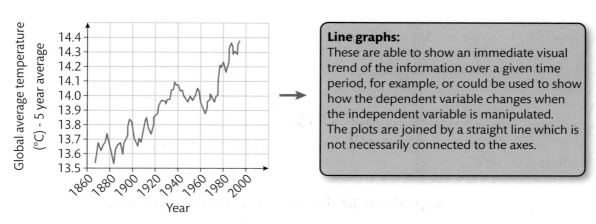

Line graphs:
These are able to show an immediate visual trend of the information over a given time period, for example, or could be used to show how the dependent variable changes when the independent variable is manipulated. The plots are joined by a straight line which is not necessarily connected to the axes.

▶ **Figure 7.15** Line graph

Pie charts:
Numerical data is represented as a proportion of a circle. The segment (central angle, area and arc length) is proportional to the amount it is representing.

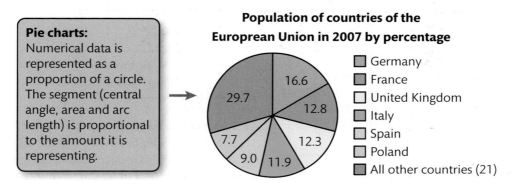

▶ **Figure 7.16** Pie chart

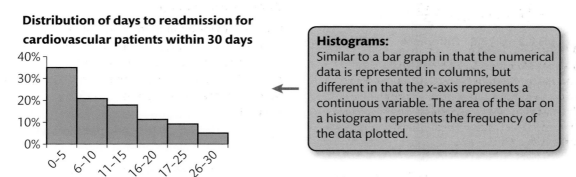

Histograms:
Similar to a bar graph in that the numerical data is represented in columns, but different in that the x-axis represents a continuous variable. The area of the bar on a histogram represents the frequency of the data plotted.

▶ **Figure 7.17** Histogram

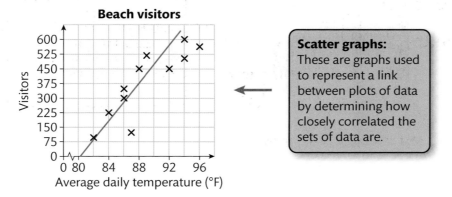

Scatter graphs:
These are graphs used to represent a link between plots of data by determining how closely correlated the sets of data are.

▶ **Figure 7.18** Scatter graph

When producing tables to display numerical information, most organisations adhere to rigid principles to ensure that the values included are well understood by the majority of readers. This generalised summary of basic principles in table layout is not necessarily used for more academic studies.

▶ Rows and columns – arrange them so that patterns in data can be identified down a column, if possible.

▶ Present numbers in the simplest format, e.g. £6.8 billion, not £6,800,000,000 or 5.2×10^{-3}, not 0.0052.

▶ Use more than one table if values are complex but of equal importance.

▶ Identify clearly what the table is showing in a heading and use a reference if needed.

Theory into practice

Table 7.7 is taken from research into the rate at which a coral reef is able to produce calcium carbonate ($CaCO_3$) for two pH levels. Are patterns identifiable? Are the numbers shown in a simple form? Does the heading convey what the table is showing?

▶ **Table 7.7** Calcification rates (mg h^{-1} cm^{-2}) at pH 8.2 and 7.8 calculated for 5–30 per cent dead area. Rates for 0 per cent and 100 per cent dead area for a coral with a low calcification rate were measured, other values by interpolation.

% dead area	pH 8.2	pH 7.8	Difference	% Difference
0	0.092	0.120	0.028	+30.4
5	0.100	0.119	0.019	+18.6
10	0.108	0.117	0.009	+0.08
15	0.116	0.116	0.000	0.0
20	0.124	0.114	0.100	−0.08
25	0.132	0.113	0.019	−14.4
30	0.141	0.111	0.030	−21.5
100	0.250	0.090	0.160	−64

Source: http://www.scielo.sa.cr/pdf/rbt/v60s1/a10v60s1.pdf

Statistics

Once collected, numerical data will need to be processed. You can obtain meaning from the data using the following methods of statistics:

▶ standard deviation – this indicates how closely a set of numerical data values are to the mean value

▶ frequency – how often a particular value occurs in a set of values

▶ t-test – a method of statistics used to compare the mean values of two data sets to determine the overall significance of the mean value difference

▶ chi-squared test – a statistical method which is used to compare the observed frequencies to expected values.

Link

Unit 3: Scientific Investigation Skills gives more detail on how these statistics are calculated.

In science reports, the type of statistical method used should be evident in the text. Data which has a **normal distribution** has a calculated mean and, subsequently, standard deviation. The calculations are then represented with the appropriate number of decimal places. Error bars are lines or shaded areas shown on graphs, bar charts and histograms. They represent the variation in the results and so give you a good idea about the reliability of the data.

Key term

Normal distribution – a set of real value data that, when plotted on a histogram, produces a bell shaped curve. This is called a normal curve and its peak is the mean.

 **PAUSE POINT**   Use the information provided in Unit 3, pages 152 and 153, to find the standard deviation of the following set of data:

Recorded 12.00pm temperatures taken from an amateur weather station in the UK on 31st May, between 2006 and 2016 – 15.0, 18.2, 17.2, 21.4, 16.2, 22.0, 17.0, 23.2, 19.2, 19.5 (°C)

Hint   Ensure that your answer is given to the same decimal numbers as the list of figures.

Extend   Produce a frequency/standard deviation curve of the data based on your standard deviation calculation.

Evaluation of scientific information

Evaluation is the process in science which we use to assess or determine the overall quality of a scientific experiment, investigation or finding. Not everyone is able to fully understand and then evaluate another piece of work or publication. We must break down the process to help identify key areas which should be our focus when attempting to judge the quality of someone else's science report:

▶ The report must have its foundation in strong, historic scientific knowledge and principles.

▶ The research used must be correctly grounded in current scientific knowledge and understanding to inform the objective of the investigation.

▶ A logical approach to the investigation must be clearly shown in the reporting structure.

▶ The method used must demonstrate a complete unbiased technique.

▶ Have the writers assessed how well their data supports their hypothesis using statistical methods?

▶ There should be a method to provide useful feedback on the work produced so as to help drive progress and discussion.

The flow diagram in Figure 7.19 outlines six stages which should form the basis when evaluating a scientific report.

Validity and reliability of data

In general terms, information or data from a scientific study is deemed to be reliable if:

▶ the study is repeatable, providing the same or very similar results when the method is repeated

▶ the results are repeatable by others

▶ a pattern in the data can be identified either in tabulated or graphical form

▶ research, secondary data supports your primary data.

Validity is different. A thermometer may be falsely providing a reading of temperature by 1°C above the actual value. It may be giving this reading every time it is used and so is providing reliability. However, the readings are always incorrect by 1°C and so the measurement is not valid. A scientific study can be deemed to be valid if:

▶ the investigation actually measures what it sets out to measure

▶ the study provides some corroboration of other similar studies

▶ the outcomes provide information which may improve the area of study

▶ a panel of individuals reviews the investigation.

> **Key term**
>
> **Validity** – how well a particular investigation measures what it claims to measure.

Sample size

The **population** from which a **sample** is taken may be of any given size, comprised of the whole population or just a proportion of it. Samples may be either random (tests on a newly developed pharmaceutical, for example) or non-random (similar pharmaceutical tests but aimed specifically at a group of individuals with a common condition).

> **Key terms**
>
> **Population** – all items, people or other individual components which have the characteristic which is to be investigated.
>
> **Sample** – a smaller selection taken from the population which has the characteristics under investigation.

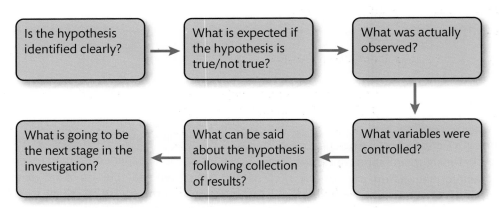

▶ **Figure 7.19** Evaluating a scientific report

Generally, the larger the sample, the more precise the outcome. In clinical trials when developing a pharmaceutical, for example, the sample size must be large enough to ensure that the results are meaningful but not so large that extra financial burden is placed on the trial, making the investigation uneconomical.

It is also important to establish the statistical probability that the findings of the sampling are not due to random chance. In clinical trials this probability is usually set at less than 5 per cent, meaning that there is a less than 5 per cent probability that the results are due to chance alone. The eventual sample sizes are calculated using some complex statistical mathematics to ensure that the outcome of the investigation is not obscured by false or misleading results based on the samples chosen. How representative the sample is of the population must also be taken into account, eliminating any bias and increasing the reliability of the study.

Number of references/authenticity of information

When you are assessing the quality of a study or producing one yourself, an important question to ask is: 'Are the number of references sufficient?' While extensive lists of sources used do not necessarily provide evidence that the findings in a study are valid or reliable, they should give a good indication of the research supporting the report.

It is fair to state that the relevant expertise within a given report will become noticeable, especially if the references used are detailed and can stand up to scrutiny. Peer review, an evaluation of the scientific work by others working in the same field of study, is a final aspect which would indicate that the information is worthy of scientific publication.

Case study

How confident can you be in the findings of a report?

61 Alvarez, S., et al., A revealed preference approach to valuing non-market recreational fishing losses from the Deepwater Horizon oil spill, Journal of Environmental Management, Vol. 145, pp. 199–209, 2014.

62 Sumaila, U.R., et al., Impact of the Deepwater Horizon well blowout on the economics of U.S. Gulf fisheries. Canadian Journal of Fisheries and Aquatic Sciences, Vol. 69(3), pp. 499–510, 2012, www.nrcresearchpress.com/doi/ full/10.1139/f2011-171#.VKL_D14DxA.

63 The Knowland Group, The Gulf Coast Oil Spill and the Hospitality Industry, The Knowland Group, August 2010, www.knowland.com/data/ caseStudy/06a17a78-f1fd-47dc-b8db-be6e9379601b.pdf.

64 Tourism Economics, The Impact of the BP Oil Spill on Visitor Spending in Louisiana: Revised estimates based on data through 2010 Q4, Prepared for the Louisiana Office of Tourism, June 2011,www.ustravel.org/sites/default/files/page/2009/11/Gulf_Oil_Spill_Analysis_Oxford_Economics_710.pdf.

65 Oxford Economics, Potential Impact of the Gulf Oil Spill on Tourism, Prepared for the U.S. Travel Association, www.ustravel.org/sites/default/files/ page/2009/11/Gulf_Oil_Spill_Analysis_Oxford_Economics_710.pdf. (Accessed March 25, 2015.)

Source: http://textlab.io/doc/1995064/summary-of-information-concerning-the-ecological-and-econ...

Check your understanding

This extract has been taken from a report which summarises the economic and ecological impact of the BP Deepwater Horizon oil spill disaster which happened in 2010. There are 65 acknowledgements/references. Would you have confidence in the findings of this report?

Case study

Sampling for fat, salt and sugar content in foods

Wilhelmine Sieben is a German born researcher based in the UK who works for an organisation which provides workers who gather information from populations in many countries of the world. The information is generally obtained from direct street questionnaires in selected cities throughout the world and is used by companies and organisations to assess the 'habits, knowledge, likes and dislikes' of the population.

Wilhelmine is an experienced market researcher who has been given a sample size of the general public to question over the course of a single weekend in a small town of population 250,000.

Details of the questionnaire and sample:

- A total of 100 people are to be questioned.
- The questions are:
  a) Do you know the recommended daily amount of salt in your diet?
  b) Do you know the recommended daily amount of sugar in your diet?
  c) Do you read food labels for fat content?
  d) How much exercise do you carry out in hours per day?
- The age range is over 15 years.

Check your understanding

1 Is the sample size representative of the population of the town?

2 Do the questions need to be more specific or should they provide a range of suggested figures?

3 Is the question on exercise relevant? Does it tell you anything about the person answering the questions?

4 Is the sampling random or is there some bias involved?

5 Are the days of the sampling relevant?

6 Could the age range be changed to produce more conclusive results?

Use and misuse of data

In the game 'Chinese whispers', the original message relayed by the instigator of the message becomes highly distorted after passing between a number of other individuals. It is sometimes very surprising to witness the full extent of the difference in the exchange of information between just two people and this aspect of misunderstanding or even misuse can happen in scientific study.

> 'UN health body says bacon, sausages and ham among most carcinogenic substances along with cigarettes, alcohol, asbestos and arsenic.'

> Source: the *Guardian*, 26 October 2015

This quotation relates to WHO research on certain processed and red meats, carried out by the International Agency for Research on Cancer. Misquotes like this cause a high degree of confusion and, in some cases, panic, when the general public become quite alarmed at the claims made.

There is no scientific basis from studies to support the claim and, in fact, there is scientific evidence to show that lean red meat is beneficial to long-term health.

In science, it is vital that we are able to use and quote information and numerical data accurately and with integrity, ensuring that whatever the outcome of an investigation, the facts are not distorted. To do this effectively, it is worth considering the following four key points:

- Make sure that the figures or information presented can be verified by others or by trusted scientific principles and publications.
- Read all points. Sometimes, the way that data has been calculated, such as in averages, may not be suitable for the sample or investigation, especially if the sample size is very small.

- Be convinced by the way that data has been collected – does it fit with scientific data collection methods or has it been taken from known sources?
- Check the figures (extreme highs and lows, missing numbers) and carry out a spot check as a way of validating a small sample.

Potential areas for further research and development

There are many areas of science which have a lot of potential for research and development (R&D). R&D takes pure scientific research and moves it towards practical applications. It implies the commercial development of an idea or a new technology so that it is both useful and makes money. R&D is mainly funded by companies and by governments, which may also fund the original scientific research or which may pick up some findings from original research and move it into R&D. There are many different factors to look for when considering whether a particular piece of research is suitable for R&D funding. These include projects which:

- **move the technology forward** – most R&D is at the very cutting edge of science, e.g. therapeutic cloning, the development of a new Very Large Hadron Collider
- **are targeted at a particular need**, e.g. more efficient, smaller batteries for smart phones, new antimicrobials, a vaccine against HIV and malaria
- **have impact and status** – R&D funding is often only targeted at large projects which will be influential and bring in a lot of money rather than small projects or projects for areas where there is little money
- **have potential for wide use** – the potential to make money from new discoveries drives most R&D so the focus is on developments which would affect a lot of people. This is also true for medical R&D, which

is largely active for diseases and conditions which affect a lot of people. This has both social benefits (the maximum number of people are helped) and economic benefits – the company sells a lot of the treatment or cure.

▸ **can be patented** – which means that any technological developments arising from the research can be used exclusively by the company paying for the R&D for a given amount of time.

Some R&D is not funded by companies or governments – individual philanthropists may pay for particular R&D projects, e.g. the Bill and Melinda Gates Foundation paying for R&D into malaria vaccines.

Evidence to support conclusions and claims made

Scientific studies do not always arrive at the correct conclusions for many different reasons. Sometimes there is considerable pressure placed on scientists by large companies or politicians, for example. In other studies, the conclusions are incorrect for quite legitimate reasons. Many scientific conclusions which find their way into the realms of the media are at the 'cutting edge' of research and ideas may be changing too quickly to produce the most up to date and accurate information. Medical claims fall into this category and, while there may be many incidences of incorrect or changing conclusions based on

evidence, it is a vital and necessary part of the scientific process to continue with experimentation until the ideas become established. This, of course, can take many years.

In all scientific reporting, the conclusions or claims made must be supported by the evidence, which should be clearly presented and must stand up to scrutiny by the scientific community. The investigation itself may be well planned, producing good results which can be used for analysis, but the investigation will not be accepted in its entirety if the conclusions are weak or do not link well with the evidence.

In many scientific studies, the final conclusions may not necessarily bring the study to a close. Good scientific practice would almost certainly ensure that the results are repeated to provide additional evidence to support the conclusions made in the initial study. In many aspects of science, the method may be slightly changed for a similar investigation. If the results obtained are also similar to the original investigation, then there is further support given to the conclusions. The 'validity' of the conclusions is strengthened.

The scientific peer review process ensures that the claims made in scientific reports are not cleared for publication until the work has been evaluated by other scientific professionals with an expertise in the same or very similar field of study. In most cases, the work may be carried out again by others who may arrive at the same set of results and final conclusions.

Assessment practice 7.2

Swap one of your scientific investigations from either Unit 2 or Unit 4 with another member of your group.

Use the information in this section to assess the overall quality of their investigation in terms of:
- references to established sources of information (where relevant)
- processing and analysis of numerical data
- validity and reliability of data
- identifying any areas where further work could be carried out
- strength of the evidence in support of the conclusions.

Plan
- What is the task? What am I being asked to do?
- How confident do I feel in my own abilities to complete this task? Are there any areas I think I may struggle with?

Do
- I know what it is I'm doing and what I want to achieve.
- I can identify when I've gone wrong and adjust my thinking/approach to get myself back on course.

Review
- I can explain the results obtained from the task.
- I can apply the activity to other situations.

C Science reporting

The way in which science is reported varies to fit the forms of reporting medium in which it is presented. For example, journals and magazines specialising in specific areas of science will contain reports which are more detailed than

newspaper or television articles. The language used by the media is very important in portraying the message and will have an immediate and also long lasting effect. The overall content of a science topic report will be tailored to the target audience and so the amount of technical information, complexity of wording and focus of the report will depend upon the audience that the medium is intended for.

Know how science is reported in different media and for different audiences

Reporting medium

We are all different and absorb information in different ways and in a variety of forms. It is not surprising therefore, that there is a range of media which report science in order to meet the needs of a variety of readers.

Specialist journals

These are publications for scientists which are valuable sources of current scientific advances and development. They are peer reviewed to ensure that the science reported is at the forefront of understanding and knowledge and is correct in terms of the science involved and the claims made. *Nature*, for example, has been in circulation since 1869 and helps to promote new scientific information to a public audience, offering a platform for discussion and presentation relating to education and matters in everyday life. Scientific breakthroughs reported include: the neutron, DNA, the ozone hole and nuclear fission.

Science magazines

These are periodic publications for non-scientists aimed at providing up to date accounts of scientific development to interested readers. Many magazines are aimed at specific areas of science such as *The Sky at Night* or *Laboratory News*, but publications such as *New Scientist* provide information on all scientific current developments, reviews and discussion. *New Scientist* is also available online.

Television

This is an easily accessible form of media which has large viewing figures from all backgrounds in society. Science programmes are tailored to the level of education background. Tight controls are in place from the regulators – Ofcom, for example, is the British communications regulator and operates under Acts of Parliament to develop media competition and maintain its commitment to high-quality television programmes for the general public. Science information is presented in stimulating graphics and by popular presenters with up to date scientific issues and breakthroughs reported immediately in news sections. Clear language and technical facilities allow for both visually and hearing impaired public audiences. This established form of media also attempts to make difficult concepts visual, for example, showing the polar ice sheet melting to illustrate the issues faced with CO_2 level rise. However, at times the science reported is simplified to the point where it is no longer accurate.

Internet and social media

This form of media has an extremely large and diverse target audience. It is used as a resource for information and is in some cases replacing written forms of communication, research and entertainment. Dependent on the provider, science articles can be sensationalist and include generalised information, although the availability of instant additional research can allow for interested readers/listeners to gather their own information. Social media in the form of Facebook Inc, Twitter Inc, YouTube™, blogs etc. are a valuable means of sharing information between people, organisations or companies, but the science relayed is often limited or inaccurate in its coverage. In some cases the science presented, especially around medical topics, is dangerously wrong.

National/local newspapers

These are cheap and popular source of general news and information with a more relaxed approach to the style of reporting they use. The regulation is more flexible and there is more political bias attached. Science is not an important feature in this media, although current events are highlighted, usually with a disturbing or eye-catching headline. (Refer back to the case study at the beginning of the unit for an example of this.) The information relayed is purposely kept to a minimum, with a clear sentence structure and focus. Scientific background is not really provided in any depth and topics vary in the manner in which they are presented. Space science, for example, is generally kept short and factual while reports on health appear sensationalist in headings and are expected to be better understood by the general public.

❚❚ PAUSE POINT   Write a journalistic report on a scientifically difficult topic which is current. The report is for a national newspaper, which has a target audience of the general public: including labourers, shop workers, delivery drivers, roofers, and others.

Hint   Focus on bridging the gap between science and the public. Keep the language basic and don't write anything which you do not understand yourself. Outline the science in precise and accurate terms, but simply. Do not sensationalise the information.

Extend   Identify how your article could possibly affect the lives of the readers.

The target audience

General public

Scientific information, news or updates on existing knowledge are of general interest to the majority of the general public because of the inherent link to our lives. At least one of the three scientific disciplines (Biology, Chemistry, Physics) will be of more interest to an individual than the others, although it is recognised that many have a general interest in how science affects their lives.

The term 'general public' is meant to include all those who do not work directly with science, nor have specific qualifications to use science in their work or undertake scientific research, study or development. People with professional qualifications not linked to science, with technical expertise or in other areas of work or study are also included as general public and may read or research science to satisfy a curiosity, enhance their understanding or simply to gain further knowledge. The average reading age of the UK general public is 9 years old!

This type of target audience is assumed to have little or no significant scientific knowledge in a given area of science and so, when reporting science to this group:

▸ Easily accessible information on the background of the science involved should be given.

▸ Examples and analogies should help to explain certain ideas and concepts.

▸ Technical words and phrases should be used only where absolutely necessary.

▸ Where possible, diagrams and other figures should be included.

▸ Include expert testimony and information to increase the level of interest and add credibility.

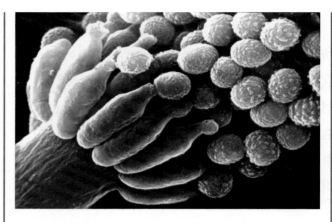

▸ *Aspergillus fumigatus,* one of the most common aspergillus species to cause disease in individuals with an immunodeficiency.

Scientists have warned that potentially deadly fungal infections are acquiring resistance to many of the medicines currently used to combat them. More than a million people die of fungal infections every year, including about 7,000 in the UK, and deaths are likely to increase as resistance continues to rise.

Researchers say the widespread use of fungicides on crops is one of the main causes of the rise in fungal resistance, which mirrors the rise of resistance to antibiotics used to treat bacterial infections in humans.

Robin McKie Science Editor

Saturday 27 August 2016 21.00 BST Last modified on Sunday 28 August 2016 00.02 BST

Scientific community

Reporting of information, findings, new research and scientific issues to the science community is necessarily detailed and contains an abundance of scientific terminology which is not understood by the general public. Sections of text may be very long and contain no further explanation of the scientific terms included, since the target audience is expected to fully grasp the meaning of the report and scientific words used in the text. The readers are assumed to have a good knowledge of the concepts and science involved.

This kind of science reporting is used in media which will be accessible to scientists or other science professionals and specific to the area and level of research carried out by the individual. As a result, television, newspaper and social media networks are unlikely to be used as forms of media to relate this kind of information.

When reports or information are presented to specific sections of the scientific community, the procedure followed needs to ensure that the report stands up to structural, language and intellectual scrutiny:

Discussion

Read the following passage taken from The Guardian online newspaper www.theguardian.com 28/08/2016. Discuss within your group whether it follows the guidelines listed above for a science report to the 'general public'.

Millions at risk as deadly fungal infections acquire drug resistance

Researchers believe widespread use of fungicides on crops is reducing effectiveness of frontline medicines.

▸ Is the report concise, well organised and well presented?

▸ Are the results or findings presented clearly and accurately?

▸ Is the language appropriate with no grammatical errors?

▸ Is the material of the report supported by other work?

▸ Can the method be replicated by others in the field of study?

▸ Are diagrams, graphs, tables etc. clear, well designed and adequate for the report?

▸ Are the conclusions supported by data and is the discussion relevant?

▸ Are all references correctly cited?

▸ Is the report within the defined scope of the publication used?

An important point to remember is the need for the scientific community to engage with other forms of media such as television, newspapers and the internet in order to ensure that the general aspects of their work are communicated to the wider public, and adding justification to their research and development. The public, of course, will ultimately use or be affected by many scientific innovations and discoveries and so should be given the chance to engage. The information provided by scientific research must be easily spread to the general population and relies to a great extent, now more than in previous centuries, on media presentation, to help in understanding, present the pros and cons, and establish some dialogue between scientists and the people. There have been many public science forums developed over the last decade, and some programmes related to wildlife, in particular, now incorporate a 'behind the scenes' section showing how many difficult aspects of filming have been dealt with.

Pressure groups

These are organisations which are funded both publicly and privately to attempt to influence the public and politicians in the hope of affecting decisions made by government on a variety of issues (see Figure 7.20). Pressure groups (or Voluntary Pressure Groups) are independent of the political system and are able to freely voice their opinions on issues which affect the population. By raising the profile of relevant and topical scientific issues, those with authority in society are forced to take note and debate the concerns openly.

In order to affect change and draw attention to points of concern, a **lobbyist** is used to approach an official within a government on behalf of the pressure group and attempt to influence the legislation or regulation relating to the concern raised. In 2014, for example, Friends of the Earth successfully lobbied MPs in the UK, who subsequently introduced an action plan to save British bees following their earlier development of the National Pollinator Strategy (NPS).

Key term

Lobbyist – a person who may or may not be a professional, acting on behalf of businesses or organisations to influence decisions made by government by approaching relevant government officials.

Campaign for Nuclear Disarmament (CND): Set up to draw attention to the threat of nuclear war.

Greenpeace: Direct intervention into global environmental sustainability issues.

Privacy International: Campaigns for freedom of information, ID cards.

World Wildlife Fund (WWF): Campaigns for natural world preservation and reduction of human damage.

Friends of the Earth: Largest environmental pressure group.

British Medical Association (BMA): An 'insider group' providing consultation and information to UK government on medical matters.

▸ **Figure 7.20** Some well-known pressure groups and common areas of campaigning

The need to develop the communication pathways between scientists, journalists, pressure groups, government and the public is a recognised and ongoing area of concern and focus. At present, pressure groups can be seen to be facilitators in communicating uncertain and developing scientific study to the public, in embryo research and environmental damage caused by shale gas mining practices - 'fracking', for example. The level of this language in communicating key scientific points needs to be a compromise between clear, but relatively basic science and detailed concepts, a balancing act which helps the general public understand the fundamental areas of concern and useful scientific information which allows an understanding of both the implications of an activity and the science involved. There have been many cases which have shown that, without this further identification of the issues on the part of the pressure groups concerned, large organisations and governments would not easily accept the scientific arguments. However, it has to be remembered that lobbyists and pressure groups are aiming to get across a very specific point of view and their use of data can be very selective. They tend not to present data which does not support their point of view!

Case study

Climate change

Attempting to communicate a difficult cause and effect argument in light of the complexities of climate modelling and the uncertainties of historical climate science is not easy. Add to this the lack of scientific information available to be able to predict the change in global and regional climate in the future, and the problem becomes even more difficult.

When faced with stubborn and very wealthy and powerful organisations and governments, the acceptance of climatic change needed to be firmly embedded in the minds of the general public in order that science could be used to force big business to agree on ways to manage and eventually reduce CO_2 emissions through government legislation.

Key events in the climate change evidence:

- Dr. Charles Keeling records CO_2 levels accurately from 1958 at the Mauna Loa Observatory, Hawaii and continues to notice a rising trend over the next 4 decades until his death in 2005.

- The Arctic Climate Impact Assessment (2004) is produced and links the rise in global temperatures with rising CO_2 levels, suggesting that human activity is the main cause.

- The Intergovernmental Panel on Climate Change (IPCC) produces its Fourth Assessment Report in 2007 stating that global warming was 90 per cent plus likely as a result of human activity.

- The G8 plus five nations release a statement that climate change is happening even faster than previously thought.

- Geological Society of London produces a position statement in 2010 which fully accepts the global warming science and identifies geological parallels, relating the warming episodes to a significant rise in sea levels and species extinctions.

- The American Meteorological Society confirms in 2012 the findings that the Earth's lower atmosphere, ocean and land surface are warming and that sea level is rising, and snow and ice cover is shrinking.

There was (and still is) much to consider for both governments and large-scale organisations, such as economic future, social changes necessary and political drive to accept alternative energy sources. Humans still need to consider possible implications of increased surface temperatures such as: release of toxic waste from military bases under the Greenland ice exposed by melting ice, emergence of deadly diseases as a result of warmer climate, species extinctions – especially where the natural habitat is very specialist.

Check your understanding

1 From the above passage, outline what you think the main difficulties are when trying to convince people that climate change is happening.

2 Of the six key events in the climate change evidence listed, identify TWO which you believe are the most compelling. Give your reasons why.

3 If scientists and politicians are not able to bring about significant change to reduce and even reverse the trend of global warming as a result of human technological development, detail the possible implications to the planet and its inhabitants over the course of the next 70 years. (You will need to research information from future projections.)

Political representatives

Science, engineering technology and research are covered by most governments by a specialist department. In the UK, for example, responsibility for this area is given to the Government Office for Science (GO-Science) – part of the Department for Business, Innovation and Skills (BIS). Unusually, perhaps, this office is further checked in its decisions and practice by the Commons Science and Technology Committee which ensures that relevant government policy is made using sound scientific evidence and professional advice.

In order for Members of Parliament (MPs) to represent our interests in science, they need either a good understanding of science itself or access to departmental organisations that can offer high-quality, detailed scientific information and professional advice. The government has therefore made sure that suitable resources are available to MPs relating to science, including:

▶ science committee and all party medical science group

▶ office of science and technology discussion

▶ Business, Innovation and Skills departmental access

▶ Foundation for Science and Technology group meetings

▶ pairing scheme between MPs and the Royal Society, as well as the Royal Society of Biology, the Royal Society of Chemistry and the Institute of Physics.

▶ engagement of MPs with societies, academies, charities and research groups.

The science community has taken steps to ensure that those providing our representation in government, local and national, are given scientific information and specialist terms in a manner which will be understood by the MPs so that they can best serve the public interest.

Our political representatives require scientific communication to be:

▶ relevant and interesting in content

▶ aimed at their specific level of science knowledge

▶ headline capturing (can the issues be delivered in a quick message?)

▶ specific but not overly detailed

▶ factual information, not speculative

▶ impartial, to ensure that the science is all that matters.

Understand the presentation of science reporting and its relationship with the reporting medium and target audience

The presentation of science reports varies considerably, dependent on both the reporting medium used and the target audience. Standardised science reports, following investigation, will be presented in a manner which has been firmly established in the industry over many years. We should all be very familiar with the structure as listed here.

▶ Title

▶ Abstract

▶ Introduction

▶ Method

▶ Results

▶ Accuracy

▶ Discussion

▶ Conclusions

▶ References and Bibliography

This structure may not always be used when science is communicated, especially since the target audience differs greatly between the reporting mediums.

Scientific articles can be published in science journals. These are the main way in which scientists publish their findings for other scientists to read. Most scientific journal articles have these headings as part of their structure but there might be slight differences as different journals have different styles of publishing.

The following two science reports demonstrate clearly the differences in the way that the science involved is communicated to two slightly different target audiences by two different reporting media; the general public (online newspaper report) and the scientific community (extract from peer reviewed journal article).

Example 1: Online newspaper/internet report (non-peer reviewed)

First ever stem cell trials on babies in the womb set to begin in January

The pioneering trial could help lessen the painful symptoms of brittle bone disease

Doug Bolton @DougieBolton *Monday 12 October 2015* (www.independent.co.uk)

Babies in the womb who have brittle bone disease will soon have their illness treated with **stem cell injections**, in a series of pioneering clinical trials set to begin in the UK in January.

The trials will be the first time stem cells have ever been used to help babies in the womb, and could potentially see the **effects of the painful brittle bone disease** made much less debilitating.

The disease affects around one in every 25,000 births, and is caused by errors in the child's DNA. These errors can mean that the collagen inside the body, which gives bones strength and structure, is either of poor quality or missing altogether.

Children born with the disease, which is officially called osteogenesis imperfecta, can be born with multiple broken bones. After birth, affected children often face growth problems and regular bone fractures, as well as hearing issues.

By injecting these children with stem cells, which can transform themselves to grow into different kinds of tissue, proper bone growth can be sparked and the negative effects of the disease can be lessened significantly.

As reported by the **BBC**, the trial will start in January, and will be led by doctors at the Karolinska Institute medical school in Sweden.

Famed children's hospital Great Ormond Street will lead the study in the UK, using stem cells from terminated pregnancies to try to help unborn children improve their chances in life. The disease won't be completely cured even if the experimental treatment is successful, but it will make the bones stronger and make affected children less susceptible to constant fractures.

In the trials, 15 babies will be treated with stem cells while in the womb, and again after they are born. Another group of 15 will be treated after birth, and the health problems of each group will be compared with those of untreated children.

Example 2: Scientific article extract (peer reviewed)

The induction and identification of novel Colistin resistance mutations in *Acinetobacter baumannii* and their implications...
Published online: 22 June 2016

Abstract:

Acinetobacter baumannii is a significant cause of opportunistic hospital acquired infection and has been identified as an important emerging infection due to its high levels of antimicrobial resistance. Multidrug resistant *A. baumannii* has risen rapidly in Vietnam, where colistin is becoming the drug of last resort for many infections. In this study we generated spontaneous colistin resistant progeny (up to >256 µg/µl) from four colistin susceptible Vietnamese isolates and one susceptible reference strain (MIC <1.5 µg/µl)...

Introduction:

Acinetobacter baumannii is emerging worldwide as a hospital acquired pathogen infecting critically ill patients. In susceptible individuals *A. baumannii* can cause a range of infections including pneumonia, bacteraemia, meningitis, blood stream infections and urinary tract infections...

Results:

Whole genome analysis of four Vietnamese A. baumannii clinical isolates

To elucidate how *de novo* mutations play a role in colistin resistance in *A. baumannii* we selected four colistin susceptible clinical isolates from patients with culture confirmed VAP in the Hospital for Tropical Diseases in Ho Chi Minh City between 2009 and 2019 (Table 1)...

Analysis of mutations in colistin resistant progeny strains

A series of longitudinal (five days) sub-culturing experiments was performed to induce and increase the MIC against colistin in the four selected Vietnamese clinical *A. baumannii* isolates and ATCC19606...

Discussion:

The rise of *A. baumannii* in Vietnam has been accompanied by the rapid emergence of antimicrobial resistance in this problematic environmental bacteria resulting in colistin being a last-resort antibiotic...

Additional Information:

How to cite this article: Thi Khanh Nhu, N. et al. The induction and identification of novel Colistin resistance mutations in *Acinetobacter baumannii* and their implications. *Sci. Rep.* 6, 28291; doi: 10.1038/srep28291 (2016).

References:

Note – full article is approximately 17 pages in length.

http://www.nature.com/articles/srep28291

Detail and accuracy

It is essential that writers of any form of science report take into account the knowledge, experiences and varying degrees of understanding of its readers. Clearly there is a major distinction between a scientific article, aimed at individuals and groups who work or have an interest in the field of study, and a newspaper report of a scientific advancement or discovery, aimed at the general public who may not understand the depth of the science involved or have any background knowledge in science whatsoever.

On comparison of the two examples shown, it is evident that the amount of detail and accuracy of information is necessarily different.

▶ **Table 7.8** Detail and accuracy of information

Example 1	Example 2
• Some detail in the information, names, procedures etc. • No further explanations of phrases such as 'stem cell injections'. • Generalising of the process, no real accuracy in terms of figures, effects or implications of the research or procedure.	• Detailed information, providing further numbered reference where appropriate. • Numerical accuracy ensured in figures and units. • Detailed analysis of experimental work.

Level of language used

The language of a science report will differ significantly in relation to the target audience. Terms used in science have developed over hundreds of years in many cases and are formed from meaningful joining of some ancient languages such as Latin. It is not surprising to see many words and phrases which are only recognisable to those working in the science discipline.

In tabloid reporting of science, there is a need for key terms to convey the science behind the report and so some scientific words must be included. These terms will have a description of their meaning included in this type of reporting. In peer reviewed articles, the inclusion of an array of scientific terms is necessary because of the detail involved in aspects such as nomenclature of biological organisms. Table 7.9 compares the language level in the two examples of science reporting.

▶ **Table 7.9** Level of language

Example 1	Example 2
• Level of the language is acceptable to general public audience, science terms used only where necessary. • Language used is appropriate to engage non-scientists and provide a summary of the processes and the topic. • Vocabulary is necessarily limited.	• The level of language is high, aimed at an audience who understands many science terms. • No attempt to explain science phrasing further or use of abbreviations. • Vocabulary is extensive, including words in the language which are not in general use.

Style of writing and correct use of terminology

When presenting science information to a mass audience of non-scientists, the style of writing needs to follow a pattern which is immediately recognisable. The overall structure of the reporting should be clear and presented in small sections which each have a particular point to make or information to convey. Sentencing is kept to a suitable length to ensure that the meaning is maintained.

Science articles should follow the standard scientific convention as shown above, beginning with a title and ending with referencing. The sentencing structure is generally clear and referencing is absolutely fundamental. Terminology used in good science reports is always directly related to the subject material and not understandable by the general public. It is always specific and accurate. Table 7.10 shows the different writing styles in the two examples.

▶ **Table 7.10** Writing style

Example 1	Example 2
• Writing style is flowing and easily followed, using shorter sentences and clear paragraphing. • Terminology is correct but few science terms are shown. • Referencing is only in relation to other reports made and not specific.	• The report is not clearly set into paragraphs and sentencing is very long with few grammatical breaks. • Structure of the report follows standard science format appropriately. • Terminology is correct and technical language is abundant. • Many referencing points included using suitable numbered format.

Visuals

The use of photographs, tables, diagrams, charts or graphs in a science report adds much to the detail in the text and in many cases helps the reader to clearly visualise the meaning of the results obtained. Using visual representation also ensures that people of different learning abilities are engaged in the reporting and that there is a break in the sometimes tedious nature of the detail provided.

Data can be represented using visuals but care must be taken to ensure that the information is converted to visual form accurately. Where a visual representation is used, a written explanation should be provided, linking the meaning of the visual information to the task and highlighting key areas. Reference to the visual representation should also be made within the main body of the report.

Using correct and clear visual forms to present the data in science enhances the information being communicated to an audience and demonstrates that specific data analysis has been adequately undertaken. When presenting to the general public, pictograms, pie charts and cartoon style diagrams may help to engage the audience and link them with the subject material better. If the audience are scientists, line graphs with error bars may be the most effective in communicating the data, for example.

Biased viewpoint

In science, presenting the findings of an investigation is the point at which clear points are made and where questions relating to the investigation are raised in order to establish if the investigation method and the conclusions drawn are complementary.

Scientists may have an underlying idea as to the outcome of an investigation before starting but must not allow this to influence their experiment as it progresses.

In many incidences throughout history, a level of bias has been shown to have been introduced by both the scientists themselves and by others who have wished for a particular outcome to further their own agenda.

Many large and influential organisations have been able to change the eventual outcome of an investigation because of the fact that they may have funded the scientific project or the scientists carrying out the research.

Bias in science can be identified in various forms:

▶ Media – the presentation of a science report may contain views which are not supported by the evidence. Certain stories may be exaggerated or sensationalised for maximum effect and the organisation may be very

selective about which aspects to emphasise during editing to have maximum impact.

▶ Funding – Influential establishments and organisations who finance investigations can put pressure on the science outcome to favour their needs, or can simply influence the type of science which gets funding.

▶ Poor science – this aspect can surface in a number of ways. A scientist who accepts certain data but rejects others because of expectation, may be influencing the final conclusions. There may be rejection of logical arguments because the data does not fit with expected outcomes. Poor and inadequate sampling technique with samples not representative of the population under study is another problem. And in some cases, scientists are simply fraudulent and present data which will make them famous or make them money.

Quantity and quality of scientific information

The two examples shown for stem cell trials and colistin-resistant bacteria are quite indicative of the variation of science reporting in general and illustrate the important differences between the contents of the two types of report.

The online newspaper report is not intended to provide an in depth scientific study and findings of the subject in question, but rather produce a useful summary of the main aspects and attempt to put this information across to non-scientists in a manner that will enable them to gain a grasp of the message and the general science involved.

As a result, these types of reports are very short, giving relevant information using some scientific terminology but not so much that readers are deterred from finishing the report. Sensational headlines are used to entice the reader into the report and may not necessarily be true. Images are used to illustrate the essence of the report and may not be given an explanation. Information quality is also tailored to the readership, highlighting key science aspects, defining some important terms but also ensuring that the fundamental science is accurately communicated.

The science journal report is not restricted to one or two pages since the important element is in answering questions raised by, or providing sound information on, the initial study. Reports of this nature may be very extensive (in this case, approximately 17 pages) to ensure that all possible avenues have been explained and investigated within the confines of the study. Communication is vital and referencing forms a very important part of this type of report to enable further information to be checked against the findings. The scope of the study may only be to touch on one important aspect of the subject material.

Assessment practice 7.3

Use one of your most recent science reports of an investigation into a physical, chemical or biological aspect.

Adapt the report to make it suitable for use on:

1 television

2 social media.

Your target audience will be the general public and you will need to address:

- detail/accuracy/quality
- language used
- style and terms used
- use of visual displays.

Plan
- What is the task? What am I being asked to do?
- How confident do I feel in my own abilities to complete this task? Are there any areas I think I may struggle with?

Do
- I know what it is I'm doing and what I want to achieve.
- I can identify when I've gone wrong and adjust my thinking/approach to get myself back on course.

Review
- I can explain the results obtained from the task.
- I can apply the activity to other situations.

Further reading and resources

www.nasa.gov: National Aeronautics and Space Administration, USA. Website provides information on research in nanotechnology and space exploration

www.nhs.uk: National Health Service official website providing information on medical treatments

www.rsc.org: Royal Society of Chemistry website with links to useful scientific journals

www.telegraph.co.uk: website of *The Telegraph* newspaper providing useful reporting on pharmaceutical companies

www.bbc.co.uk: website of BBC providing information and reports on all science topics including GM crops

www.newscientist.com: popular and informative website with reports and discussion on most science topics

www.energy.gov: UK based Department of Energy, information on energy sources

www.nationalgeographic.com: information site for the National Geographic publication covering many environmental issues

www.who.int: official World Health Organization website

www.un.org: website of the United Nations providing updated information on current concerns and processes

www.worldwildlife.org: World Wildlife Fund pressure group information website

www.nature.com: website for *Nature* magazines and scientific journals on all science related topics

THINK ▶▶FUTURE

Eiryl Fear

Science journalist

I began writing science material as a teacher and completed my Ph.D. during part-time study. I was fully aware of the different ways in which science is reported in different media but soon developed an in depth understanding of the subtle methods used to direct my reporting to a specific audience. My job involves writing and editing stories and articles which have just been released and to establish the facts within the story. This involves a lot of time researching the science behind the claims. My writing, as a freelance journalist, may be for professional publications, specialist publications, the general media or specialist scientific journals and magazines. The media used (my clients) expect my science knowledge to be excellent and my articles to be clear, accurate, concise and able to target those people with good science understanding and also the general public

I am currently performing research for a particular story on the problems associated with cosmetic microbeads, very small plastic particles used in the cosmetics industry in products that the public uses for washing and personal hygiene. These particles make up about 4 per cent of the microplastics which enter our oceans. I have also discovered that other particles such as those from broken pieces of food and drinks containers or synthetic fibres from clothing are also found in large quantities in our rivers. This is one of many diverse research facts which continually make my work interesting. Communicating science information to a varied audience is both demanding and exciting because some of my work features in other articles written by other science journalists and one story actually found its way on to prime-time television.

Focus on skills

Using your understanding of science reporting

The skills that you have developed from the study of this unit will help you to understanding the need for detail in your writing and the amount of accurate research necessary before an article can be considered for publication.

Put yourself in the place of a science journalist, finding out information from a particular story which has come to light recently on the television. You must carry out research using many different sources and also enlist the help of a trustworthy partner to help develop reliable science links and to provide assistance in the editing of written articles:

1 Think clearly about areas of research you will need to explore and establish a working plan.
2 What attributes would you want in a colleague to help in the activity?
3 How do you record information that you have found?
4 How reliable is the source and the science information?
5 How can you check the accuracy of the information – what sources of study could you use?

Working individually and with others

Science journalists work alone for much of the time but also need to enlist the help of other people in the research and editing process. Discuss the following important points with a partner in your group to determine what you need to consider and why another person can provide a secondary valuable perspective.

1 The style of the article, length of time needed on research, deadlines and the target audience.
2 Who will provide science information – doctors, industry experts etc?
3 Do you need to make a visit to a university, industry or research establishment?
4 What format will the report take? Research into other scientific journals and reports.
5 Will you need to timetable regular update and editing meetings with other colleagues?
6 Selection of suitable visuals such as photographs, tables, graphs, etc.

Getting ready for assessment

This section has been written to help you do your best when you take your final assessment test. Read through it carefully and ask your tutor if there is anything you are not sure about.

About the test

Here is a brief summary of the Unit 7 assessment. It is divided into two parts.

Part A

You will be given your source booklet before your assessment. This will contain articles all relating to a contemporary scientific issue. You should complete the following tasks. Plan your time carefully!

- Read the articles several times. Analyse them in the context of how the scientific issue is being tackled, the intended audience of the article and the validity of the scientific content, including organisations and people who may be mentioned. You can highlight or annotate your articles. Here are some examples of how you might approach doing this.
- You must do further research. You need to find out about any organisations or scientists mentioned in the articles and decide how valid their opinions might be and how influential they are. You can also look for reliable evidence which confirms or refutes the ideas put across in the articles provided.

Part B

You complete your taskbook which contains copies of the articles. It must be completed independently and under supervised conditions. You write your answers directly into the taskbook, which is then submitted for assessment. The questions are all based on the articles provided. **Different questions are given different numbers of marks. Make sure you spend most of your time on the questions where you can gain most marks.**

The areas most likely to be highlighted in the questions include:
- **The implications of the scientific issue or issues in the articles**
- The organisations/individuals mentioned in the articles and their potential influence on the scientific issue
- **The validity of the judgements in one or more of the articles**
- Possible future areas for development or research linked to the main scientific issue
- **Your understanding of the scientific issues and ability to explain them to a given audience.**

Questions covering these issues will carry the most marks.

Annotating the Part A Articles

Here are some sections from three articles on treating malaria. Read them through and see how they have been highlighted and annotated. This gives you an idea of the task you will have to complete.

Article 1: What is Malaria?

Every 2 minutes a child dies from malaria - a disease spread by a single mosquito bite. *Nothing But Nets* works with supporters and partners around the world to raise funds and awareness about the disease and advocate for malaria prevention.

> This student has used yellow highlighting to show the organisations they want to look up.

Malaria is a disease caused by the blood parasite *Plasmodium*, which is transmitted by mosquitoes. Each year, an estimated 219 million people are infected with malaria, causing approximately 600,000 deaths – mostly children under the age of five.

> Here the student has underlined the main science issues explained in the text.

.... Every 2 minutes, a child in Africa dies from a malaria infection and 90% of all malaria deaths occur in the region. When combined with HIV/AIDS, malaria is even more deadly, particularly for pregnant women and children.

Malaria is a big problem – and the disease has big consequences for families, communities, and countries. Fortunately, there are small things that can help make a huge impact in the fight against malaria. It's easy to help: *Nothing But Nets* works with our UN partners to prevent malaria in Sub-Saharan Africa by sending nets to save lives and raising voices to let policymakers know that the fight against malaria is important. Learn about the many ways you can take action now!

Did you know?

Four Nobel Prizes have been awarded for work associated with malaria to Sir Ronald Ross (1902), Charles Louis Alphonse Laveran (1907), Julius Wagner-Jauregg (1927), and Paul Hermann Müller (1948).

> The student is using a different coloured highlighter to show up individuals involved in malaria research.

The two most effective and potent anti-malaria drugs come from plants with medicinal values recognized for centuries: artemisinin from the Qinghao plant and quinine from the cinchona tree.

How Is Malaria Prevented?

In the poorest parts of the world, window screens are lacking, anti-malaria drugs are expensive, and so far an effective malaria vaccine does not exist. … Infections can be prevented by sleeping under long-lasting insecticide-treated bed nets (LLINs). These nets, which are designed to last at least three years, work by creating a protective barrier against deadly malaria-carrying mosquitoes that typically bite at night.

The benefits of these bed nets extend even further than protecting those sleeping underneath them. The insecticide woven into each net makes entire communities safer – killing mosquitoes so that they can't go on to bite others who may not be protected by a net. Bed nets can reduce malaria transmissions by as much as 90 percent in areas with high coverage rates. It costs just $10 to send a bed net to the families who need them. According to the 2012 World Malaria Report, 90 percent of people with a bed net use it. Join us now. Send a net and save a life.

Other Prevention Measures and Treatment

In addition to bed nets, malaria can be prevented by applying insecticide to the inside walls of individual homes. Mosquitoes that land on treated walls are killed, preventing the transmission of malaria....

Source: Nothing but nets, charity involved in reducing and preventing the spread of malaria: http://www.nothingbutnets.net/new/saving-lives/what-is-malaria.html

Article 2: Drug-resistant malaria an 'enormous threat' — vigorous international effort needed to contain it

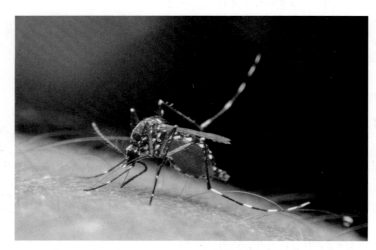

A drug-resistant form of malaria is spreading in southeast Asia and represents an "enormous threat" to the world.

The resilient form of the *mosquito-spread parasite* has been found in many parts of Myanmar (Burma) as well as Cambodia, Thailand, Laos and Vietnam.

The drug artemisinin is normally given as part of a combination therapy to battle the disease, but now a strain of the parasite is not responding to it.

The alarming discovery is a major blow to global health efforts to reduce the number of deaths from the mosquito-spread parasite.

Although initially other drugs given in the combination treatment could keep malaria at bay, the parasite is likely to develop resistance to the partner drugs as well - and there's evidence that this is happening already.

Health workers fear that the strain could soon spread to India, where thousands more lives would be at risk.

More underlining shows up a different aspect of the malaria story.

"It's hugely worrying," says Professor David Conway from the London School of Hygiene & Tropical Medicine, "both for people in southeast Asia and the rest of the world."

"Should these [drugs] fail today, there's nothing waiting in the wings that's going to be affordable and adequately tested in time."

Something similar happened in the 1950s, when malaria became resistant to a drug called chloroquine. It spread across the world and eventually reached Africa.

"The global spread of chloroquine resistance resulted in the loss of millions of lives in Africa and, clearly, Myanmar is an important part of the frontline in the battle to contain artemisinin resistance," say the authors of the study, published in the *Lancet Infectious Diseases.*

"A vigorous international effort to contain this enormous threat is needed," they say.

Conway adds: "It's not too late, but action needs to be taken now to stop the spread."

Healthcare workers should make sure that malaria isn't being treated with artemisinin on its own (which is happening in some places "against all advice"), and efforts should be made to eliminate counterfeit medicines.

Continued molecular testing of the strain in real-time and monitoring patients in other territories, particularly Africa, for resistance is vital.

Source: http://www.mirror.co.uk/news/technology-science/science/drug-resistant-malaria-enormous-threat---5197706

The London School of Hygiene and Tropical Medicine has an excellent reputation.

The student highlights a publication quoted in the article to look up later.

Article 3: The ongoing battle against drug-resistant malaria

Artemisinin-based combination therapy

Artemisinin-based combination therapy (ACT) has been integral to the recent successes in global malaria control. The main idea behind ACT was to provide an inexpensive, short-course treatment that would also help protect against the development of drug resistance…

Here the student is using a different pattern of underlining to show up where one article reinforces another.

Is history repeating itself?

In 2009, researchers reported concerns that artemisinin was taking longer to clear parasites from patients infected with *Plasmodium falciparum* along the Thailand-Cambodia border — a worrying sign of emerging drug resistance. Since then, researchers have reported slow parasite clearance in four countries in the Greater Mekong Subregion…

What's more, if artemisinin resistance were to arise in Africa or emerge independently elsewhere, as has happened with other antimalarial drugs, the public health consequences would be catastrophic…

In response to this threat, the World Health Organization (WHO) launched an emergency plan of action to tackle artemisinin resistance in the Greater Mekong Subregion covering 2013–2015. They proposed an immediate and coordinated increase in efforts to tackle malaria in Cambodia, Laos, Myanmar, Thailand and Vietnam. Currently the WHO's goal is to initiate elimination activities by 2020 in order to remove malaria completely from Greater Mekong Subregion countries by 2030.

The World Health Organization needs looking up, then make notes about it.

But how are we going to stop the spread of drug resistance if we haven't been able to in the past? Well, now we have one more weapon in our arsenal that we didn't have before – genome sequencing!

Genomics vs. malaria – the fight is on

At the time that artemisinin resistance was first discovered in early 2009, no one knew which genetic changes were responsible, and pinpointing those changes proved more challenging than expected. However, faster and cheaper genome sequencing techniques have enabled us to learn a lot more about the underlying genetic changes responsible. … By finding these genetic changes scientists are hoping that they may eventually be able to track and then prevent the spread of artemisinin resistance.

Clues on chromosome 13

In 2012, and then again in 2013, a couple of genome-wide association studies (GWAS) looking at the *P. falciparum* genome pointed towards two regions next to each other on chromosome 13 as potential sites of the mutations associated with artemisinin resistance. However, they needed to find out for sure if these mutations were directly involved in resistance. A year or so later, a collaboration led by scientists at the Institut Pasteur in Paris came up with an experiment that pointed them in the right direction.

> The Institut Pasteur needs looking up, then make notes about it.

Over a five year period, the scientists grew and nurtured a strain of *Plasmodium* parasite that they knew did not have any resistance to artemisinin. Every so often during this period they gave the colony of parasites a small amount of artemisinin. They hypothesised that sooner or later an artemisinin resistant parasite would emerge because of the selection pressure of the drug (the pressure to adapt in order to survive!). Sure enough, after four years of exposure to the drug, artemisinin resistant parasites were seen. With DNA sequencing they were then able to study the genome of the resistant parasites and compare them to the genome of the original, non-resistant strain of *Plasmodium.*

They found several genetic changes in the resistant parasite genome but the most significant one occurred bang in the middle of the previously-identified regions on chromosome 13, in a gene called *kelch13*.

Like spies in an enemy country, genomics can provide us with the intelligence to track drug resistance emerging in the malaria parasite. This gives us more time to plan our counterattack before drug resistance becomes more widespread.

Source: http://www.yourgenome.org/stories/the-ongoing-battle-against-drug-resistant-malaria

> Is this website reliable? More research is needed!

Revising for your test

In most exams the questions you have to answer depend on you learning about a particular aspect of science, and being able to answer questions on it. So before the exam, you have to revise the topic thoroughly. Unit 7 is rather different. The questions in the test are based on the skills which you have developed in researching and identifying key points of information from an extensive passage of text. They do not significantly test your knowledge of the subject material, so you will not be expected to recall information.

The best way to revise for the test is to work through the research and discussion sections in this unit, making sure that you understand the key differences in scientific reporting and issues involved in scientific development including:

- both sides of the issues involved – the positives and the negatives
- the overall quality and balance of reports including any coverage of data analysis, sampling and supporting evidence
- the reporting medium used and corresponding target audience
- the presentation of science reports.

Making notes

You may not have to learn all the science in Unit 7 by heart – but you *do* need to practise doing research and making notes. Always bear in mind the questions you are likely to be asked. You need to link all the information to the specific article it comes from. Here is a sample of some early notes made by a student….

Notes for Unit 7 assessment

1   The main science issue is the problem of malaria and the development of drug-resistant strains of the parasite which causes the disease e.g. In Article 1: an estimated 219 million people are infected with malaria, causing approximately 600,000 deaths – mostly children under the age of five.

In Article 2: A drug-resistant form of malaria is spreading in southeast Asia and represents an "enormous threat" to the world. In Article 3: Resistance to antimalarial drugs is one of the biggest problems currently facing malaria control.

The secondary issues include how do we treat or prevent malaria and how can we deal with the problem of the drug resistant parasites …

2   Scientific organisations mentioned

Article 1

- World Health Organization (WHO): WHO is a specialised part of the United Nations and is very influential. – it collects data on diseases from all over the world and issues guidelines for treatments, flags up epidemics and pandemics etc. It put together the 2012 World Malaria Report based on data from 104 countries with endemic malaria. VERY influential globally (WHO publications website, Wikipedia).

Article 2
- London School of Hygiene and Tropical Medicine: world-leading centre for research into tropical medicine and public health. Part of the University of London. Founded 1899. Research carried out here has big impact all over the world (2016 CWTS Leiden Ranking for research impact).

Article 3
- World Health Organization (WHO): see Article 1
- Institut Pasteur: in Paris, internationally renowned for research into microbiology and disease — based on the work on microbiology, vaccines etc of Louis Pasteur.

Have a go and see if you can build up your own notes based on the articles.

Sample questions and answers

Remember: These examples are much shorter than you would write in your exam, because they are only based on single-page articles.

1 Discuss the implications of the scientific issue described in the article.

12 marks

Use your notes to remind yourself about the main issues in the three articles. Make sure you refer to all three of the articles and to other reading you may have done. This question carries a lot of marks so plan your timing and your answer carefully.

To score well you need to draw a wide range of links to and between the ethical, social, economic and/or environmental implications of the science. You are assessed on the structure of your answer which should be clear, coherent and logical so plan carefully.

The main scientific issue identified in these three articles is the spread of drug-resistant malaria in many parts of the world. There are two aspects to this issue. One is the problem of malaria as a global disease. Every year around 219 million people are infected with malaria, causing approximately 600,000 deaths. Malaria is a disease caused by the blood parasite Plasmodium, which is transmitted by mosquitoes. Most of the deaths are in children under 5 years old. The level of disease in populations affects society because people are too ill to work effectively and look after their children. Economically this affects families and countries. Individuals can't make enough money to support their families, and countries cannot build secure economies.

> Student has read and summarised the article main points effectively.

The other problem is treating malaria successfully. It is common in poor countries for example on the continent of Africa and areas such as Cambodia and Myanmar. One line of defence is to try and stop the spread of mosquitoes as effectively and cheaply as possible — for example using insecticide impregnated mosquito nets. Other methods include general insecticides and anti-malarial drugs. Although these

> Student has looked at the problem and identified issues, avoiding bias.

can be expensive, combined anti-malarial therapies can be very effective. Unfortunately, forms of the malaria parasite are evolving which are resistant to our best anti-malarial drugs, and this is a big problem going forward. Solutions will depend on hi-tech genome analysis, which is expensive. This raises environmental issues because insecticides can have damaging effects on the environment. It also raises ethical issues because the countries which most need the solutions are also very poor.

> Student recognises some of the environmental and ethical issues of potential solutions.

2 Identify two organisations and two scientists mentioned in the articles and suggest how they may have had an influence on the main scientific issues. **6 marks**

Look back at your notes to remind yourself about the organisations and individuals mentioned in the three articles.

If you have made good notes, you will have these organisations and individuals already identified, which will make answering this type of question much easier. To score highly you need to make clear links between the institutions and people mentioned and the original articles.

Two examples of influential organisations mentioned in the articles are:

The World Health Organization, known as the WHO, is a specialised part of the United Nations and is very influential. It collects data on diseases from all over the world and issues guidelines for treatments, flags up epidemics and pandemics etc. It put together the 2012 World Malaria Report based on data from 104 countries with endemic malaria. VERY influential globally (WHO publications website, Wikipedia)

The London School of Hygiene and Tropical Medicine is mentioned in Article 2. It is a world-leading centre for research into tropical medicine and public health. It is part of the University of London and it was founded in 1899. It specialises in tropical diseases such as malaria and specialist scientists come here from many different countries to study these diseases and how to treat them. Research carried out here has a big impact all over the world.

Two examples of influential scientists mentioned in the articles are: Sir Ronald Ross (1902), the doctor who demonstrated that malaria is spread by mosquitos. His work allowed people to begin to work out how to prevent the spread of the disease. (Wikipedia, Nobel prize web site)

Professor David Conway is very active in anti-malaria research. He is professor at the London School of Hygiene and Tropical Medicine, works in UK and African countries, has published over 170 research articles, and is well known in his field (LSHTM website).

3 Discuss whether the articles are expressing valid concerns about the problems linked to treating malaria. In your answer you should consider:
- How the articles have interpreted and analysed the scientific information to support the conclusions/judgements being made
- The validity and reliability of data
- References to other sources of information.   **12 marks**

Use your notes to remind yourself about the validity of the articles. What does validity mean? What will you be looking for? Make sure you refer to the points made in the bullet points given to guide your answer.

Look at your comparison of the three articles to help you highlight the main points you need to answer this question well. You will be assessed on the structure of your discussion as well as the content. It must be clear, coherent and logical. Make sure you take the time to plan carefully.

If results are valid they measure what they are supposed to measure. If results are reliable, the investigation produces stable, consistent results which other people can replicate. It is important to be sure that an article is valid and reliable before you take any notice of it. There is considerable agreement between the three articles, which in itself suggests that the conclusions are probably reliable.

> Student explains what valid data is and why it is important.

Article 1 was published on a website by a charity which works with the UN. They are trying to raise money to supply insecticide impregnated mosquito nets in Africa to help prevent the spread of malaria. Charities should use reliable data but they are trying to persuade you to give money so it is important to double check their sources as they may be biased in their choice of evidence. When the data is published in reliable journals or on websites it suggests the content will be valid. Article 1 refers to the 2012 Malaria Report from the WHO – a very reliable source because it collects valid data from all over the world. However, much of the information it gives about malaria is supported by the other two articles, which increases its validity.

> Student shows they are away of the possibility of bias in an charity.

Article 2 is taken from the website of a popular newspaper. Not all science articles in newspapers are reliable so it is important to check the sources used and the way they are interpreted. This article focuses on the drug-resistant strains of the malaria parasite spreading from Myanmar. It quotes at least two reliable sources:

- report based on work by Professor David Conway – very reputable, LSTM

- report from a study published in the Lancet Infectious Diseases, a very reputable scientific journal which only publishes peer-reviewed valid work.

> Student highlights a number of reputable scientists and sources in all three articles.

Although the article is for a popular newspaper website, it uses the highly reliable sources it has chosen as evidence for its conclusions - that there is a global risk as a result of the emergence of artemisinin resistance in malarial parasites. These conclusions are also backed up in Article 3, suggesting that although they are part of the popular media aimed at a non-scientific readership, they are nevertheless valid and reliable. The article finishes by stating that the continued molecular testing of the malarial parasites, especially in Africa, is vital. This is not supported or explained in this article, but there is a lot more about it in Article 3 with good backing evidence.

Article 3 is published on the www.yourgenome.org website developed by the education team at the Wellcome Trust Genome Campus. This is the top institution globally for analysis of the genomes of different organisms, and it has many internationally renowned scientists working there on various projects. That alone suggests that the content will be both valid and reliable, and that any conclusions drawn will be supported by evidence. Reading through the article, evidence and data are presented from a variety of sources including the Worldwide Antimalarial Resistance Network of the WHO and Nature Genetics, a highly reputable peer reviewed journal.

The sources quoted in this article are very reliable. The content also confirms the impact of malaria, some of the problems in preventing malaria and the problems of drug-resistant malaria parasites described in Articles 1 and 2. For example, Article 3 quotes vector management as well as effective antimalarial drugs as an important reason why death rates from malaria have fallen by 47% since the year 2000. This reinforces the message in Article 1 about the importance of simple methods of vector control such as insecticide impregnated mosquito nets. Later in Article 3, it confirms the conclusion in Article 2 that malaria parasites have become resistant to various drugs and that the current drug resistance has emerged from South east Asia. Article 3 gives us a lot more information on the genetic basis for the development of drug resistance in malarial parasites. Considering where the article was developed, and the quoted sources, we can be fairly certain that the judgement made in the article that genome analysis of the malaria parasites is a vital tool in the ongoing battle against drug-resistant malaria is completely justified.

> Student makes judgements on the sources quoted in the articles based on good evidence.

4 **Suggest possible areas for further development or research related to the scientific issue covered by the articles.** **5 marks**

Look back at your notes – when you were reading the articles and making notes you will have thought about potential future development and research. In this answer you can suggest more than one direction for future R&D. You will need to justify your choices/explain why that research is needed. You are expected to link your ideas for further development/research to science from all three articles.

The big problems highlighted in these articles are the issues of malaria – the toll it takes on human lives and the economies of affected countries – and the growing problem of drug resistance in the malarial parasite which causes the disease.

Here are some potential areas for further research and/or development:

New drugs to cure malaria which work in a different way to the current medicines. This would mean that even the malarial parasites resistant to the current drugs would be wiped out as they would not have resistance to the new drug mechanism. A new drug like this could be used in combination with a current anti-malarial drug so it continues to be effective. This would be similar to the use of piperaquine with artemisinin.

Genome analysis to find out where the drug resistance genes are found.

Genome analysis reveals the DNA sequences of the malarial parasites. We can use this information to show us when the malarial parasite begins to develop resistance to a drug and change the treatments used in this area. Then scientists can use this information to help develop potential medicines and vaccines.

Development of new insecticides to be used on water or on mosquito nets. We need new insecticides which are safe in the environment and safe for people but effective at killing mosquitoes.

> Student has highlighted the problem and suggested three interesting and different areas for further research and development, showing evidence of researching the subject and understanding the issues.

5 **Write an article for a tabloid newspaper on the growing problems of malaria based on the information contained in these articles. You will need to identify:**
   - **the target audience**
   - **the level of language to be used**
   - **amount of detail and accuracy**
   - **what titles or captions may appeal to your target audience.**   **15 marks**

Think carefully about your target audience. Plan what you want them to know. You will need a good headline to catch your reader's attention.

Decide how you can include good, valid, reliable science but still make your article interesting – you want people to keep reading to the end. Use information from all three articles – show the examiners that you have read and taken in information from all your sources and understand where they agree and where they take different approaches. If you can, include some extra information from your reading around.

You MUST keep your tone, style and level of scientific terminology the same throughout the article – don't start off all chatty and then end up writing as if you are delivering a piece for a scientific journal!

You will get credit for a well-organised article, with a clear, logical, coherent structure – so make your planning time count.

A malarial time-bomb – and it could affect YOU!

Choose a big, bold headline if you are writing a newspaper article – this one uses emotive terms like 'time-bomb' and makes it feel personal to people with the use of the emphasised 'YOU'.

Every year about 600,000 people die of malaria. Most of them are children under 5 years old. The World Health Organization estimates that around 219 million people are infected with malaria. It is caused by a tiny parasite, spread by the bites of infected mosquitos. Malaria wrecks individual lives and destroys economies when many of the working population are infected with this dreadful disease.

So what? Malaria doesn't affect us here in the UK – but only because it is too cold for the mosquitos which carry the malaria parasite to survive. But in Shakespeare's times malaria (the ague) was common. As global warming increases, our old enemy could return – with a vengeance.

This is only the start of this answer. It is important to keep going in the same tabloid style!

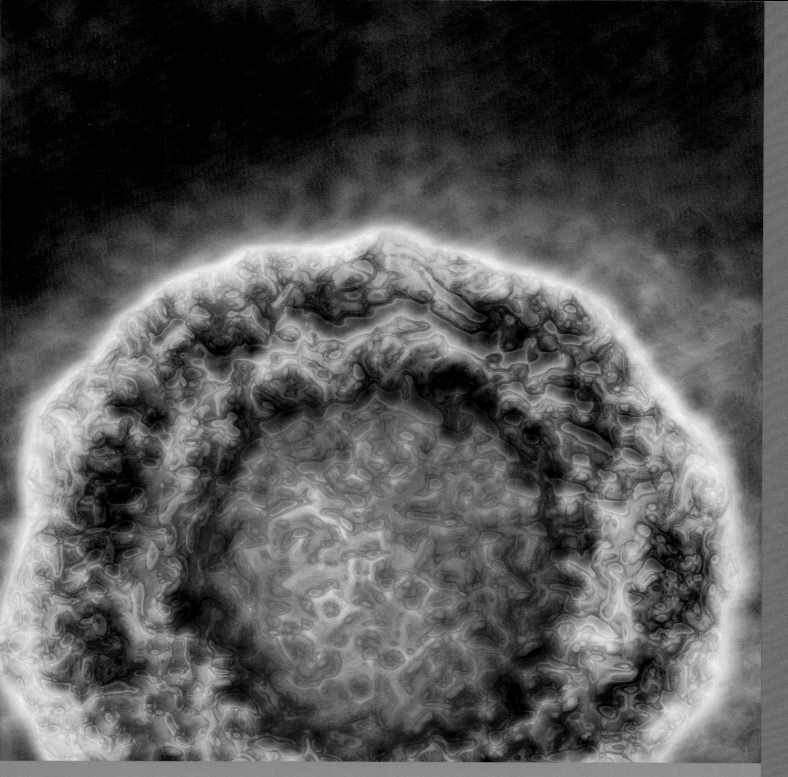

Microbiology and Microbiological Techniques

17

Getting to know your unit

Microorganisms are essential to life on Earth. The last universal common ancestor (LUCA) of all life forms on Earth gave rise to bacteria and archaea about 2700 million years ago. These single-celled life forms were probably the only life forms on Earth until about 1500 million years ago, when single-celled eukaryotic cells (cells with a proper nucleus and membrane-bound organelles) evolved; multicellular life forms, including protists, plants, fungi and animals, started to evolve around 700 million years ago. Today, microorganisms occur in many different habitats on Earth. Many types of bacteria live inside animals and plants, including those that live in human digestive tracts, many of which are essential to our wellbeing. A few types of microorganisms cause diseases, but most are harmless. Many microorganisms are essential for agriculture, food production, medicines and biotechnology.

Microbiology is the branch of biology that studies organisms too small to be seen with the naked eye, including bacteria, archaea, viruses, some fungi and some protists. Microbiologists also study abnormal disease-causing proteins called prions, although these are not organisms. In this unit, you will study the characteristics of different types of microorganisms and the factors they need for growth. You will use aseptic techniques, prepare and use different types of growth media, use different inoculation techniques and develop skills to use microscopes.

The skills associated with microbiological techniques are essential to other areas of biological sciences, such as crop health and food production, genetic modification, biochemical and biomedical research, livestock health and forensic science. This unit will help you to progress towards a wide range of related higher education courses, including medical microbiology and biomedical sciences.

How you will be assessed

In this unit, you will be assessed by a series of internally assessed tasks set by your tutor. Throughout the unit, you will find assessment activities that may help you work towards your assessment. Completing these activities will not mean you have achieved a particular grade, but the research you carry out for them will be relevant and useful when you come to carry out your final assessment.

It is important to check that you have met the Pass grading criteria, shown in the table below, as you work your way through the assignments.

To achieve a Merit or Distinction, you need to present your work in such a way that you meet the criteria for those grades.

To achieve Merit, you need to analyse and discuss. For Distinction, you need to assess and evaluate.

The assignments set by your tutor will consist of a number of tasks designed to meet the criteria in the table below. Some tasks will be written and some will be lab-based practicals with written reports. Tasks may also involve reviewing and analysing case studies.

Assessment criteria

This table shows what you must do in order to achieve a **Pass**, **Merit** or **Distinction** grade and where you can find activities to help you.

Pass	Merit	Distinction
Learning aim A Understand the importance of microbial classification to medicine and industry		
A.P1 Explain how the structures and characteristics of microorganisms are used to classify them.	**A.M1** Compare the characteristics of microorganisms used for classification.	**AB.D1** Evaluate the use of microscopy techniques to observe structures and classify microorganisms.
Learning aim B Undertake microscopy for specimen examination in laboratories		
B.P2 Set up and use a light microscope and oil immersion lens to observe structures of microorganisms under magnification.	**B.M2** Compare the use of different microscopy techniques to observe the structures of microorganisms.	
B.P3 Illustrate, with accuracy, the structures of microorganisms observed using a light microscope and an oil immersion lens.		
Learning aim C Undertake aseptic techniques to culture microorganisms		
C.P4 Prepare and inoculate growth media and measure microbial growth using aseptic techniques.	**C.M3** Demonstrate skilful application of aseptic techniques in inoculation and preparation of growth media and in measuring microbial growth.	**CD.D2** Evaluate the aseptic techniques used to culture microorganisms with specific reference to the type of media, methods of inoculation and the biocontainment procedures carried out.
C.P5 Explain biocontainment procedures used in your centre laboratory and within industrial laboratories.	**C.M4** Compare biocontainment procedures in your centre laboratory to those used in industrial laboratories.	
Learning aim D Explore factors controlling microbial growth in industrial, medical and domestic applications		
D.P6 Carry out investigations into the effect of growth requirements on microorganisms.	**D.M5** Analyse how the growth of microorganisms is affected by changing environmental factors.	
D.P7 Explain how growth inhibitors affect microorganisms.		

Getting started

How many different types of microorganism can you think of? Are all microorganisms harmful? How do scientists characterise microorganisms? Why is hygiene in the home, food processing plants and in hospitals important? Discuss these questions with a partner and think about them as you go through this unit.

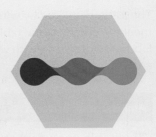

A Understand the importance of microbial classification to medicine and industry

There are many different types of **microorganisms**. Some are not made of cells, some are **prokaryotes** and some are **eukaryotes**. Organisms are classified following the principles of taxonomy developed in the eighteenth century by the Swedish botanist, Carolus Linnaeus. He developed a hierarchy of groups. Organisms within a particular group share similar structures, biochemistry and physiology. They therefore have some enzymes and **genes** in common with other members of the group. Carolus Linnaeus developed the binomial system, in which all living organisms have two names, a generic and specific name – for example, the bacterium, *Escherichia coli (E. coli)* that lives in the digestive tract. The generic name always starts with an upper case letter and the specific name begins with a lower case letter. The scientific names of living organisms are written in italics. Classification of organisms can also help identify evolutionary relationships between organisms and groups of organisms. Comparing the genes of different groups of organisms can verify or disprove these relationships. Many thousands of new species or living organisms are discovered each year and, sometimes, existing species have to be reclassified as we learn more about them.

All living things on Earth are classified by first placing each in one of three domains – Archaea, Bacteria and Eukarya (eukaryotes) (Figure 17.1). Within the Eukarya domain, there are different kingdoms – protists (Protista), fungi (Fungi), plants (Plantae) and animals (Animalia).

Key terms

Microorganisms – microscopic organisms that include viruses, bacteria, fungi and some protists.

Prokaryotes – organisms made of cells that have cell surface membranes, cytoplasm (see below) and a cell wall, but do not have a proper nucleus containing DNA. Their DNA floats free in the cell's cytoplasm.

Eukaryotes – organisms made of cells that contain a nucleus and other specialised structures (membrane-bound organelles; see below).

Gene – length of DNA that codes for one or more proteins, or that codes for a regulatory length of RNA.

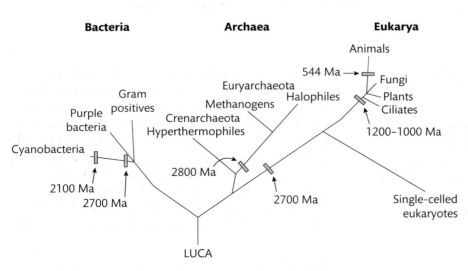

▶ **Figure 17.1** A universal tree of life showing the three domains. The tree is based on sequence comparisons of ribosomal RNA, analysed by the scientist Carl Woese and his team. There is less variation in ribosomal RNA within the entire animal kingdom than there is between different groups of methane-producing archaea

▸ **Table 17.1** Some organisms classified according to the three-domain classification system

	Humans	Plasmodium	Yeast	E. coli
Domain	Eukarya	Eukarya	Eukarya	Bacteria
Kingdom	Animalia	Protista	Fungi	Prokaryotae
Phylum	Chordata	Apicomplexa	Ascomycota	Proteobacteria
Class	Mammalia	Aconoidasida	Saccharomycetes	Gammaproteobacteria
Order	Primates	Haemosporida	Saccharomycetales	Enterobacteriales
Family	Hominidae	Plasmodiidae	Saccharomycetaceae	Enterobacteriaceae
Genus	*Homo*	*Plasmodium*	*Saccharomyces*	*Escherichia*
Species	*sapiens*	*malariae*	*cerevisiae*	*coli*

Within each kingdom, there are subgroups called phyla, within these are subgroups called classes, then orders, families, genera and species. Table 17.1 shows the classification of some organisms.

Viruses are difficult to classify as they have no cellular structure and are considered borderline as to whether they are living. Scientists are not sure of the origin of viruses and have not traced viruses to the last universal common ancestor, although viruses have probably been around for a very long time. Viruses are classified according to their shape, type of nucleic acid, mode of replication, **host** organisms and type of disease they cause.

Key terms

Virus – submicroscopic infective agent consisting of a protein coat and nucleic acid – either DNA or RNA, but not both.

Host – organism inhabited by an infecting agent or parasite.

 PAUSE POINT What is the binomial system used in biological classification?

Hint Think about the contribution of Linnaeus.

Extend Why are viruses not part of the three-domain classification system?

Microorganisms and infectious agents

In this section you will learn about the differences in relative sizes, structural features and means of reproduction of various microorganisms, including bacteria, fungi, protozoa (a subgroup of protists), viruses and viroids, as well as infectious proteins, called prions. You will read about pathogenic (disease-causing) and non-pathogenic examples.

Bacteria

Bacteria are single-celled (unicellular) prokaryotic organisms. They range in size from 0.2 to 2 μm in diameter and 1–8 μm in length (one micrometre, μm, is one millionth of a metre). Some bacteria live in extreme environments, such as thermal oceanic vents, hot sulfur springs and in the Earth's crust. Many types of bacteria inhabit the intestines of animals, including humans. The relationships between the bacteria and animals are usually **symbiotic** and, in those cases, the bacteria are essential to us. However, some types of bacteria are pathogenic. Bacterial diseases include bubonic plague, anthrax, meningococcal meningitis, cholera, syphilis, tuberculosis (TB), leprosy, bacterial pneumonia and whooping cough.

Key term

Symbiotic – an interaction between two organisms where both organisms benefit.

Distinct communities of up to 30 000 different species of bacteria also live in the roots, on leaves and inside the flowers of plants. Those bacteria help plants obtain nutrients and resist diseases.

Many bacteria are useful and some are used in the making of foods, such as sourdough bread, yoghurt and cheese.

Bacterial cells are prokaryotic. They have:

- a cell surface membrane
- **cytoplasm**
- a cell wall made of peptidoglycan
- no membrane-bound **organelles**
- no proper nucleus – their DNA floats free in the cytoplasm
- small rings of DNA, called plasmids, and a large circular loop of DNA
- smaller ribosomes than those of eukaryotic cells.

> **Key terms**
>
> **Cytoplasm** – gel-like substance enclosed by the cell surface membrane; about 80% water; medium in which many metabolic reactions take place; organelles are suspended in the cytoplasm.
>
> **Organelle** – an organised and specialised structure within a cell; some, e.g. mitochondria, are membrane-bound and are found only in eukaryotic cells. Ribosomes are organelles that are not bound by a membrane and occur in prokaryotic and eukaryotic cells.

Some bacteria also have:

- an outer capsule
- hair-like structures called pili (singular: pilus) for adhering to surfaces or to host cells
- one or more flagella (singular: flagellum; whip-like structures) to propel them.

Figure 17.2 shows the generalised structure of a bacterial cell.

Bacteria reproduce asexually by binary fission (Figure 17.3).

- Their DNA uncoils and replicates.
- Their plasmid DNA also replicates.
- The DNA molecules are pulled to opposite ends of the cell and then a wall forms between them.

Fungi

Fungi are eukaryotic organisms. Some, such as yeasts, are unicellular and some, such as mushrooms, are multicellular. Fungi are abundant in all parts of the world and, along with bacteria, are the main decomposers, which means that they feed on and break down dead matter in ecosystems, helping to recycle the nutrients. Fungi feed heterotrophically (use organic carbon for growth); they secrete enzymes and acids into the environment to digest food, which they then absorb through their surface. Mould fungi may spoil food. Some fungi are parasites and can

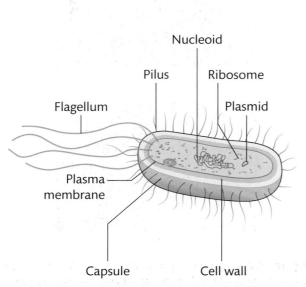

▶ **Figure 17.2** Generalised structure of a bacterial cell

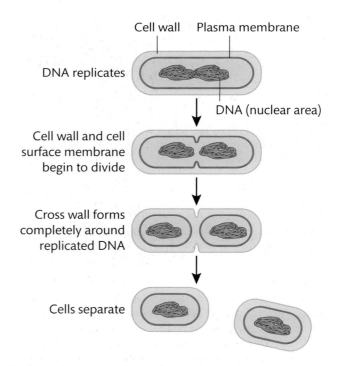

▶ **Figure 17.3** Bacterial reproduction by binary fission

infect other organisms. These fungi cause diseases and are also described as pathogenic. Of the fungi that infect humans, some cause:

▶ candidiasis (thrush) (Figure 17.4)
▶ ringworm and athlete's foot (*Tinea* spp.)
▶ farmers' lung (aspergillosis)
▶ opportunistic infections, such as a type of pneumonia in immunosuppressed people.

Other fungi cause disease in livestock, fish and crop plants.

However, fungi are also of great economic importance, as sources of:

▶ food (e.g. mushrooms)
▶ antibiotics, such as penicillin
▶ fusidic acid, which is used against MRSA (methicillin resistant *Staphylococcus aureus*)
▶ substances that inhibit viruses or cancer cells
▶ enzymes
▶ the immunosuppressant, cyclosporine.

Fungi are also used in:

▶ brewing and baking
▶ making soy sauce, Quorn, tempeh and blue cheeses.

Lichens are symbiotic associations between a fungus and either a cyanobacterium (photosynthetic prokaryotic microorganisms) or a green alga (photosynthetic algae). More than 100 000 species of fungi have been identified so far. Scientists who study fungi are called mycologists.

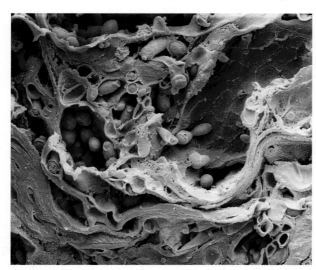

▶ **Figure 17.4** Scanning electron micrograph (SEM) with colour added of the surface of a human tongue infected with *Candida* fungus, shown as orange. *Candida* does not become a problem until there is a change in the oral microbiota favouring *Candida* over other species; this may be caused by antibiotics, chemotherapy or conditions such as diabetes, malnutrition or immunodeficiency. Mag ×1600

Fungi:

▶ are eukaryotic. Their DNA is in the form of linear chromosomes that are housed inside a nucleus, bound by a double membrane.
▶ can reproduce asexually by spores or by fragmentation. This involves mitosis (a type of cell division). They can also reproduce sexually by meiosis (another type of cell division).
▶ contain membrane-bound organelles, such as mitochondria. Most respire aerobically although some, such as yeast, are facultative anaerobes.
▶ have cell walls that contain chitin.
▶ store carbohydrate in the form of glycogen.
▶ have **hyphae**, which grow at their tips.
▶ are more closely related to animals than to plants.

> **Key term**
>
> **Hyphae** – cells that grow as long tubular thread-like filaments. A network of hyphae is called a mycelium.

Yeast cells reproduce by *budding* (Figure 17.5).

▶ The parent cell forms a protuberance or bud.
▶ As the bud elongates, the DNA in the cell nucleus replicates.
▶ The cell nucleus divides into two and one nucleus migrates into the bud.
▶ A wall forms between the bud and the parent cell, and the bud breaks away.

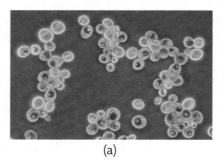

(a)

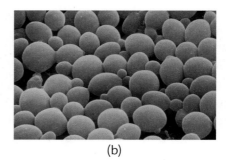

(b)

▶ **Figure 17.5** (a) Light micrograph of yeast cells, *Saccharomyces cerevisiae*. Mag ×340 (b) Colour enhanced SEM of yeast cells, *Saccharomyces cerevisiae*, budding. Mag ×3000

Moulds reproduce by forming spores that may be sexual or asexual.

▶ Asexual spores form in sporangia at the tips of aerial hyphae from one organism (Figure 17.6).
▶ Sexual spores form when cells of two organisms fuse.
▶ Sexual spores are not produced as often as asexual spores. Sexual spores often have to survive adverse conditions before they germinate.

Protozoa

Protozoa are a subgroup of the kingdom Protista (sometimes called Protoctista), which contains organisms that do not fit into the other four kingdoms (prokaryotes, fungi, animals and plants). Protozoan infecting agents cause malaria (Figure 17.7), sleeping sickness and amoebic dysentery.

Protozoa:

▶ grow best within a temperature range of 10°C–45°C.
▶ are unicellular eukaryotic cells.
▶ can reproduce *asexually* by fission.
▶ can reproduce *sexually*.
  • Two fertilised cells are produced when haploid nuclei (nuclei with half the normal number of chromosomes) from different cells fuse. Each fertilised cell now contains DNA from another individual.
  • When these cells divide, their daughter cells will not be genetically identical to the original parent cell.
▶ can be parasitic and cause disease, such as amoebic dysentery, giardiasis and toxoplasmosis. The protozoa causing these infections are transmitted via faeces or faecal contamination of food or water.

Parasites can be spread by insects that transmit the protozoa when they bite humans (or other animals). The insects are the vectors (but not the cause) of the diseases.

  ▶ Female *Anopheles* mosquitoes are vectors (carriers) for *Plasmodium*, which causes malaria, and spread malaria.
  ▶ Sand flies are vectors for *Leishmania* and spread leishmaniasis.
  ▶ Tsetse flies are vectors for trypanosomes and spread African sleeping sickness.

Viruses

Viruses occur in all types of ecosystem and can infect all types of living organisms: bacteria, protists, fungi, plants and animals. Viruses are the most abundant entities on Earth and are far smaller, but even more numerous than bacteria. Scientists estimate that there are several hundred thousand species of viruses.

▶ **Figure 17.6** Light micrograph of the fungus, *Aspergillus*, with sporangia that house spores at the end of the aerial thread-like hyphae.

▶ **Figure 17.7** Coloured transmission electron micrograph (TEM) of human red blood cells infected with the malaria parasite, *Plasmodium gambiae*, which is dividing within each cell and is released when the red blood cells burst. Mag ×2800

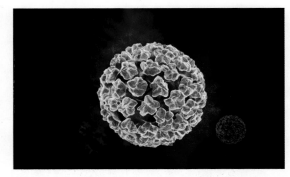

▶ **Figure 17.8** Transmission electron micrograph with colour added of human papilloma virus (HPV). The diameter of this virus is about 50 nm. Can you work out the magnification factor?

Among the diseases viruses cause in humans are HIV/AIDS, hepatitis A, B and C, measles, mumps, influenza, colds, polio, glandular fever, herpes, hepatitis, viral meningitis, chicken pox and Ebola. Some viruses cause cancers, such as leukaemia; cervical cancer is caused by the human papilloma virus (HPV) (Figure 17.8). Smallpox is a viral disease that was eradicated by 1980, but there are samples of the virus in high security laboratories.

Viruses are described as **akaryotic**. They:

▸ are not made of cells, but consist of a capsid (protein coat) made of smaller units, called capsomeres, surrounding some nucleic acid.

▸ have nucleic acid that is *either* DNA *or* RNA, but not both, at the same stage of the life cycle. The nucleic acid contains genes.

▸ can have a **lipid** envelope around the protein coat.

▸ have no cytoplasm, membranes or organelles.

▸ can only reproduce when inside a host cell.

▸ can be crystallised and kept for long periods of time.

▸ exist as independent particles called virions when not inside a host cell.

> **Key terms**
>
> **Akaryotic** – having no cell structure, no cytoplasm and no organelles. Consist of nucleic acid and a protein coat.
>
> **Lipid** – fats or their derivatives, including fatty acids, oils, waxes and steroids; lipids are insoluble in water, but soluble in organic solvents, such as ethanol.

There are many different types of virus, so they display a wide variety of shapes and sizes. Most are much smaller than bacteria, with diameters of 20–300 nm (one nanometre, nm, is one millionth of a millimetre, mm). Some filoviruses, such as the Ebola virus, are much longer, at up to 1300 nm (1.3 μm), with a width of 80 nm.

Because viruses are not cellular, they cannot reproduce by binary fission. The replication of viruses takes place inside a host cell. They use the host cell's organelles to build new virus particles that then rupture and kill the host cell as they burst out of it. Each new virus particle can then infect another host cell. Figure 17.9 shows the structure of an influenza virus.

Viruses are nature's genetic engineers and, because many of them can infect many different hosts, they have transferred

▸ **Figure 17.9** The influenza H1N1 virus. The protein spikes on the surface are haemagglutinin (H), which helps attach the virus to host cells, and neuraminidase (N), which helps release virus particles from infected cells. Inside the virus is its RNA

genes between species of living organisms. About 8%–10% of the human **genome** is viral in origin. They have therefore played a crucial role in the evolution of organisms on Earth.

> **Key term**
>
> **Genome** – all the genes within a cell/organism.

Bacteriophages

Bacteriophages are viruses that infect bacteria. They attach to the bacterial cell surface and inject their nucleic acid into the bacterium. The genes in the viral nucleic acid can then be expressed. The viral nucleic acid is replicated and the structures inside the bacterial cell, such as ribosomes, are 'hijacked' and used to assemble many new viral protein coats. When several hundred new virus particles have been made inside the bacterium, the cell ruptures, releasing the new viruses (Figure 17.10).

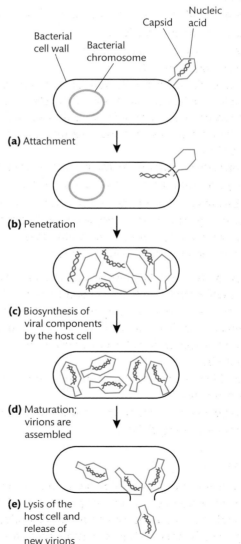

▸ **Figure 17.10** Replication cycle of a bacteriophage virus

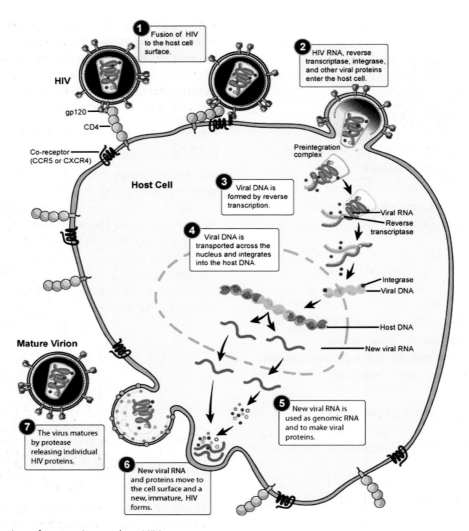

Figure 1: Diagram of HIV replication showing numbered steps:

1. Fusion of HIV to the host cell surface.
2. HIV RNA, reverse transcriptase, integrase, and other viral proteins enter the host cell.
3. Viral DNA is formed by reverse transcription.
4. Viral DNA is transported across the nucleus and integrates into the host DNA.
5. New viral RNA is used as genomic RNA and to make viral proteins.
6. New viral RNA and proteins move to the cell surface and a new, immature, HIV forms.
7. The virus matures by protease releasing individual HIV proteins.

Labels: HIV, gp120, CD4, Co-receptor (CCR5 or CXCR4), Host Cell, Mature Virion, Preintegration complex, Viral RNA, Reverse transcriptase, Integrase, Viral DNA, Host DNA, New viral RNA

▶ **Figure 17.11** Replication of a retrovirus such as HIV

Retroviruses

Retroviruses have RNA genomes and they also contain the enzyme, reverse transcriptase. During their replication cycle inside a host cell retroviruses use the enzyme to produce copy DNA, using RNA as the template. This is the opposite of transcription. The DNA may then be incorporated into the host genome (host DNA in the cell nucleus) and this may remain inactive for several years. The double-stranded copy DNA is then used as a template to make single-stranded RNA that is packaged into the new virus particles, which can burst out of the infected cell.

Figure 17.11 shows the replication of a retrovirus, such as HIV. The glycoprotein spikes on the virus envelope attach to the antigens on the surface of a T-lymphocyte white blood cell. The RNA viral genome (red) and reverse transcriptase enzyme (blue dots) are injected into the cell. Reverse transcriptase catalyses the transcripiton of RNA to DNA, which then integrates into the host genome and may remain dormant there for several years. When activated, the virus genes direct the synthesis of new virus particles that bud off from the cell, taking part of the host cell's surface membrane to form their envelopes.

Most retroviruses infect vertebrates. When they incorporate their gene into the host genome, they may cause cancer, because the insertion of viral genes can disrupt the action of host genes that control normal cell division. Human T-cell leukaemia is caused by a retrovirus. HIV is also a retrovirus. Not all RNA viruses are retroviruses. The influenza virus contains RNA but it is *not* a retrovirus.

Viroids

Viroids are smaller than viruses. They are infectious pathogens and consist only of short strands of circular, single-stranded RNA (Figure 17.12). They do not have protein coats. Viroids do not code for any proteins and they replicate using the enzyme RNA polymerase. The first viroid was discovered in 1971.

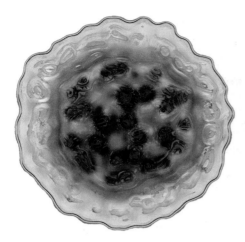

▶ **Figure 17.12** The viroid that causes hepatitis D. Its diameter is 35 nm. Can you work out the magnification factor?

Prions

During the 1960s research scientists discovered that infectious proteins cause forms of transmissible spongiform encephalopathy, such as scrapie in sheep, and kuru (laughing disease) and new variant Creutzfeldt-Jakob disease (vCJD) in humans.

Prions (pronounced *pree-ons*) is a term that was first used in 1982 and means proteinaceous (made of protein) infectious particle. It is derived from the words **pr**otein and infec**tion**. Prions:

▶ are not living organisms but are misfolded proteins (proteins that are incorrectly folded and consequently misshapen)

▶ are stable and not denatured by heat or digested by protease enzymes.

When transmitted to another organism, prions can cause some of the proteins of that organism to misfold and form aggregates (clumps). The aggregates accumulate in the infected tissue (brain and nerve tissue), causing tissue damage and cell death. The spongiform diseases are not treatable and are usually fatal.

Figure 17.13 shows prion proteins taken from an infected hamster.

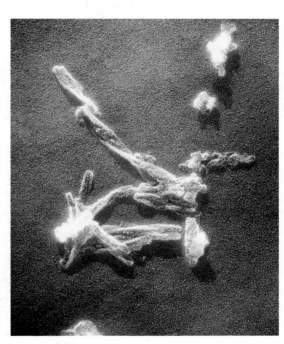

▶ **Figure 17.13** Scanning electron micrograph (SEM) with colour added of prion proteins taken from the brain of an infected hamster. Mag ×135 000

⏸ **PAUSE POINT**   Compare the features of bacteria, fungi and protozoa.

Hint   Think of similarities and differences between the three groups of organisms.

Extend   Why are prions not regarded as organisms?

Classification

Here, you will find out about some of the characteristics used to classify bacteria, viruses and fungi.

Bacteria

Bacteria share many common features, as given above. However, there are differences between them that can be used to classify and identify them. Some of the differences are due to their shapes, cell wall structure (and subsequent Gram stain) and their oxygen requirements. In 1923 David Hendricks Bergey published a manual used to classify

bacteria into species, families and orders, based on their structural and functional characteristics. *Bergey's Manual of Systematic Bacteriology*, initially a set of four books, is still in use and has been amended regularly, to include evolutionary relationships between the bacteria as they have been discovered. There are currently five volumes of the manual.

▶ Volume 1 includes information on Archaea and phototrophic (photosynthetic) bacteria.

▶ Volume 2 is sub-divided into three books and deals with the Proteobacteria (a major phylum of Gram-negative bacteria).

- Volume 3 deals with the Firmicutes (a phylum of bacteria, most of which are Gram-positive and many of which produce spores).

- Volume 4 includes information on the phyla Bacteroidetes, Spirochaetes, Tenericutes, Acidobacteria, Fibrobacteres, Fusobacteria, Dictyoglomi, Gemmatimonadetes, Lentisphaerae, Verrucomicrobia, Chlamydiae and Planctomycetes.

- Volume 5 deals with the Actinobacteria (a phylum of Gram-positive bacteria, many of which are soil dwelling).

Gram staining is a microbiological laboratory technique that is almost always used as the first step in preliminary identification of a bacterial species. Having carried out Gram staining, further tests are needed to identify the species. These include examining their shapes and finding their oxygen requirements. Biochemical tests are also used to see which enzymes are present.

Gram staining

In 1884, the Danish bacteriologist Hans Christian Gram developed a staining method to make bacteria in infected lung tissue appear more visible. He found that some bacteria, for example, *Rickettsia* bacteria that cause typhus fever, did not retain a stain, called crystal violet, at all. Gram staining (see 'Gram staining', under 'Specimen and slide preparation',) differentiates bacteria according to their cell wall structure. Gram-positive bacteria retain the purple crystal violet stain because they have thick

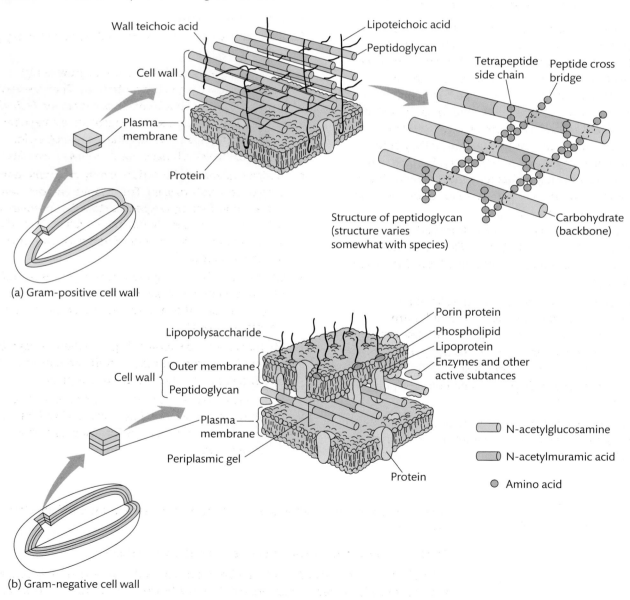

(a) Gram-positive cell wall

(b) Gram-negative cell wall

▶ **Figure 17.14** (a) Gram-positive bacterial cell wall and (b) Gram-negative bacterial cell wall

walls made of peptidoglycan. Gram-negative bacteria have thinner peptidoglycan walls and an inner and outer membrane; this means they lose the purple dye and appear pink, due to a counterstain, which is usually safranin (Figure 17.14).

Most species are Gram-positive or Gram-negative, but some are Gram-variable or Gram-indeterminate. The bacteria that cause tuberculosis (TB) do not stain readily with Gram staining and are described as acid-fast.

Phenotypic classification – observing the bacterial shapes

Phenotype refers to the visible characteristics of an organism. Once bacteria have been stained and viewed under a microscope, their shapes can be observed.

Round-shaped bacteria are described as cocci (singular: coccus). If the cocci bacteria occur in chains, they are described as streptococci; in pairs, they are diplococci; and if in clusters, they are staphylococci. The names of some bacteria indicate their shape, for example, *Staphylococcus aureus* are round-shaped bacteria that appear as clusters and live harmlessly on the skin. However, *S. aureus* can cause infections if they enter a deep wound, for example, after surgery. *Streptococcus mutans* are round-shaped bacteria that occur in chains, live in human mouths and can cause tooth decay. Other shapes of bacteria include:

▶ rod-shaped bacteria, which are described as bacillus (plural: bacilli) (Figure 17.15). *Bacillus cereus* in rice causes food poisoning. *Lactobacillus bulgarica* is used in making yoghurt.

▶ corkscrew-shaped, or spiral, bacteria, known as spirilla (singular: spirillum). Examples are *Spirillum minus*, which causes rat-bite fever, and the spirochetes, such as *Helicobacter pylori* (which causes peptic ulcers), *Borrelia* spp. (which cause Lyme disease) and *Treponema pallidum* (which causes syphilis).

▶ curved rod-shaped, or comma-shaped, bacteria, which are described as vibrio. An example is *Vibrio cholerae*, which causes cholera.

▶ **Figure 17.15** Bacterial shapes: spirilla, bacilli and cocci

Oxygen requirements

Bacteria are also grouped based on their need for oxygen to grow (Figure 17.16).

▶ Facultatively anaerobic bacteria can grow in high oxygen or low oxygen concentrations. They are very versatile. The bacteria *Staphylococcus aureus, Escherichia coli* and *Streptococcus* spp. are facultative anaerobes. (Some fungi can also be facultative anaerobes, for example, the yeast fungus, *Saccharomyces cervisiae*.)

▶ Obligate anaerobic bacteria will only grow where there is no or very little oxygen. They respire without oxygen and are poisoned by oxygen. *Bacteroides*, found in the colon, are anaerobes. (Some fungi, such as those that live symbiotically in the guts of ruminants, are also obligate anaerobes.)

▶ Obligate aerobes only grow where oxygen is plentiful. *Mycobacterium tuberculosis*, which causes TB, is an obligate aerobe. (Most fungi, other than yeasts, are obligate aerobes.)

▶ Microaerophilic bacteria only grow where oxygen is present in low concentrations. *Neisseria gonorrhoeae*, which cause gonorrhoea, is microaerophilic.

▶ Aerotolerant bacteria do not require oxygen to respire, but they are not poisoned by oxygen. Most *Clostridium* spp. are aerotolerant, as they form spores that are resistant to oxygen.

Ⅱ PAUSE POINT   How are Gram staining, cell shapes and oxygen requirements used to help classify bacteria?

(Hint)   Think about how the cell wall structure influences Gram staining.

(Extend)   When bacteria are inoculated into a tube of special broth, some grow only at the top of the broth whereas others grow only at the bottom. What do you think are the oxygen requirements of each of these bacteria?

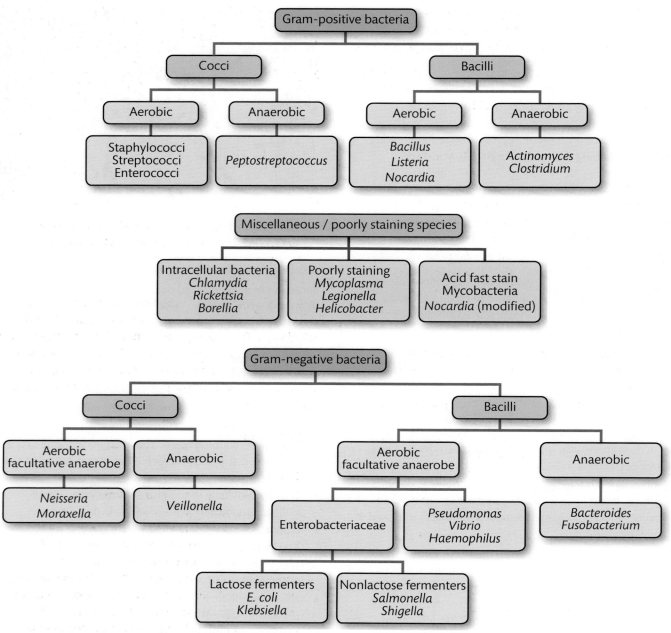

Figure 17.16 Classification of some bacteria

Fungi (Mycota)

The main phyla (usually called divisions) of fungi are:

▶ Chytridiomycota ▶ Basidiomycota

▶ Zygomycota ▶ Glomeromycota.

▶ Ascomycota

Chytridiomycota

The Chytridiomycota are also called chytrids (Figure 17.17). They do not have mitochondria in their cells and produce motile spores with flagella, which move through water. They used to be classified as protists, but modern DNA analysis techniques have redefined them as fungi. Some of them are

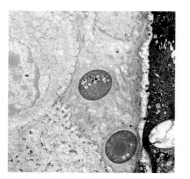

▶ **Figure 17.17** A chytrid fungus, *Batrachochytrium dendrobatidis*, on the skin of an amphibian

anaerobes and live symbiotically inside the digestive tracts of large herbivores. Some of the anaerobic chytrids also dwell in landfill sites, where they help to decompose matter.

The oldest known living organism is a large chytrid fungus, whose underground mycelium covers over 900 hectares; the whole organism is 9000 years old.

Zygomycota

Zygomycota are also called pin moulds. The moulds that you may find growing on fruit, such as raspberries, or cheese kept for too long in the fridge, are pin moulds. Some are zygote-forming mycorrhizae (fungi that live symbiotically inside plant roots and play an important role in aiding uptake of minerals by the plants). Zygomycota are mostly microscopic. Their spores form in round-shaped spore cases called sporangia (singular: sporangium). Examples include *Mucor* spp. and *Rhizopus* spp. (Figure 17.18).

Ascomycota

The sac fungi produce spores for reproduction in a sac-like structure, called an ascus. Cup fungi include morels, truffles and some mushrooms. The ascomycete *Botrytis cinerea* (grey mould) infects many crop plants, including grapes and other soft fruit such as strawberries and raspberries. *Aspergillus niger* causes black mould of onions and some ornamental plants, but is also used within the food and biotechnology industries as a source of enzymes. *Aspergillus* species can infect humans, causing ear infections and lung infections (Figure 17.6). *Penicillium* spp. are a source of the antibiotic penicillin. *Fusarium graminearum* infects cereal crop plants.

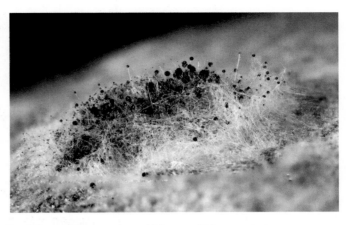

▶ **Figure 17.18** Bread mould fungus, *Rhizopus nigricans*, growing on bread. Mycelia, consisting of hyphae and sporangia containing spores, are visible.

Basidiomycota

Basidiomycota are also called club fungi. They include many types of mushrooms, toadstools, puffballs and stinkhorns. The visible parts of these fungi are the fruiting bodies that appear above ground. They produce spores in a club-shaped spore case called a basidium (plural: basidia). Rust fungi, major pathogens of cereal crops, are basidiomycetes.

Case study

Ergotism

The fungus *Claviceps purpurea* infects rye, particularly during wet summers. Bread made from infected rye may contain alkaloids produced by the fungus. These chemicals are similar to the chemical LSD and can result in ergotism. They cause convulsions, mania, psychosis, hallucinations, delirium, headaches, nausea and gangrene. Scientists think that ergotism may have caused historical incidents of people being 'bewitched'; for example, the Salem witch trials in New England in the early 1600s, which are the subject of Arthur Miller's play *The Crucible*. In the Middle Ages, the gangrenous disease 'St Anthony's Fire' was ergotism. In 1951 there was a small outbreak of ergotism, due to infected bread, in a French village. Five people died and several were committed to an asylum.

Check your knowledge

1 Why do you think ergotism was attributed to being bewitched in the Middle Ages?
2 Why do you think people in the 1951 outbreak of ergotism were committed to an asylum?

Case study

The 'curse' of Tutankhamun

Six of the 16 people present when Howard Carter found the tomb of Tutankhamun in 1922 died before 1932. Howard Carter died in 1939. The myth of the 'curse of the mummy' was reinforced by these deaths. Lab studies have shown that some ancient mummies carry mould such as *Aspergillus niger* and *Aspergillus flavus*, which, if breathed in, can cause bleeding in the lungs. Lung-infecting bacteria such as *Pseudomonas* and *Staphylococcus* also grow on Egyptian tomb walls.

Check your knowledge

1 Is there sufficient evidence to suggest that Carter and his team were infected by fungi or bacteria present in Tutankhamun's tomb?

Glomeromycota

Members of this division of fungi form mycorrhizae in the roots of land plants. These mycorrhizae increase the uptake of minerals by the plants. This symbiotic association has been around for about 400 million years.

Viruses

There are more than 200 000 identified species of virus and many more waiting to be discovered. Scientists do not know the origin of all viruses, so they cannot be placed within the 'tree of life' (Figure 17.1) as other organisms are. The classification of viruses is based on their:

- size
- capsid morphology (shape: round, long, with or without an envelope)
- type of nucleic acid
- mode of replication
- host organism (bacterium, fungi, protist, animal or plant)
- pathology (type of disease they cause).

The International Committee on Taxonomy of Viruses (ICTV) began its classification of viruses in the 1970s. It continues to develop and refine this classification. The taxonomy system for viruses shares certain features of the taxonomic system for living organisms. However, instead of the sequence 'kingdom, phylum, class, order, family, genus, species', virus classification begins with the level 'order'. Each taxon (group), apart from species, has a specific suffix:

- order (-virales)
- family (-viridae)
- subfamily (-virinae)
- genus (plural: genera) (-virus)
- species.

The Baltimore classification of viruses places viruses into seven groups (I–VII), depending on their type of nucleic acid. Nucleic acids, DNA and RNA, exist in the following forms:

- double-stranded DNA with a coding and template strand (ds DNA)
- single-stranded coding strand of DNA (ss +DNA)
- single-stranded template strand of DNA (ss −DNA)
- positive-sense single strand RNA, similar to messenger RNA (mRNA; a copy of the coding strand of DNA) (ss +RNA)
- negative-sense single strand RNA, complementary to mRNA (ss −RNA).
- double-stranded RNA; one strand is mRNA and the other strand is complementary to it (ds +/− RNA)

Some viruses (Group VII) have gapped nucleic acid. Table 17.2 shows examples of viruses from each group, their genomes and examples of diseases caused.

> **Link**
>
> See Unit 11: *Genetics and Genetic Engineering* for more information about nucleic acids.

▶ **Table 17.2** Examples of viruses from each group

Group	Genome	Examples of viruses	Examples of diseases caused
I	ds DNA	adenoviruses, herpesviruses, poxviruses	meningitis, chickenpox, smallpox
II	ss +DNA	parvoviruses	parvovirus infection
III	ss −DNA	reoviruses	gastroenteritis
IV	ss +RNA	picornoviruses, togaviruses	hepatitis A, polio, SARS, foot and mouth disease, yellow fever, hepatitis C, rubella
V	ss −RNA	orthomyxoviruses, paramyxoviruses, rhabdovirus	influenza, measles, mumps, Ebola, Marburg disease, rabies
VI	ds +/−RNA	retroviruses	HIV/AIDS
VII	gapped nucleic acids	hepadnaviruses	hepatitis B

The ICTV has established six orders of viruses. Table 17.3 shows these orders with some examples of diseases caused and hosts infected.

▶ **Table 17.3** Examples of viruses from each order

Order	Genome	Hosts	Examples of disease-causing viruses
Caudovirales	ds DNA	bacteria	T$_2$ phage virus (causes bacteria to lyse (split))
Herpesvirales	ds DNA	eukaryotes	*Herpes simplex* (causes cold sores in humans)
Mononegavirales	ss –RNA	plants and animals	Ebola (causes haemorrhagic fever)
Nidovirales	ss +RNA	vertebrate animals	SARS (severe acute respiratory syndrome)
Picornavirales	ss +RNA	plants, insects and some other animals	strawberry mottle virus
Tymovirales	ss +RNA	plants	potato virus X

Case study

Giant viruses with cell-like features

In 1957, a scientific paper defined viruses as potentially pathogenic, having only one type of nucleic acid, DNA or RNA, unable to grow and replicate by binary fission, and lacking their own metabolic machinery. In 1992, in Bradford, UK, researchers trying to find the organism responsible for an outbreak of pneumonia found an organism similar to a Gram-positive coccus bacterium. They isolated it from a culture of *Acanthamoeba polyphaga* in a water sample from a hospital cooling tower. The researchers named it *Bradfordcoccus*, but they found it impossible to get meaningful results from the usual tests to identify a bacterium. The organism was stored and in 2003 it was taken to the University of Marseilles, France. In France, a team used transmission electron microscopy to examine the organism. To their great surprise, the French researchers found it was an enormous virus and renamed it *Acanthamoeba polyphaga mimivirus* (APMV), in reference to the host organism from which it was first isolated and because it could mimic a bacterium. APMV is about the same size as the bacterium *Staphylococcus aureus*. Since then, other giant viruses have been discovered and a new viral family, *Mimiviridae*, has been defined. *Mimiviridae* are generally nucleocytoplasmic large DNA viruses.

Inside the capsid of APMV is a lipid membrane and the virus contains fibrils of peptidoglycan. It has double-stranded DNA and contains 1.2 million base pairs, encoding about 1000 genes, which include some genes for protein translation apparatus, and enzymes related to DNA repair, RNA modification and carbohydrate metabolism. These large viruses can also be infected by other smaller viruses.

The discovery of these giant viruses has reopened the debate 'are viruses living organisms?' In the future the entire classification system for living organisms could well change and all the living organisms on earth could be placed into one of two groups: ribosome-encoding organisms (cellular organisms) and capsid-encoding organisms (viral organisms), dispensing with the idea of domains.

Check your knowledge

1 Which aspect of the APMV's structure was responsible for it being thought of as a Gram-positive bacterium?

2 In what ways is APMV different from other viruses?

3 How have viruses played a key role in evolution on Earth?

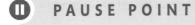

PAUSE POINT   How are viruses classified?

Hint   Think about their nucleic acids.

Extend   Discuss whether viruses should be classified as living or non-living.

Microorganisms in medicine and industry

Microorganisms have many uses in medicine and industry.

Identification of causes of disease

If a patient has symptoms of an infectious disease (one caused by an infecting organism), clinical tests on their blood, urine and faeces samples can be used to identify the causative organism. Microscopic examination of blood samples may reveal eukaryote infections, such as malaria, caused by *Plasmodium* spp. Serological blood tests can identify antibodies, produced by the body against particular antigens (foreign bodies), and identify a viral infection, such as HIV. Tests on urine can help identify the cause of kidney and bladder infections. Tests on faeces can identify bacterial infections, such as those causing gastroenteritis, typhoid, cholera, salmonella, and food poisoning (the latter is often caused by *Campylobacter* spp.).

Specialised growth media are widely used for the isolation and identification of microorganisms and/or testing their sensitivity to various antibiotics. Different types of bacteria have different nutritional requirements. In addition to nutrients needed for the growth of all bacteria, special-purpose media contain one or more chemical compounds that are essential for the growth of specific types of bacteria.

There are many special-purpose media, used for:

▶ isolation of bacterial types from a mixed population of organisms

▶ differentiation among closely related groups of bacteria, on the basis of the appearance of their colonies and biochemical reactions (leading to colour changes) within the medium

▶ finding the number of bacteria per unit volume of a sample of water, sewage or food and dairy products

▶ identification of bacteria by their ability to produce characteristic chemical changes in different media.

If the pathogen is a bacterium then a Gram stain is carried out to see if the organism is Gram-positive or Gram-negative (see 'Gram staining' sections under 'Classification', above, and 'Specimen and slide preparation'). Other biochemical tests are carried out and then differential, selective or enriched media are used to identify the growth requirements of the pathogen.

Selective media support the growth of one group of organisms, while suppressing the growth of others. For example, if the only type of sugar in a growth medium is lactose, then only bacteria that can utilise lactose (lactose fermenters) will be able to grow. Selective media are used to isolate specific groups of bacteria. Nutrient agar with added penicillin will select Gram-negative bacteria, as only Gram-positive bacteria are susceptible to penicillin. Media with added potassium tellurite, sodium azide or thallium acetate will inhibit the growth of Gram-negative bacteria, as these chemicals inhibit enzymes that only Gram-negative bacteria possess, and select Gram-positive bacteria.

Differential media can distinguish among closely related groups of organisms. Specific dyes or chemicals in a medium can either change the growth of organisms or change the colour of the medium around the organisms' colonies.

Enriched media are supplemented with highly nutritious materials, such as blood, serum or yeast extract, to encourage the growth of organisms that will not grow on other types of media.

Table 17.4 and Figure 17.19 show some examples of special-purpose media and the organisms which each selects.

▶ **Table 17.4** Some special-purpose media

Medium	Classification	Supplementary ingredients	Type of organisms isolated
Mannitol salt agar (MSA)	selective and differential	7.5% NaCl (table salt), mannitol (a carbohydrate) and phenol red (a pH indicator)	*Staphylococci* spp.; it can differentiate between *S. aureus*, which ferments mannitol to produce acid, giving yellow zones around the colonies, and *S. epidermidis*, which does not produce acid and has red zones around the colonies.
MacConkey agar	selective and differential	Lactose (a sugar), bile salts, crystal violet, neutral pH red indicator and peptone	Gram-negative enteric (gut-dwelling) bacilli; *E. coli*, which ferments lactose, to produce acid, and produces pink or red colonies. *Salmonella* and *Shigella* (cause of dysentery) do not ferment lactose, but use peptone to produce ammonia, which raises the pH and gives white colonies.
Eosin methylene blue agar (EMB)	selective and differential	Lactose, eosin and methylene blue	Gram-negative intestinal pathogens; *E. coli* produces metallic green colonies, *Salmonella* and *Shigella* produce colourless colonies.

▸ **Table 17.4** *Continued*

Medium	Classification	Supplementary ingredients	Type of organisms isolated
Blood agar	enriched and differential	Sheep or ox blood	Most bacteria will grow, but haemolytic organisms (which destroy red blood cells) can be distinguished; these bacteria cause blood cells to lyse, releasing haemoglobin. *S. aureus* produces a light/golden yellow pigment and *S. epidermidis* produces a white pigment.
Choc agar	enriched	Heated red blood cells	Pathogens such as *Neisseria* spp. (cause of meningococcal meningitis, gonorrhoea) and *Haemophilus influenza* (cause of pneumonia, bacterial meningitis, and bone infections).
Potato dextrose agar	selective	Potato extract (starch), dextrose (a type of glucose)	General purpose medium for mould and yeast fungi. It can be supplemented with antibiotics or acid to suppress bacterial growth. It is also used for growing clinically significant (pathogenic) yeasts and moulds; it encourages spore formation and pigment production in some dermatophytes (fungi that infect skin, such as ringworm and athlete's foot). It is used for plate count methods in the food and dairy industry, and for testing cosmetics.

❚❚ PAUSE POINT   What are differential and selective media?

> **Hint**   These are both special types of growth media for microorganisms.
>
> **Extend**   What sort of growth medium would you use to grow fungi in the lab?

▸ **Figure 17.19** Selective and differential media

Bio-waste processing

Many tons of bio-waste are produced every day in hospitals, health clinics, care homes, nursing homes, medical research laboratories, dentists, vets, GP surgeries, schools and colleges. This waste is distinct from normal household rubbish and it has to be properly disposed of to prevent infections from spreading. There are different means for disposing of different types of bio-waste.

▸ Incineration is used for pathological waste, such as used wound dressings, body parts, blood and body fluids and used sharps. All that is left after incineration is ash.

▸ Autoclaving is for used sharps, discarded Petri dishes, aprons and gloves. An autoclave uses steam generated under pressure and reaches a temperature of about 120°C. This reduces the microbiological load to a level where it can be disposed of safely. The same autoclaves must not be used to sterilise waste and to sterilise supplies.

▸ Microwave radiation is used for sharps, including needles, lancets and scalpels.

▸ Chemical treatment is used for chemicals that have been used to clean labs, operating theatres or wards. Bleach can also be used to disinfect bio-waste.

Bio-waste has to be properly managed to protect workers, the general public and the environment. Large hospitals or university research labs may have on-site disposal facilities. However, as these are expensive, smaller establishments may package and transport their waste for treatment. To do this, they may employ a bio-waste disposal service, whose employees are trained to safely take away the waste in special containers that are leak-proof and clearly marked with biohazard symbols. Workers handling the waste must observe standard precautions.

Regulations

In the UK, clinical waste and the way it is handled is closely regulated. Legislation includes the:

▸ Environmental Protection Act 1990 (Part II)
▸ Waste Management Licensing Regulations 1994
▸ Hazardous Waste Regulations (England and Wales) 2005
▸ Special Waste Regulations in Scotland.

Sometimes, robots are used to handle clinical waste (Figure 17.20).

▶ **Figure 17.20** A robot handling infectious waste that is labelled with a biohazard warning sign

Food and beverage production

Fermentation carried out by microorganisms changes the taste and texture of food and may preserve it. For thousands of years, humans have used yeasts, moulds and bacteria to make food products, such as beer, wine, yoghurt, bread, cheese, sausage, soy sauce, and fermented meat, fish and vegetables. During the mid eighteenth century, scientists realised that microorganisms were responsible for these fermentation products. After the Second World War, the food and drinks industry began to develop biotechnology techniques to use microorganisms in food and drink production.

Many microorganisms are used in the production of food and drinks. Lactic acid bacteria (LAB) are a very important group, as they produce metabolites, including lactic acid, that inhibit the growth of other microorganisms. LAB are used for making sausage, cheese, kimchi (pickled vegetables) and yoghurt. However, they can spoil beer and wine, because they lower the pH.

Key term

Fermentation – a metabolic process that occurs in yeasts and bacteria to convert sugar to acids, gases or ethanol.

Wine

Harvested grapes are crushed to form must. This is exposed to carbon dioxide to kill the natural yeasts and a specific strain of yeast, *Saccharomyces cerevisiae*, is added. As the yeast ferments the sugar in the must, ethanol is produced, as well as carbon dioxide. Sweet (dessert) wines are given a pre-fermentation microbial treatment. They are infected with a fungus, *Botrytis cinerea*, which causes water loss. This increases the sugar content of the grapes and also destroys the malic acid in the grapes. Special glucophilic (glucose-loving) yeasts are used for the fermentation; they utilise the glucose, leaving fructose and producing a sweet wine.

Beer

Beer is made from malted barley; the grains are allowed to germinate so that they produce amylase enzymes, which digest the starch in the grains, forming the sugar maltose. The grains are then heated to denature enzymes and stop germination. The grains are crushed and soaked to produce a sugary liquid called wort. This is then boiled with hops (for flavour). When cooled, yeast, either *Saccharomyces cerevisiae* (for beer) or *S. carlsbergensis* (for lager), is added. As the yeast respires anaerobically, it converts sugars to ethanol and carbon dioxide. The spent yeast is then separated from the liquid and may be used to make Marmite. The beer is filtered and **pasteurised** before being canned or bottled.

Bread

Flour (from wheat, rye or corn) is mixed with water and the yeast, *S. cerevisiae*. Amylase enzymes from the cereal grains are present, or fungal amylases may be added to the mix. The enzymes hydrolyse the starch in flour to sugars that the yeast respires, producing carbon dioxide (which makes the dough rise). Any ethanol made is evaporated during the baking process. Proteolytic (protein-digesting) enzymes in the yeast change the texture of the flour.

Yoghurt

Milk is heated to about 90°C to kill any harmful bacteria and drive out dissolved air, giving a low oxygen environment. The milk is then cooled to about 45°C and a mixture of microaerophilic bacteria (*Lactobacillus bulgaricus* and *Streptococcus thermophilus*) is added. *L. bulgaricus* breaks down the milk proteins to peptides, on which the streptococci feed, producing methanoic acid, which stimulates the growth of *L. bulgaricus*, producing lactic acid and lowering the pH to 4.4. Live yoghurts also contain the bacteria *Lactobacillus acidophilus* and *Bifidobacterium bifidum*.

Fruit juice

To extract more juice from crushed fruit, pectinase and cellulose enzymes are added. These enzymes break down the pectin between plant cell walls and the cellulose in the cell walls, increasing the volume of juice extracted. As fruit juice has a low pH, the enzymes added must work well at low pH values. Pectinase is obtained from the fungus, *Aspergillus niger*, which secretes pectinase when it infects plants.

Theory into practice

Microorganisms can be used to increase agricultural production. One example is silage, made from grass harvested in the summer and used for winter feed. A strain of *Lactobacillus plantarum* is added to cut grass, which is then stored in large polythene bales or silage pits. The bacterium competes with other bacteria naturally present, and helps the silage to ferment quickly and reach a stable pH. Cows find silage palatable and easy to digest. Other bacteria, *Serratia rubidaea* and *Bacillus subtilis*, are added to the cut grass, to inhibit the growth of moulds.

Check your knowledge

1 Why do you think the growth of moulds in silage is inhibited?

 PAUSE POINT   Explain how microorganisms are useful in the food and drinks industries.

Hint   Include reference to bacteria and yeasts.

Extend   Why are beers and wines pasteurised?

Nitrogen fixation

The growth of plants depends upon not only photosynthesis, but on mineral availability. Nitrogen is an essential component of proteins, nucleic acids and other important chemicals, such as certain coenzymes crucial for photosynthesis and respiration. Molecular nitrogen, N_2, makes up 79% of the Earth's atmosphere, but the triple bond between the two nitrogen atoms make this molecule almost inert and plants cannot utilise it. For plants to utilise nitrogen, it has to be combined with hydrogen in the form of ammonium (NH_4) or combined with oxygen as nitrate (NO_3^-).

Microorganisms play a key role in recycling nitrogen. Soil-dwelling bacteria break down dead matter, faeces

and urine, to release ammonia (NH_3), which other bacteria change to nitrites (NO_2^-) and then nitrates (NO_3^-). In cold, poorly aerated soils, bacteria may utilise nitrate as a source of oxygen for respiration and release molecular nitrogen (N_2) back into the air. Some bacteria contain the enzyme nitrogen reductase and can combine molecular nitrogen with hydrogen, in a process called nitrogen fixation, to produce ammonia (NH_3). Some nitrogen-fixing bacteria are free-living in soil; others form symbiotic relationships with certain plants, protozoa or termites. Lightning also produces a small amount of ammonia. During the Haber–Bosch process, developed in the 1940s, nitrogen and hydrogen are heated under high pressure and in the presence of a catalyst, which makes ammonia that can be used to make fertiliser.

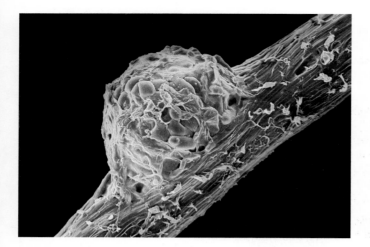

▶ **Figure 17.21** Coloured SEM of a root nodule on a pea plant, *Pisum sativum*, containing the bacteria *Rhizobium leguminosarum*. Mag ×105

Symbiotic relationship

Leguminous plants, for example, peas, beans, clover, lupins and alfafa, have symbiotic nitrogen-fixing bacteria, *Rhizobium* spp., encased in nodules in their roots (Figure 17.21). The plants produce hydrogen, which the bacteria use, with nitrogen reductase and ATP, to make ammonia, as outlined in the equation below.

$$N_2 + 8H^+ + 8e^- + 16\ ATP \rightarrow 2NH_3 + H_2 + 16\ ADP + 16P_i$$

The enzyme nitrogen reductase consists of two proteins, one joined to iron and one containing molybdenum and iron. Some free-living nitrogen-fixing bacteria, such as *Clostridium* and *Desulfovibrio*, are anaerobes. Other free-living nitrogen fixers, such as *Azotobacter*, *Klebsiella* and some Cyanobacteria, are aerobic. However, oxygen inhibits nitrogen reductase, because it reacts with the iron components of the proteins. Legumes have leghaemoglobin in their roots, which combines with oxygen, providing a low oxygen environment for *Rhizobium* to carry out nitrogen fixation. *Azotobacter* has a high rate of respiration and uses up oxygen; it also builds a slimy capsule around itself, reducing the rate of diffusion of oxygen into the cell. Cyanobacteria are photosynthetic, but they only fix nitrogen in special structures, called heterocysts, which do not contain the enzymes that split water to generate free oxygen.

Alder trees (*Alnus* spp.) and sea buckthorn, a sand-dune dwelling plant, have root nodules containing symbiotic nitrogen-fixing bacteria of the genus *Frankia* that are actinomycete bacteria. As a result, alder trees can grow in soils poor in nitrate content.

Discussion

Scientists are researching how to genetically modify cereal plants to contain the nitrogenase enzyme. This would mean the plants can fix nitrogen and would reduce the need for fertilisers. Fertilisers can lead to environmental problems, including leaching of nitrates and **eutrophication** of watercourses. However, nitrogen fixation to produce ammonia has high energy costs and would use some of the plants' carbohydrate reserves and hydrogen atoms.

Discuss the pros and cons of genetically modifying cereal crops for nitrogen fixation.

Key term

Eutrophication – a form of water pollution caused when excess fertilisers leach into lakes and rivers. This excess encourages the growth of algae, which then cover the surface, preventing oxygen reaching deeper aquatic plants. The plants die and aerobic bacteria decompose them, using up the oxygen in the water. Animals die or leave the area. Anaerobic bacteria produce methane and hydrogen sulfide.

Antibiotic and hormone production

Antibiotics

Antibiotics are chemical substances produced by microorganisms to kill or inhibit the growth of other, competing, microorganisms. They are widely used to treat infectious diseases.

During the Second World War, penicillin, an antibiotic discovered accidentally in 1929, was produced from the mould fungus, *Penicillium rubens*, on a commercial scale (Figure 17.22). The process was inefficient at first,

▶ **Figure 17.22** Production of an antibiotic by culturing a *Penicillium* sp. under carefully controlled conditions in vats (fermenters)

▶ **Table 17.5** Source, use and mode of action of some antibiotics

Antibiotic	Source	Use	Mode of action in pathogen
Penicillins (penicillin, amoxycillin, methicillin)	Mould fungus *Penicillium* spp.; some are semi-synthetic and derived from penicillin	Streptococcal infections, syphilis, Lyme disease	Interferes with cell wall synthesis.
B-lactams (cephalosporins)	Semi-synthetic, derived from mould fungus, *Acremonium*	Wide range of bacterial infections	
Bacitracin	Bacterium: *Bacillus subtilis*	Eye and ear infections (usually as an ointment)	
Teicoplanin	Bacterium: *Actinoplanes teichomyceticus*	MRSA, endocarditis	
Vancomycin	Soil bacterium: *Amycolatopsis orientalis*	MRSA, colitis, *C. difficile*, pneumonia, skin infections, endocarditis (usually given intravenously)	
Colistin	Bacterium: *Paenibacillus polymyxa*	Eye, ear and bladder infections	Interferes with cell membrane structure and function.
Ciproflaxacin	A fluoroquinolone	Urinary tract infections, bacterial diarrhoea, gonorrhoea, joint infections	Inhibits replication of DNA and prevents cell division.
Norfloxacin	Synthetic fluoroquinolone		
Rifampicin	Soil bacterium: *Amycolatopsis rifamycinica*	Meningococcal meningitis, typhus fever, cholera, MRSA, TB, leprosy, legionella, bubonic plague	Inhibits the first stage (transcription) of protein synthesis.
Chloraphenicol	Filamentous bacterium: *Streptomyces venezuelae*		
Oxytetracycline Tetracycline Doxycycline	Filamentous soil bacteria: *Streptomyces rimosus*, *Streptomyces aureofaciens*, *Streptomyces* spp.	Syphilis, chlamydia, Lyme disease, malaria, acne, pneumonia, chronic bronchitis, campylobacter, anthrax, bubonic plague	Inhibits the translation stage of protein synthesis at ribosomes.
Clarythromycin	Semi-synthetic, derived from erythromycin	Streptococcal infections, stomach ulcers, Lyme disease, syphilis	
Erythromycin	Bacterium: *Saccharopolyspora erythraea*		

because shallow pans were needed for the mould to grow on the surface of the nutrient medium and receive enough oxygen for aerobic respiration. Later, a strain was developed that grows submerged in the medium. Modern production techniques use a mutant strain of the fungus that yields better results.

There are many antibiotics besides penicillin. Many soil-dwelling microorganisms, such as *Streptomyces*, filamentous actinomycete bacteria, have been screened and found to produce antibiotics. Once a microorganism that produces an antibiotic is found, further investigations identify:

▶ its optimum growth conditions (temperature, pH, oxygen and nutrient requirements)

▶ whether it produces one useful compound or more

▶ whether the antibiotic produced is new or the same as one already in use

▶ if it is new, whether it is more effective than one already in use.

The antibiotic-producing microorganism is cultured, under its optimum growth conditions, in large fermenters. As the microorganism grows it secretes the antibiotic, which can then be extracted from the medium. Table 17.5 shows the source, use and mode of action of some antibiotics.

Discussion

Antibiotics are often prescribed when not needed. In some countries, they can be bought over the counter. Antibiotics have also been added to animal feed. These practices have selected for antibiotic-resistant strains of bacteria. Antibiotic resistance is now widespread, because bacteria can pass DNA to each other. There is a £10 million prize for scientists to find new bactericidal drugs to which the bacteria will not become resistant.

Discuss the pros and cons of selling antibiotics over the counter. Why should patients always complete their course of antibiotics?

Hormones

Bacteria can be genetically modified to produce useful proteins such as human insulin and human growth hormone. Insulin helps regulate blood glucose levels; it causes excess glucose (absorbed from the gut into the blood), to leave the blood and enter liver, muscle and other cells, where it is converted to glycogen for storage, or for respiration. Insulin is a small protein hormone and used to be obtained from frozen pig pancreases. However, this could not provide enough insulin to treat all those suffering with diabetes.

By the 1970s scientists had found out that insulin is a small protein made of 51 amino acids. Recombinant DNA technology (genetic engineering) is now used to make insulin with genetically modified bacteria, usually *E. coli*. Such insulin-producing bacteria were first commercialised in 1982 (Figure 17.23). The gene coding for insulin is small (in line with the small protein) and difficult to find within

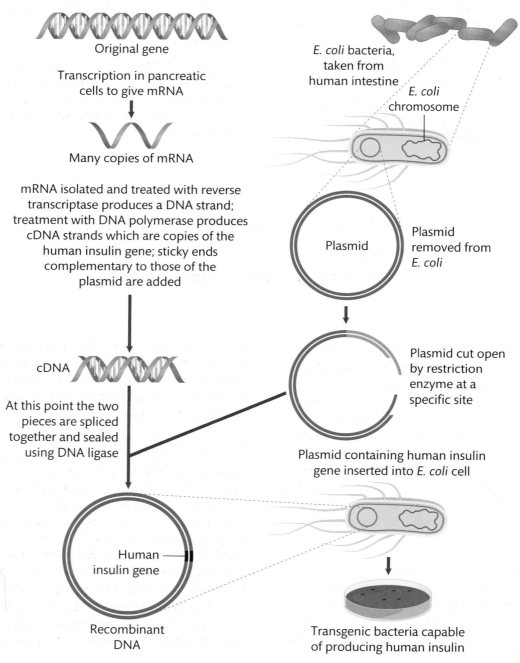

Original gene

Transcription in pancreatic cells to give mRNA

Many copies of mRNA

mRNA isolated and treated with reverse transcriptase produces a DNA strand; treatment with DNA polymerase produces cDNA strands which are copies of the human insulin gene; sticky ends complementary to those of the plasmid are added

cDNA

At this point the two pieces are spliced together and sealed using DNA ligase

Human insulin gene

Recombinant DNA

E. coli bacteria, taken from human intestine

E. coli chromosome

Plasmid

Plasmid removed from *E. coli*

Plasmid cut open by restriction enzyme at a specific site

Plasmid containing human insulin gene inserted into *E. coli* cell

Transgenic bacteria capable of producing human insulin

▶ **Figure 17.23** Production of bacteria containing the human insulin gene

the human chromosome. Therefore, the steps used to make the bacteria produce insulin start with obtaining mRNA, as shown below.

1   mRNA transcribed from the gene for human insulin can be obtained from beta cells of islets of Langerhans, in the human pancreas, where insulin is made.

2   The enzyme reverse transcriptase is added to the mRNA and this catalyses the process of making a single complementary strand of DNA.

3   To make double-stranded DNA, the enzyme catalyst, DNA polymerase, is added, which uses the single strand of DNA as a template; the double-stranded DNA is the gene for human insulin.

4   Three unpaired nucleotide bases, sticky ends, are added at each end of the gene.

5   Plasmids are obtained from *E. coli* bacteria and treated with restriction endonuclease enzymes to cut them open at a specific DNA base sequence, leaving sticky ends.

6   The insulin genes are mixed with the plasmids and some of the genes fit into the opened plasmids; then, DNA ligase enzymes seal the genes into the plasmids.

7   Such plasmids, containing a new piece of DNA, are called recombinant plasmids.

8   The recombinant plasmids are mixed with *E. coli* bacteria and calcium chloride solution, which makes the walls of the bacteria more porous.

9   The mixture is subjected to heat shock, with one minute at 0°C followed by just under a minute at 40°C; this also helps the bacteria take up the plasmids.

10 The modified *E. coli* bacteria are grown on agar plates to form colonies. These are then grown in large fermenters. The human insulin the bacteria produce, due to the plasmids, is harvested and purified, and used to treat diabetics.

 PAUSE POINT   Which groups of microorganisms are important sources of antibiotics?

Hint   Study Table 17.5.

Extend   Why do these microorganisms produce antibiotics?

Flora and fauna of the human digestive tract

When bacteria were first identified, they were classified as plants, because they had cell walls and, at the time, all living organisms were classed as either plants or animals. The term flora referred to plants in an area and the bacteria living in and on our bodies were called the human body flora. Scientists have since found out a lot more about living organisms and use the classification system with five kingdoms. The microorganisms (bacteria, viruses, fungi and protists) in and on our bodies are now called the microbiota.

Symbiotic relationship

The several thousand species of microorganisms that normally colonise (live in) our gastrointestinal (GI) tract (gut), saliva, mouth lining, throat, upper and lower respiratory tract, conjunctiva of the eye, vagina and on our skin surface have symbiotic relationships with us. They obtain shelter and nutrients from us (the hosts) and in turn supply us with some nutrients (e.g., vitamin K), hormones

(some help to regulate appetite) and many other products of their genes (the microbiome). They also help reduce infections from other invading microorganisms by competing with them for space and food or by secreting chemicals that kill them.

The organisms of the gut microbiota are essential for our wellbeing. The GI tracts of newborn babies are colonised by bacteria from the mother's anus and vagina during birth, and the mother's skin after birth. Each of us has between 10 and 100 trillion microorganism cells, far more than the number of cells in our bodies, weighing about 1.5–3.0 kg in total. Imbalances in the range of species present in our bodies, caused by prolonged exposure to antibiotics or from eating a diet lacking fruit and vegetables, can contribute to health problems, such as obesity. Certain live yoghurts can help to colonise the gut, but you need to eat the right foods, including plenty of vegetables, to keep the bacteria alive in your gut.

A persistent infection

John contracted an infection of *Clostridium difficile* while he was in hospital for several weeks, being treated for endocarditis with intravenous antibiotics. Everyone has these bacteria inside their digestive tract but the other bacteria usually keep *C. difficile* in check, so that it does no harm. However, if you are on a long course of antibiotics, some of your gut bacteria can be killed and *C. difficile*, also known as C. diff, can multiply and release toxins that cause swelling and irritation of the colon. This inflammation is known as colitis and the symptoms are diarrhoea, fever and abdominal cramps.

This infection can be difficult to treat and John had tried various antibiotics (vancomycin and fidaxomicin), but with no positive result. He suffered from recurring bouts of C. diff, which was very debilitating and could eventually prove fatal. John had read about the gut microbiota – the 1000 or so different species of bacteria and other types of microbes that live in our guts. These bacteria help digest some of our food and they make certain vitamins (e.g., vitamin K)

that we can use. They also make hormones that help us regulate our appetites, as well as keeping some infectious bacteria at bay. He also learned that C. diff flourishes when this balance of gut microbes is upset (e.g., after long exposure to antibiotics) and that putting this bacterial balance back to normal can cure C. diff. He discussed this with his GP who recommended that he have a faecal transplant. This involves a doctor or nurse placing a sample of faeces (taken from a healthy donor and screened) into his colon, using a catheter. It worked (as it does in over 90% of cases) and he is now well again. His GP has advised him to maintain a healthy diet to encourage the growth of the good bacteria in his colon; this diet involves plenty of fruit and vegetables, especially leeks and onions.

Check your knowledge

1 Why do you think the donor of the faeces has to be healthy?
2 Which diseases do you think are being looked for when the donated faeces sample is screened?

Assessment practice 17.1 · A.P1 · A.M1

1 Produce a large poster showing how microorganisms fit into the biological classification system.
2 Explain, by means of annotated diagrams, how the structures and characteristics of microorganisms (bacteria, fungi and some protists) are used to classify them.
3 Produce another poster to describe and explain, with the help of annotated diagrams, how viruses are classified. Discuss why viruses are classified in a separate system and are not part of the three domain/five kingdom classification system.
4 Compare the characteristics of microorganisms that are used for their classification.

Plan
• Do I know what I am being asked to do?
• Do I have enough information or should I extend my research?

Do
• I know what I want to achieve.
• I can check my work to see where I have gone off task and can make changes to put this right.

Review
• I can instruct someone else on how to complete the task more efficiently.
• I know what I would do differently next time and the approach I would take with the parts that I found difficult this time.

B Undertake microscopy for specimen examination in laboratories

Microscopes

There are many different types of microscope that can be used for examining different types of microbiological specimens. The purpose of all microscopes is to enable scientists to see small specimens, which cannot be seen with the naked eye. They all magnify objects and the **image** should be clear. The clarity depends on the resolution of the microscope.

> **Key term**
>
> **Image** – a real image is an image located where light rays from an object are focussed. A real image can be seen on a screen placed at the focal point. A virtual image is one where light rays appear to, but do not actually, come from that image.

Some useful definitions

Image

When lenses bend light rays coming from an object so that these light rays now enter your eye, the image is where the light rays you see 'seem' to be coming from. The image you see when using an optical microscope is a real image. It is magnified and inverted, so when you move the slide left to right, the image appears to move right to left (Figure 17.24).

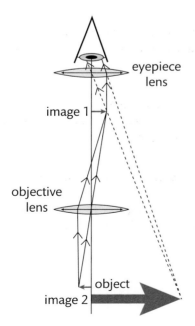

> **Figure 17.24** Virtual image formation by a compound light microscope

Magnification

Magnification describes how much bigger an image appears, compared with the original object. Microscopes produce linear magnification, which means that if a specimen is seen magnified ×1000, the image appears 1000 times longer and 1000 times wider than the object's true size.

The objective lens in a microscope magnifies the object and forms a virtual image. The eyepiece lens magnifies this virtual image more, forming a real image. Table 17.6 shows the total magnification, with various combinations of objective and eyepiece lenses.

> ▶ **Table 17.6** Total magnification values with combinations of eyepiece and objective lenses

Objective lens magnification	Eyepiece lens magnification	Total magnification
×4	×10	×40
×10	×15	×150
×40	×10	×400
×100	×10	×1000

Resolution

Resolution is the ability of an optical instrument to see or produce an image that shows fine detail clearly. Electron microscopes have much higher resolutions than light microscopes.

Focus

Most light microscopes are parfocal. This means that when one lens is in focus, other lenses will have the same focal length and can be rotated into position, without further major adjustments. The image is focussed using lenses and, when in focus, gives a clear image. The focal point of a lens is the point at which the light rays it bends meet. The distance between the lens and the focal point is the focal length.

Stereomicroscopes

Stereomicroscopes (dissection microscopes) (Figure 17.25) are optical (light) microscopes that use light illumination and low magnification, allowing inspection or dissection of small objects (e.g., microorganisms). Light is reflected off the surface of the object being viewed, rather than passing through it. There are two objectives and two eyepieces, which provide slightly different viewing angles

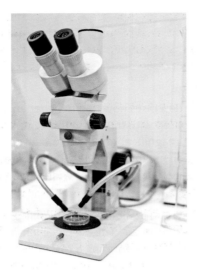

▶ **Figure 17.25** A stereomicroscope

to each eye. This gives three-dimensional visualisation of the object being viewed. Stereomicroscopes are often used for microsurgery. Some stereomicroscopes have video dual CCD camera pickups attached that display the images onto a high-resolution LCD monitor. Special software can convert the two images into an integrated three-dimensional image that viewers can see, if they wear special glasses.

Compound light microscopes

Light microscopes are also called optical microscopes (Figure 17.26). Compound microscopes have an objective lens close to the object being viewed, which focusses a real image of the object inside the microscope (see Figure 17.24). That image is then magnified by a second lens (the eyepiece lens), which produces an inverted virtual image of the object.

Some compound microscopes have a special objective lens, called an oil immersion lens. To use this, a drop of Canada balsam oil, which has a **refractive index** very close to that of glass, is placed on the stained specimen and the objective lens is positioned so that it is in the oil. This gives a greater resolution and higher magnification. You may be able to use oil immersion objectives to view stained specimens of bacteria that you prepare.

> **Key term**
>
> **Refractive index** – light changes speed as it passes form one medium, such as air, to another, such as glass. This change in speed can lead to a change in direction of the wave, which is called refraction. Refractive index is the ratio of speed of light in vacuum/speed of light in medium.

The wavelength of visible light is 400–700 nm. This is too large to pass between some of the smallest organelles within cells, so they cannot be seen. However, optical microscopes do have several benefits, including that they are:

▶ inexpensive

▶ easy to use

▶ portable and can be used in the field

▶ able to be used to examine living specimens as well as stained sections.

A photograph of a specimen seen under a light microscope is called a **photomicrograph**.

> **Key term**
>
> **Photomicrograph** – photograph taken through a microscope, to show a magnified image of an object.

Ocular tube

1. The specimen on a slide is placed here on the stage and clipped into place.

Condenser

4. Whilst viewing the image adjust the iris diaphragm for optimum light.

Light source

5. Make sure that the object you wish to view is directly over the hole in the stage. Now rotate the nosepiece and bring the ×10 objective into place over the specimen. Look down the ocular tube and use the fine focus knob to focus the image.

Arm

2. By rotating the nosepiece, the lowest power (smallest) objective lens is placed over the specimen.

3. Adjust the coarse focus knob, while looking into the eyepiece, until the image you see is clear and in focus.

Fine focus knob (see step 5)

6. Repeat Step 5 using the ×40 objective lens.

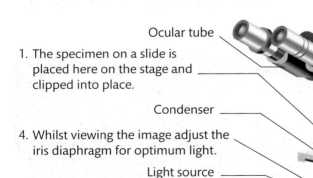
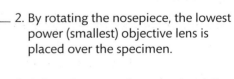

▶ **Figure 17.26** Annotated diagram showing how to use a light microscope

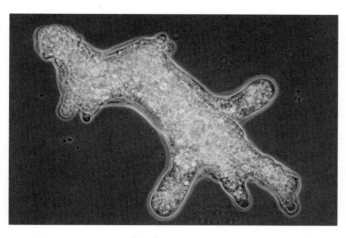

▶ **Figure 17.27** *Amoeba proteus*, a free-living protozoa viewed under phase contrast. Mag ×140

Phase-contrast microscopes

Phase-contrast microscopes are very useful for observing unstained specimens (Figure 17.27). This may include living and moving protists, such as amoebae. The microscopes have special phase-contrast objectives and a condenser (see 'Condenser', under 'Setting up and using a compound light microscope') that make different cellular components, which each have slight differences in their refractive indices, visible. As light passes through a specimen, some of this light is refracted (bent), due to slight differences in density and thickness of the various parts of the specimen's cells; that is, the refractive index of the specimen is different from that of the surrounding medium. The special optics of the microscope convert the differences between the transmitted light and refracted rays to a significant variation in light intensity (brightness), giving a clear image that appears dark against a light background.

Research

Find out about laser scanning (confocal) microscopes. Visit:

www.bristol.ac.uk/synaptic/research/techniques/confocal.html

www.physics.emory.edu/faculty/weeks/confocal

Electron microscopes

Electron microscopes use a beam of fast-travelling electrons, which has a wavelength of about 0.004 nm. As this is a much shorter wavelength than that of visible light (around 400–700 nm) these microscopes have a much higher resolution than optical microscopes (so organelles, for example, can be seen). Electron microscopes produce very clear, highly magnified images.

In an electron microscope, electrons are fired from a cathode and focussed by magnets, rather than glass lenses, onto a screen or photographic plate. There are two types: transmission electron microscopes and scanning electron microscopes.

Transmission electron microscopes

Transmission electron microscopes were developed in the 1930s.

▶ The specimen, which can be a very thin section of a larger object, has to be chemically fixed by being dehydrated and stained with metal salts.

▶ The beam of electrons passes through the stained specimen. Some electrons pass through and are focussed on the screen or photographic plate.

▶ The electrons form a two dimensional black and white (grey-scale) image. When photographed, this is called an **electron micrograph**.

Key term

Electron micrograph – photograph of an image seen with an electron microscope.

Transmission electron microscopes can currently produce a magnification of up to ×2 000 000. A new generation of microscopes is being developed and those microscopes can magnify a specimen up to ×50 000 000.

Scanning electron microscopes

Scanning electron microscopes were developed in the 1960s (Figure 17.28). Electrons do not pass through the specimen, which is whole, but cause secondary electrons to bounce off the specimen's surface and be focussed on a screen. This results in a 3D image, with a magnification

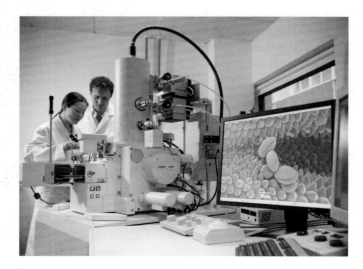

▶ **Figure 17.28** A scanning electron microscope

from ×15 up to ×200 000. The image is black and white, but computer software programmes can add false colour.

Both types of electron microscopes:
▶ are large and very expensive
▶ need a great deal of skill and training to use
▶ can only observe dead specimens, as the specimen has to be in a vacuum.

The metallic salts used to stain specimens may be hazardous to the user, so you need to follow safety guidelines carefully when using electron microscopes.

Ⅱ PAUSE POINT Distinguish between the terms 'resolution' and 'magnification'.

Hint Think about clarity and size of image.

Extend Discuss the advantages and disadvantages of light microscopes and electron microscopes.

Specimen and slide preparation

During your course you will need to prepare slides of microorganisms for observation under a light microscope.

Cover slips

When you mount a sample on a slide, you may want to use a cover slip. This will protect the sample and the objective of your microscope.

Flat slides

To make a slide of fungal hyphae, place a strip of sticky tape, sticky side down, onto a fungal mycelium (which may be on an agar plate) and then position the sticky tape face down onto a slide. A cover slip is not necessary. For most specimens you should use a flat microscope slide, but for some specimens you may use a concave slide.

Concave slides

Many species of bacteria are motile. This movement can be observed using a suspension of living bacteria, for example, *Pseudomonas aeruginosa*, in a hanging drop slide. To do this, place a diluted drop of a bacterial suspension onto a cover slip. Position a concave slide, with a ring of petroleum jelly around the depression, over the drop of bacterial suspension. Turn over the slide and cover

slip, so that the bacterial suspension is hanging from the underside of the cover slip into the depression on the slide. You could use a drop of oil on the cover slip and then an oil immersion objective lens, to view the bacteria under reduced light (see Figure 17.29).

Wet and dry mounts

Dry mount slides work best for specimens such as pollen and dust. Place the specimen on the slide and position a cover slip over it. As always, you should hold the slide and cover slip at the edges, so as not to make fingerprints on the slide.

To set up a wet mount slide, place liquid onto the slide and add the specimen to the liquid. Add more liquid if needed, to make sure the specimen is covered. Gently lower a cover slip (to make sure no air bubbles form) onto the specimen in the liquid. Make sure the cover slip does not float; any excess liquid can be absorbed using a fine tissue.

Air drying and heat fixing

To examine bacteria, you can make a heat-fixed **smear slide** by rubbing a loopful of bacteria from a colony into a drop of sterile water on a slide. Alternatively, you can place a drop of liquid bacterial culture directly onto a dry flat slide. Dry and heat fix the slide, either by passing it swiftly through a

a. Warming the wax-grease tube

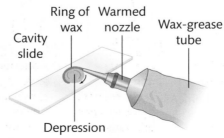

b. A circular wall of wax is made around the depression

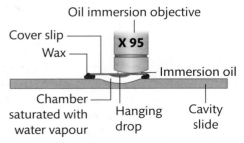

c. The slide is shown correctly set up for oil immersion viewing

▶ **Figure 17.29** The hanging drop method, with an oil immersion objective lens, used to view live, motile specimens

Bunsen flame, or by placing it onto a hot plate (Figure 17.30). In some cases you could leave the slide to dry naturally in the air. The heat-fixed smear or air-dried slide can then be stained (see below). For oil immersion, it is best to *not* use a cover slip.

> **Key term**
>
> **Smear slide** – a slide prepared when a sample (for example, blood, bacteria or sediment) is spread thinly (smeared) on the slide in preparation for examination.

Staining

The staining techniques you will most often use are simple staining, negative staining and Gram staining.

Simple staining

Once a heat-fixed or air-dried smear of a sample is made, you can add a drop of methylene blue for one minute and then wash off the dye. Dry the slide again, then add a drop of immersion oil and observe the sample using an oil immersion objective lens (Figure 17.31).

Negative staining (Indian ink)

For negative staining, you could use a stain such as eosin or nigrosin (see Figure 17.32). Indian ink is a good substitute. The negative charges on the molecules of the stain will not penetrate the cells of some microorganisms (e.g., bacteria), because these cells have a negative charge on their surface. There is no need to heat fix the cells, so there is no risk of any distortion to their shape. The microorganisms will show up as unstained against a dark background. This method is ideal for finding the true shape and size of bacterial cells. To perform negative staining of bacteria, place a loopful of

a. Passing the slide through a Bunsen flame

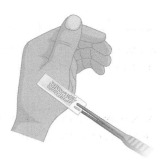

b. Testing to ensure that the slide is not overheated

▶ **Figure 17.30** Use of a Bunsen flame to dry and heat fix a slide

bacteria (from a colony or from a liquid culture) in a spot of Indian ink, positioned at one end of a microscope slide. Use another microscope slide to spread the mixture out and form a thin smear. Allow the slide to dry in the air and then place a drop of immersion oil onto the dried slide. Then you can examine the bacteria using oil immersion.

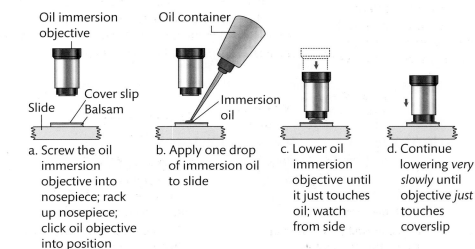

a. Screw the oil immersion objective into nosepiece; rack up nosepiece; click oil objective into position

b. Apply one drop of immersion oil to slide

c. Lower oil immersion objective until it just touches oil; watch from side

d. Continue lowering *very slowly* until objective *just* touches coverslip

e. Look down microscope and *slowly raise* objective using fine focus until specimen comes into focus

▶ **Figure 17.31** Using an oil immersion objective lens, which is spring-loaded and has six lenses inside it

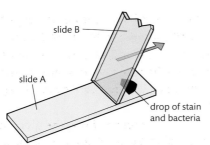

slide B

slide A

drop of stain
and bacteria

(a) place the end of slide B on the surface of
slide A and pull its slowly towards the drop

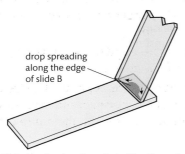

drop spreading
along the edge
of slide B

(b) when the drop is contacted it will run along
the whole width of slide B

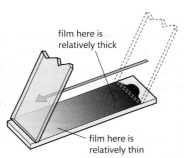

film here is
relatively thick

film here is
relatively thin

(c) to make the film: push slide B quite quickly
along slide A dragging the drop behind; do
this once only; do not attempt to push the
drop as this will result in too thin a film

▶ **Figure 17.32** How to carry out negative staining

Gram staining

Gram stain is the most widely used stain, because Gram staining is one of the first stages performed to identify a bacterium. Gram stain divides bacterial cells into two major groups: Gram-positive or Gram-negative (see 'Gram staining', under 'Classification').

Gram stain is an example of a differential stain. To perform Gram staining, apply four different chemical reagents in sequence to a heat-fixed smear of bacteria.

The first reagent, crystal violet, is called the primary stain; this colours all the cells purple.

The second reagent, Gram's iodine, is a mordant; this intensifies the purple colour. In Gram-positive cells, the crystal violet–Gram's iodine complex binds to the magnesium–RNA content of the thick bacterial cell walls (made of peptidoglycan), forming a complex that is difficult to remove.

The third reagent, 95% ethanol, is a decolouriser. It removes colour from Gram-negative cells that do not have the same structure in their cell walls (they have thin peptidoglycan walls) and do not bind to the crystal violet–Gram's iodine complex. The Gram-negative bacterial cell walls also contain more lipids (in the inner and outer membranes), which the ethanol dissolves; this leads to large pores, through which the crystal violet dye leaks out.

The fourth reagent, safranin, is a counterstain; this adds a different colour (from the crystal violet purple), pink, to the decolourised Gram-negative cells (see Figure 17.14).

⏸ PAUSE POINT   Explain the difference between a simple stain, negative stain and a differential stain.

(Hint)   Gram stain is an example of a differential stain.

(Extend)   Discuss the use of cover slips, cavity slides and oil immersion lenses.

Setting up and using a compound light microscope

Component parts of the light microscope and their functions

Figure 17.26 shows the parts of an optical microscope and how to use it.

Using the microscope

Condenser

Below the stage is a condenser. The function of this is to condense the light onto the specimen, making the specimen brighter and the image clearer. There are two lenses in the condenser and the condenser is properly adjusted when light passing through it is focussed on the specimen.

To adjust the condenser:

▶ set up the microscope for low power work

▶ place a pencil against the surface of the sub-stage lamp bulb

▶ look down the microscope and move the pencil slowly over the lamp surface until it is seen as a blur

▶ while still looking down the microscope, feel beneath the stage for the condenser control and use it to move the condenser up and down slightly, until the pencil tip comes sharply into focus. Remove the pencil.

Iris diaphragm

The iris diaphragm adjusts the amount of light that passes through the specimen. To adjust the iris diaphragm:

▶ make sure the microscope is set up for low power work and the condenser has been adjusted

▶ remove the eyepiece and place it on the bench; look down the ocular tube, where you should see a bright disc of light (this is the back lens of the low power objective)

▶ while still looking down the tube, feel for the iris control; open and close it a few times to see its effect. Set the iris diaphragm in a position where the back lens of the low power objective is two-thirds filled with light.

If you are examining a very transparent specimen and the image is too bright, you can adjust the iris diaphragm to reduce the amount of light passing through it.

Light source

Some microscopes have a built in light source. Others have a mirror that can be adjusted, to shine light from a lamp that is directed onto the mirror, up through the specimen.

Stage

The stage is where the slide is placed. Position the slide so that the specimen is over the hole on the stage and then clip the slide into place. Once you have finished examining the slide, remove it from the microscope and place it in a suitable receptacle – for example, a pot of bleach or disinfectant.

Use of objectives (lowest power lens and other lens magnifications), coarse focus and fine focus

Use the objectives (the different magnification lenses) and focus knobs (coarse and fine) to make a clear image of your specimen. Make sure the slide is clipped in place on the stage and then:

▶ rotate the nosepiece until the lowest power objective lens clicks into place (always start using the lowest power objective lens)

▶ use the coarse focus knob to lower the objective lens as low as it will go, without touching the slide

▶ look down the eyepiece and turn the coarse focus slowly to move the objective lens *away* from the slide, until the specimen comes into view (the image may not be clear)

▶ use the fine focus knob and bring the specimen into sharp focus. Adjust the iris diaphragm if the image is too bright or dim

▶ move the slide gradually until a part of the specimen you wish to view under higher power is in the centre of the field of vision

▶ rotate the nosepiece and bring the next higher magnification objective lens into place

▶ look down the eyepiece, adjust the fine focus to make a clear image of the specimen and make sure that the part you wish to view is still in the centre of the field of vision

▶ repeat with the next higher magnification objective lens, until you reach the magnification you desire.

Oil immersion lens

Oil immersion lenses are placed very close to the slide, in a drop of Canada balsam oil. They are spring-loaded, to avoid damage to the lens or to the slide. Oil immersion lenses are usually used to visualise heat-fixed and stained smears (without a cover slip). To use an oil immersion lens:

▶ make sure that the slide is dry and place one drop of immersion oil on the specimen

▶ place the slide on the stage and rotate the nosepiece so that the oil immersion lens is in place

▶ without looking down the eyepiece, adjust the focus knob and move the oil immersion objective down, as close to the slide as possible, so that it is in the oil

▶ look down the eyepiece and adjust the fine focus knob to move the lens slowly away from the slide, but keep it in the oil, until the specimen comes into view and a clear image is visible.

Care of microscopes

When you have finished using the microscope:

▶ make sure that you clean the microscope stage, using fine tissue

▶ using lens tissue moistened with a drop of xylol, clean the objective lenses and then polish them with a fresh piece of clean lens tissue. Make sure all grease, immersion oil and dust is removed

▶ using lens tissue, clean the eyepiece objectives so they are free of grease, dust and fingerprints

- cover the microscope, or place it in its box
- if the microscope needs to be moved, using one hand under the foot/base and the other hand to grip the arm, carefully carry the microscope back to its place of storage.

Calculating the actual size of specimens under the microscope

You can measure the actual size of a specimen under the microscope if the eyepiece objective has a graticule in it, which has been calibrated using a stage micrometer. You can also calculate the actual sizes of specimens from a photomicrograph or electron micrograph, provided you know the magnification.

All microscopic measurements are made in micrometres (μm). One micrometre (μm) is one thousandth of a millimetre (mm).

A small acetate eyepiece graticule is an inexpensive scale that can be inserted into the eyepiece tube of a light microscope (Figure 17.33). When you view an object, the scale of 100 divisions is superimposed on it. However, the scale is arbitrary, as it represents different lengths at different magnifications; that is, at different magnifications, the number of scale divisions covering a specimen varies. The eyepiece graticule has to be calibrated, using a stage micrometer. A stage micrometer (Figure 17.34) looks like an ordinary microscope slide, but a small scale of 1 mm divided into 100 divisions is etched on it. Each division is therefore 10 μm.

To calibrate an eyepiece graticule, follow the steps below.

- With the eyepiece graticule in place, clip the stage micrometer in position and use low power to focus on the scale on the stage micrometer.
- Align the eyepiece graticule (you can swivel the eyepiece lens) and the stage micrometer, as shown in Figure 17.35 (a). In the example in Figure 17.35 (a), you can see that 1 mm (1000 μm) of the stage micrometer corresponds to 40 eyepiece divisions (epd) on the eyepiece graticule.

$$\text{Therefore, 1 epd} = \frac{1000}{40} = 25 \ \mu\text{m}$$

- Using the next higher magnification objective lens, focus on the stage micrometer and align the scales, as shown in Figure 17.35 (b). In the example shown in Figure 17.35 (b), 100 epd now correspond to 1 mm (1000 μm).

$$\text{Therefore, 1 epd} = \frac{1000}{100} = 10 \ \mu\text{m}$$

- Continue and calibrate each objective lens, including the oil immersion lens. Table 17.7 shows the values of eyepiece divisions at different magnifications for most modern microscopes.

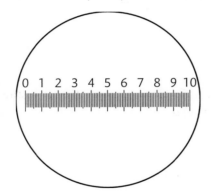

▶ **Figure 17.33** An eyepiece graticule

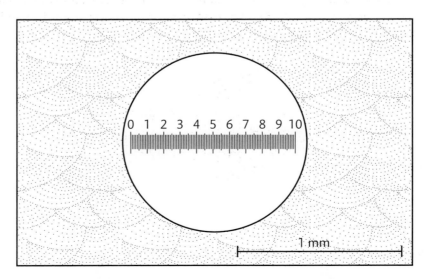

▶ **Figure 17.34** A stage micrometer

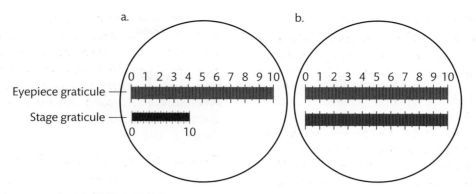

▶ **Figure 17.35** Image of an eyepiece graticule in line with a stage micrometer at (a) ×40 magnification and (b) ×100 magnification

▶ **Table 17.7** The values of eyepiece divisions at different magnifications for most modern microscopes used in schools

Magnification of eyepiece lens	Magnification of objective lens	Total magnification	Value of one eyepiece division (μm)
×10	×4	×40	25.0
×10	×10	×100	10.0
×10	×40	×400	2.5
×10	×100	×1000	1.0

Worked example

Figure 17.36 shows the image of an amoeba (a protist) seen under low magnification of an optical microscope.

Under low power, total magnification ×40, the specimen has a length of 28 epd.

At this magnification 1 epd = 25 μm.

Therefore, the true length of the amoeba is 28 × 25 = 700 μm.

At a total magnification of ×100, the same specimen would have a length of 70 epd.

Therefore, the true length of the amoeba is 70 × 10 = 700 μm.

Try this:

A large spirochete bacterium, *Borrelia burgdorferi*, which causes Lyme disease, measures 15 epd when viewed under oil immersion at a total magnification of ×1000. Another specimen of the same species of bacteria measures 30 epd at the same magnification. What are the true lengths of these specimens?

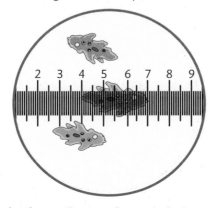

▶ **Figure 17.36** An amoeba seen under low (×40) magnification of an optical microscope

Calculating the actual size of specimens from micrographs

Photomicrographs or electron micrographs are photographs taken using a microscope. Provided you know the magnification at which the photo was taken, you can calculate the true length of a specimen from the micrograph.

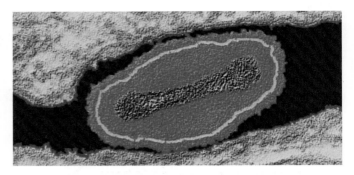

▶ **Figure 17.37** Coloured SEM of bacteria (orange) and *Vaccinia* virus particles (green) that cause cowpox.

Assessment practice 17.2

B.P2   B.P3   B.M2   AB.D1

For this assessment practice, you will need to present a portfolio of practical activities you have carried out.

1 Set up and use a light microscope with an oil immersion lens; observe and draw structures of microorganisms, such as amoeba, fungi and bacteria. Use a calibrated eyepiece graticule to measure the lengths of your specimens.

2 Make accurate drawings of the structures of the microorganisms you observe with an oil immersion lens. Use a sharp HB pencil, draw clear unbroken lines and do not shade. Label each drawing; clearly state the magnification and scale of the drawing. Calculate the true measurements of the microorganisms and some of their structures.

3 Write a short report comparing the use of light microscopes, stereomicroscopes, phase-contrast and electron microscopes to observe the structures of microorganisms.

4 Evaluate different microscopy techniques for observing structures and classifying microorganisms. Refer to ease of use, cost, training and level of detail revealed, due to resolution and magnification.

Plan
- Do I know what I am being asked to do?
- Do I have enough information or should I extend my research?

Do
- I know what I want to achieve.
- I can check my work to see where I have gone off task and can make changes to put this right.

Review
- I can instruct someone else on how to complete the task more efficiently.
- I know what I would do differently next time and the approach I would take with the parts that I found difficult this time.

C Undertake aseptic techniques to culture microorganisms

This section covers the use of aseptic techniques in the preparation of growth media and its inoculation and incubation.

Safety and prevention of contamination in microbial culturing

In microbiological laboratories there are protocols and procedures to safeguard the staff and make sure that samples and specimens are not contaminated. Specialised equipment is also used.

Classifications of biosafety (levels 1–4)

A biosafety level (BSL) is a set of containment precautions used to isolate dangerous biological agents (microorganisms) in laboratories. The lowest level is BSL-1 and the highest is BSL-4. You need to be aware of these BSLs when working with microorganisms. Table 17.8 outlines the safety protocols for BSLs 1–4.

Personal protective equipment

You should wear basic protective clothing, which may include lab coats, disposable aprons and eye protection (goggles). Gloves can be worn and nitrile gloves avoid

▶ **Table 17.8** Safety protocols of the biosafety levels and examples of their use

BSL	Risk levels	Example microorganisms	Typical lab and safety procedures to follow
1	Low risk to individuals and communities.	*B. subtilis, Staphylococcus albus, Saccharomyces cerevisiae* and non-pathogenic *E. coli* (Any organisms that do not cause disease in healthy humans.)	This is the type of lab found in schools and colleges. Hand washing on entry to lab and before exit; attention to personal hygiene; no eating or drinking or mouth pipetting; no requirement for containment cabinets; work can be carried out on open benches; lab has a lockable door, but this is not usually locked; people carrying out investigations observe aseptic techniques and wear some protective clothing, such as disposable aprons; all potentially infectious material has to be decontaminated before disposal.
2	Moderate risk to individuals and low risk to communities.	*Salmonella* spp., *Staphylococcus aureus,* hepatitis B and C viruses, adenoviruses, HIV, pathogenic strains of *E. coli, Plasmodium falciparum, Toxoplasma gondii*	Pathology and research labs. Procedures of BSL1 *plus*: personnel need more training for handling pathogens, given by a senior qualified and competent scientist; limited access to the lab while work is in progress; extreme precautions with contaminated sharp items; safety cabinets used if aerosols will be generated.
3	High risk to individuals and moderate risk to communities.	*Mycobacterium tuberculosis*, yellow fever virus, *Yersinia pestis* (cause of bubonic plague), SARS, *Chlamydia*, rabies virus	Pathology and research labs dealing with various bacteria, parasites and viruses that can cause serious or potentially lethal diseases, but for which there are treatments. Procedures of BSL 1 and 2 *plus*: specific training for lab personnel for handling pathogens; all procedures for handling microorganisms are carried out in containment cabinets; personnel wear protective clothing and may be required to remove make up and jewellery.
4	High risk to individuals and communities.	Ebola virus, smallpox virus, Herpes B virus, Marburg virus, Lassa virus, *Clostridium botulinum*	Public Health labs and some medical research labs. Procedures of BSL 1–3 *plus*: high levels of security for access – via director; double-door entry with an airlock; airflow systems and negative pressure in the lab, so air always enters the lab and does not flow outwards; air is filtered; multiple containment rooms; lab separate from other buildings; pressurised personnel suits; operators manipulate cultures of pathogens by putting their hands inside special gloves inside sealed cabinets; established protocols for all procedures; extensive training for personnel; personnel may be required to shower before entry and before leaving; own clothes not worn in lab; protective clothing decontaminated.

problems with those who are allergic to latex. However, hand washing should be sufficient in school and college labs, and sometimes wearing gloves can be dangerous. *Never* wear disposable polythene gloves, as they reduce manual dexterity and if they brush against a Bunsen flame, they will burn and melt onto your hands, causing serious burns.

Safety procedures to follow in a biosafety level 1 microbiology practical

Your school or college lab is a biosafety level 1 area. You must always wash your hands before and at the end of the practical session and observe the other procedures detailed in Table 17.8. You must observe aseptic techniques.

Step-by-step: Aseptic techniques

25 Steps

Using good aseptic techniques prevents contamination of the microbiological sample. This is called positive containment. The following steps detail good aseptic techniques.

1 If you have a heavy cold or a stomach upset check with your tutor before you begin a microbiology practical.

2 Cover any cuts or grazes with a clean waterproof dressing.

3 Do not eat or drink anything in the lab; do not do any hand to mouth operations, such as chewing pen tops; no mouth pipetting, always use mechanical pipettes and micropipettes.

4 Always wear a clean disposable plastic apron; this protects your clothing and reduces the risk of contamination of the microorganisms you are growing.

5 If possible, close windows and doors to prevent airborne contamination.

6 Wash your hands with warm water and antiseptic soap before you start. Dry them with a paper towel and place this towel in the disposal bag provided.

7 Make sure benches have smooth surfaces, so that bacteria cannot collect in crevices (you could cover any benches with cracked surfaces using a clean plastic sheet).

8 Spray your bench area with disinfectant, leave it to work for 10–15 minutes and then wipe with a paper towel. Place the paper towel in the disposal bag.

9 Work near a lit Bunsen burner (hot flame). You will need this to flame loops, but the updraft will also prevent airborne bacteria from falling onto the plates you want to inoculate. Place the Bunsen flame on yellow when not in use, as this is cooler and more visible.

10 Sterilise your equipment before use. Hold wire loops in the hottest part of a blue Bunsen flame, until they glow red hot. Allow them to cool in the air, dip them in methanol and then quickly flame the loops again to burn off the methanol.

11 Sterilise glass instruments, such as spreaders, by dipping them in 70% methanol and then flaming them lightly.

12 Always keep the methanol in a small beaker, well away from the Bunsen flame.

13 Never place bungs or caps from flasks of bacterial cultures on the bench; hold them in the crook of your little finger.

▼

14 *Before* and *after* you use a (sterilised) loop or pipette to withdraw a sample from a culture tube, pass the neck of the culture tube through the Bunsen flame.

▼

15 Before you inoculate plates or tubes label them with your name, date and the culture. Label Petri dishes on the bottom. Use a permanent marker pen. Do not lick labels.

▼

16 Always tape a Petri dish lid to the Petri dish in two places, with sticky tape. Do not seal the dish all the way round, as this excludes air and could encourage the growth of anaerobic contaminants.

▼

17 Incubate the plates upside down, so that any condensation runs onto the lids and not onto the microorganisms.

▼

18 Report any spillages immediately.

▼

19 Incubate Petri dishes at no more than 30°C, to avoid encouraging the growth of any pathogenic contaminants.

▼

20 When examining Petri dishes, do not remove the lids. In some cases a bactericidal agent can be placed within the lid of the Petri dish 24 hours before you have access to it. In this case, you may, under instruction from your tutor, take off the lid and obtain a colony of bacteria for staining.

▼

21 Place all used instruments in a pot of bleach or disinfectant after use. Place used Petri dishes in the disposal bag provided. This bag will be properly disposed of by the technicians.

▼

22 Swab your bench as you did before the session.

▼

23 Place your apron in the disposal bag.

▼

24 Wash your hands with warm water and antibacterial soap and dry with a paper towel. Place the towel in the disposal bag.

▼

25 Technicians will autoclave used Petri dishes, aprons and towels before sealing them into a bag for safe disposal.

Biosafety cabinets

Where potentially dangerous microorganisms are being handled in a laboratory, they need to be contained. This ensures that personnel are not accidentally infected and prevents accidental release of pathogens into the environment. Biosafety cabinets (BSCs; also called laminar flow cabinets) are enclosed spaces made of stainless steel with no gaps or joints where microorganisms could collect. The surface is very smooth, also preventing the harbouring of microorganisms. BSCs operate at negative pressure, so that a smooth (laminar) flow of air is constantly entering the cabinet from the outside, where the air pressure is greater, down a pressure gradient. The air from the cabinet is then filtered so no microorganisms can escape (they are stopped by the filter). BSCs may also have an ultra violet lamp and this light can be switched on to kill any microorganisms within the BSC. The lamp has to be switched off when the cabinet is in use.

> **Discussion**
>
> Why do you think the UV lamp in a laminar flow cabinet has to be switched off when the cabinet is in use?

Classes of BSCs

There are three classes of BSCs, which are defined based on the level of protection they give to personnel and the environment, and the level of protection of the product.

▶ **Class I:** These BSCs provide personnel and environment protection but no product protection. They are commonly used to house centrifuges or procedures, such as aeration of cultures, which could generate aerosols. Aerosols, containing microorganisms, travel upwards and then rain down, contaminating anything below them.

▶ **Class II:** These BSCs protect personnel, environment and products. A fan on top of the cabinet draws sterile air over the products being handled. This air is then filtered before leaving the cabinet.

▶ **Class III:** These BSCs are sometimes called glove boxes. They are used in maximum containment labs and give maximum protection, where BSL-4 pathogens are being handled. The enclosure is airtight and all material leaves through a double-door autoclave (Figure 17.38). Gloves attached to the front prevent direct contact with hazardous material.

Growth media

For most investigations into factors that affect microbial growth (see section D), you will use liquid culture media, called nutrient broth, or you will use nutrient agar poured into Petri dishes (nutrient agar plates).

Nutrient broths

Nutrient broth is made from beef extract that contains peptones (digested proteins). To make it, dissolve 8 g beef extract and peptone in sterile water and then top up the volume to 1 L. Portion out the broth into test tubes, stopper it with cotton wool and foil caps, and then sterilise it by heating it in an autoclave.

Nutrient agar plates

Mix 20 g Bovril (beef extract), 5 g sodium chloride (table salt) and 15 g agar powder (weighed out in a fume cupboard to avoid inhalation) with 100 mL water into a paste. Add more water to make the volume to 1 L. Sterilise, by heating in an autoclave, and then allow the mixture to cool to 50°C. Keep the liquid nutrient agar in a thermostatically controlled water bath set at 50°C, until you are ready to pour the plates (below).

Pouring the plates

Petri dishes are supplied sterilised in packs of ten. Spread the plates on the bench. Lift the lid of one plate, but keep it above the plate, while quickly pouring in enough liquid nutrient agar to cover the plate to a depth of about 5 mm. Flame the neck of the flask of liquid agar before and after

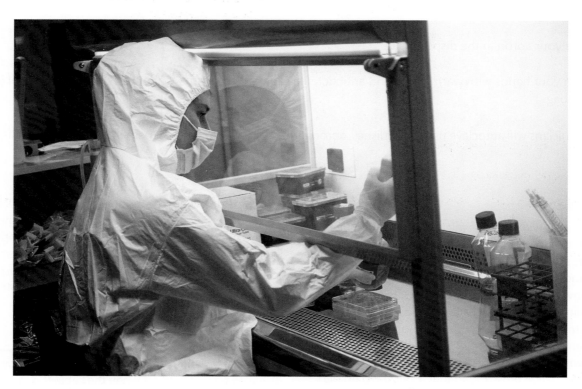

▶ **Figure 17.38** A biosafety cabinet

pouring from it (Figure 17.39). Replace the lid and gently swirl the plate to distribute the liquid agar. Allow the plate to cool and the nutrient agar to set. Repeat with all the plates. You should be able to pour about 50 plates from 1 L of medium. You can reduce the quantities pro rata, to make smaller volumes of media. These plates can be stored, upside down, in a fridge until you are ready to inoculate them with bacteria or fungi.

The same technique is used for making, sterilising and pouring plates of specialised media.

Selective media

The section 'Microorganisms in Medicine and Industry' describes the use of differential and selective media. Table 17.9 gives the recipes for some of the selective

1. Arrange sterile Petri dishes on bench. Do not open. Label on base

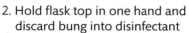

2. Hold flask top in one hand and discard bung into disinfectant

3. Flame mouth of flask

Molten nutrlent agar

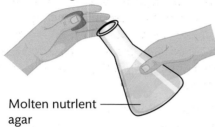

4. Remove lid of a Petri dish, mix and pour molten agar immediately

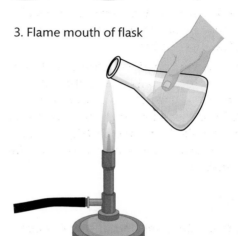

5. Replace lid on Petri dishes and allow plates to set (15 minutes approximately)

6. Discard empty flask into disinfectant

▶ **Figure 17.39** How to pour an agar plate

▶ **Table 17.9** Recipes for some selective media. Quantities are in g per L of water.

MacConkey agar (pH 7.1)		Mannitol salt agar (pH 7.4)		Blood agar (pH 7.3)		Potato dextrose agar (pH 5.6)	
Bacto peptone	17.000	Beef extract	1.000	Infusion from sheep heart	500.0	Potato infusion from 200 g	4.0
Proteose peptone	3.000			Tryptose	10.0	Dextrose	20.0
Lactose	10.000	Sodium chloride	75.000	Sodium chloride	5.0	Agar	15.0
Bile salts	1.500	D-Mannitol	10.000	Agar	15.0		
Sodium chloride	5.000	Agar	15.000				
Agar	13.500	Phenol red	0.025				
Neutral red	0.030	enzymatic digest of casein	5.000				
Crystal violet	0.001	enzymatic digest of animal tissue	5.000				

growth media. You can make these media, using aseptic techniques, and use them to inoculate with various types of bacteria (see next section).

Your school or college may have dried powder for making selective media, which you could use instead of making them from scratch using the recipes in Table 17.9. If you use the dried powder, follow the instructions and safety considerations given with those powders.

Research

Go to http://www.nuffieldfoundation.org/practical-biology/making-nutrient-agars to find more recipes for selective media.

Inoculation and incubation

You will be expected to carry out inoculation of liquid and solid media, using aseptic techniques at all stages.

Step-by-step: Inoculating liquid media

8 Steps

Make sure that your tubes of liquid media to be inoculated are labelled with your name, the date and the microorganism to be added.

1 Flame the loop and allow it to cool (if you are using a disposable, sterile plastic loop, do *not* flame it).

▼

2 Take the culture tube and shake it to distribute the bacteria. Loosen the cap and then remove the cap from the culture tube, but hold it by your little finger while holding the culture tube.

▼

3 Flame the neck of the culture tube.

▼

4 Insert the sterile loop into the culture tube and obtain a loopful of culture. Hold this loop while reflaming the neck of the tube and replacing the cap. Take care, the tube will be hot.

▼

5 Pick up the tube of liquid to be inoculated. Remove the cap (hold onto it) and flame the neck of the tube.

▼

6 Insert the loopful of bacterial culture into this tube and then withdraw it.

▼

7 Flame the neck of the newly inoculated tube again and replace the cap. Place this tube in a rack.

▼

8 Sterilise your wire loop by flaming it, or dispose of the loop into a jar of disinfectant on the bench, if it is plastic.

The inoculated tubes can be incubated (see below) and the amount of growth can be measured using turbidity.

Inoculating solid media

When one bacterium lands on a suitable solid medium at a suitable pH, temperature and oxygen level, it will divide by binary fission. Each new cell will divide and grow again. After 24–48 hours, where there was one bacterium, there will be a visible colony containing several thousand million bacteria, all genetically identical to the original bacterium.

To inoculate solid media, aseptic technique should be used throughout. Follow the instructions under 'Growth media', above, to make the solid media and pour the plates. Before you start, label the plates on the base with your name, date and the bacteria to be added.

Streak plates

Figure 17.40 shows how to prepare a streak plate. This method spreads out one loopful of bacterial culture, so that the bacteria on the last streak are very spread out. This means that the colonies that form on the last streak will be distinct and easy to see.

1. Dip inoculating loop in 70% ethanol and flame until it glows red

2. Still holding loop by handle, remove lid from broth culture, holding in little finger as shown

3. Pass neck of culture container through flame

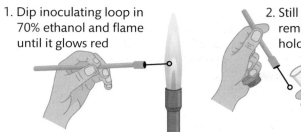

4. Dip the cool loop into the broth culture

5. Flame neck of culture container again, as in Step 3, and replace lid

6. Raise lid of Petri dish with other hand, only enough to allow loop inside. Streak surface in three parallel lines

7. Resterilise loop and streak as shown. Sterilise loop at each 'corner'

8. Seal dish with sticky tape and incubate

two pieces of tape securing lid to base

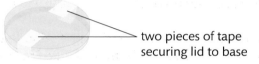

▶ **Figure 17.40** How to prepare a streak plate

Lawn plates

To prepare a lawn plate:

▶ pipette 1 mL of bacterial culture onto the middle of a nutrient agar plate. (Do *not* mouth pipette.)

▶ use a sterilised bent glass rod to spread this evenly over the jelly.

When incubated, lawn plates produce a lawn of bacteria, rather than individual colonies. This is particularly useful if you wish to investigate how effective antibiotics, or other antibacterial substances, are (see 'Growth inhibitors').

Optimising growth for examination

Storage during incubation

As already discussed in the step-by-step to 'Aseptic techniques' you should use two pieces of sticky tape to anchor the Petri dish lids into place, but do not seal the plates completely; this would exclude the air, depressing the growth of aerobic microorganisms and encouraging the growth of anaerobic contaminants. Plates are usually incubated upside down, so that any condensation runs onto the lids and not onto the growing microorganisms. If the well method (see 'Measuring microbial growth') is used to investigate the effects of antimicrobial substances, then the plates should be incubated the right way up.

Importance of incubation temperature

Some bacteria (psychrophiles) grow best at cold temperatures, between −20°C and +10°C. Thermophiles grow best at high temperatures, 41°C–122°C. Most of the bacteria you will investigate will be mesophiles; these grow best at 20°C–45°C. If you are investigating the effect of pH, antimicrobial agents or oxygen requirements, you should incubate your plates or tubes at 25°C–30°C. Unless specified by your tutor, you should not incubate plates at 37°C, as this could encourage the growth of contaminants that are pathogens.

Length of incubation

For any investigation where the independent variable is a factor *other* than length of incubation, you need to incubate all your plates for the same duration. The lower your incubation temperature, the longer it will take for visible colonies to appear on the plates, or for the liquid medium to become turbid (cloudy). A time span of 24–48 hours is usually sufficient. If your lessons are further apart than that, your tutor or technician can place your incubated plates in the fridge (to slow further growth) before you examine them. If your lessons are one day apart, incubating the plates at 30°C would be a good idea.

Assessment practice 17.3

For this assessment practice you will need to present a portfolio of practical activities you have carried out.

1 Skilfully using aseptic techniques, prepare and inoculate liquid and solid growth media and measure microbial growth.

2 Write a report explaining and comparing the biocontainment procedures used in your school/college laboratory and those used within industrial laboratories.

Plan
- Do I know what I am being asked to do?
- Do I have enough information or should I extend my research?

Do
- I know what I want to achieve.
- I can check my work to see where I have gone off task and can make changes to put this right.

Review
- I can instruct someone else on how to complete the task more efficiently.
- I know what I would do differently next time and the approach I would take with the parts that I found difficult this time.

D Explore factors controlling microbial growth in industrial, medical and domestic applications

In this section, you will find out about the growth requirements of microorganisms, how you can measure microbial growth and the effect of growth inhibitors on microorganisms.

Growth requirements

There are many factors that affect the growth of bacteria and fungi (Figure 17.41).

▶ When microorganisms are first introduced into a growth medium, they have to synthesise enzymes to utilise the nutrients in the medium. This takes time, as genes have to be switched on and transcribed (copied to mRNA) before their genetic codes can be translated into proteins like enzymes. This means the growth rate is slow at first; this is called the lag phase. Organisms are adjusting to the surrounding conditions. This may mean taking in water, cell expansion, activating genes and synthesising specific enzymes. The cells are active but not reproducing so population remains fairly constant. The length of this period depends on the growing conditions.

▶ Once they are able to utilise the nutrients in the medium, microorganisms enter an exponential growth phase, also called the log phase. The population size doubles each generation as every individual has enough space and nutrients to reproduce. In some bacteria, for example, the population can double every 20–30 minutes in these conditions. The length of this phase depends on how quickly the organisms reproduce and take up the available nutrients and space. In some species, their numbers double every 30 minutes. You would need to use log graph paper to plot the growth curve during this phase, as millions of bacterial cells are produced. The log phase lasts as long as there are plenty of nutrients, space and oxygen, and as long as levels of toxic waste do not build up enough to kill the microorganisms.

▶ When the microorganisms begin to run out of nutrients and oxygen, or start to be poisoned by their own waste products, the rate of cell division equals the rate of cell death; this is called the stationary phase. Nutrient levels decrease and waste products like carbon dioxide and other metabolites build up. Individual organisms die at the same rate at which new individuals are being produced. *Note:* in an open system, this would be the carrying capacity of the environment.

▶ When more cells die than those being produced, the microorganisms are in the decline phase or death phase. Nutrient exhaustion and increased levels of toxic waste products and metabolites lead to the death rate increasing above the reproduction rate. Eventually, all organisms will die in a closed system.

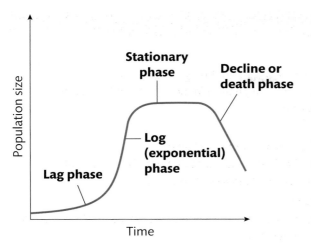

▶ **Figure 17.41** Growth curve for a population of microorganisms

Nutrients

All microorganisms require certain basic nutrients. They all need a source of carbon, as all organic molecules contain carbon. The bacteria and fungi you will culture will not be photosynthetic. Therefore, they will need organic sources of carbon, such as glucose, starch, fats or proteins. Microorganisms also need a source of nitrogen, in the form of protein, to make amino acids (proteins) and nucleic acids. They need sulfur and phosphorus, in the form of sulfates and phosphates. Nutrient agars and broth also contain traces of metallic elements, such as calcium, zinc, sodium, potassium, manganese, magnesium and iron, to aid enzyme function and other aspects of metabolism. Tiny amounts of vitamins are also added, as these organic substances are sources of coenzymes that aid enzyme action.

Water

All living cells need water, which enables low molecular mass nutrients to cross cell membranes. A large part of the cytoplasm of cells is water.

Light and temperature

Some bacteria and some protists are photosynthetic, which means they need light as an energy source and can use inorganic carbon dioxide as the source of carbon. Most bacteria and fungi that you will investigate use organic molecules to supply both their nutrition and energy and do not need light.

Most bacteria and fungi are mesophiles and grow best at temperatures of 20°C–45°C. Some psychrophilic bacteria grow best at low temperatures (–20°C to 10°C) and thermophiles grow best at high temperatures (41°C–122°C).

Oxygen requirements

The bacteria you will grow are likely to be aerobic. Yeast is a facultative anaerobe and can grow with or without oxygen.

> **Link**
>
> See the section on 'Oxygen requirements', under 'Classification', for more about the oxygen requirements of microorganisms.

pH levels

The pH of the extracellular environment affects the activity of the enzymes used by microorganisms. Enzymes work within a narrow range of pH; any deviation in pH can slow their activity. That is, both lowering the hydrogen ion concentration (increasing the pH) and raising the hydrogen ion concentration (lowering the pH) interferes with the hydrogen bonds that hold the tertiary structures (3D shapes) of the enzyme molecules. If the 3D shapes of the active sites in enzymes change, then they are no longer complementary to the shape of the substrate molecules. This slows the reactions that the enzymes catalyse. At extremes of pH, many enzymes are denatured and their active sites do not fit the shape of the substrate molecules at all; this means no metabolic reactions take place and the microorganisms cannot divide and grow. Most of the microorganisms you will investigate have enzymes that work best at a neutral pH (pH 7). However, some bacteria can live in hot sulfur springs, where the pH is equivalent to that of sulfuric acid.

Growth surfaces

Bacteria can exist in a planktonic, free-living form. They can also be attached (adhered), by their pili, to a surface or within the confines of a biofilm. In a biofilm, bacterial cells are immobilised in an organic polymer matrix, which they secrete, on a solid surface. Biofilms of bacteria form in pipes, on the rubber seals of washing machines, on teeth that are not cleaned, in catheters, drips and sometimes on implants, such as pacemaker leads and artificial hip joints. The matrix protects the bacteria from the action of antibiotics and disinfectants. The chemicals produced by the bacteria can signal to one another and also up or down regulate gene expression. Biofilms are therefore problematic.

> **Research**
>
> Go to http://www.sciencedirect.com/science/article/pii/S1002007108002049 to find out more about biofilms.

Growth inhibitors

There are many substances and procedures that can inhibit the growth of microorganisms. Table 17.10 shows some techniques that are used to control the growth of microorganisms.

▶ **Table 17.10** Techniques used to control the growth of microorganisms and their features

Control technique	Features
Disinfectants	Antimicrobial agents that can be applied to non-living objects, such as surfaces and instruments, to kill microorganisms on those surfaces. Examples: phenol, bleach, alcohol, formaldehyde, hydrogen peroxide, iodine solution, ozone gas, copper.
Antiseptics	Antimicrobial agents that can be applied to skin to reduce the risk of infection and sepsis. Antibacterials may be **bacteriostatic** (prevent bacteria from multiplying) or **bactericidal** (kill bacteria). Examples: TCP, hydrogen peroxide, ethanol, boric acid, tincture of iodine, phenol, salt water.
Antibiotics	Chemicals produced by microorganisms that can kill or inhibit the growth of other microorganisms; mainly bacteria but also some fungi and protozoans. Examples: penicillin, tetracycline, amphotericin, voriconazole, bacitracin.
Antivirals	Chemicals that either prevent viruses from entering host cells (e.g., Tamiflu®, antibodies), prevent viruses from uncoating (e.g., amantadine) or inhibit the replication of viruses inside host cells (e.g., interferon, acyclovir, AZT).
Virucides	Chemicals that can kill viruses on surfaces, such as door handles and hospital surfaces. Examples: copper, Virkon, bleach, Lysol, zidovudine.
Sterilisation procedures (heat and pressure)	At increased pressure, steam generated by heating water reaches temperatures of 115°C–120°C and can kill viruses and bacteria. Examples: pressure cookers, autoclaves.
Irradiation	Gamma rays can irradiate equipment such as Petri dishes and kill any bacteria or viruses. UV light can also be used to kill bacteria.
Antimicrobial soaps	Solid and liquid soaps reduce the surface tension and damage the cell surface membranes of microorganisms.
Controlled atmospheres for food preparation	Storing fruits and vegetables in high levels of carbon dioxide and nitrogen, low levels (8%) of oxygen and cool temperatures delays ripening during transportation. Low oxygen levels are maintained to inhibit the growth of anaerobic bacteria, which may cause food spoilage. The fruits and vegetables are subjected to a burst of ethene after transport, to promote ripening.
Adjusting the osmotic potential of food	Cells lacking water cannot carry out metabolism, as the enzyme-controlled chemical reactions need to be in solution; these cells cannot divide. Salted vegetables are preserved, because the low water potential of the salt solution causes water to leave, by osmosis, from the cells of microorganisms. Adding sugar to jams and marmalades also reduces the water potential, causing water to leave the cells of microorganisms by osmosis.
Curing food	Adding nitrates and nitrites, as well as sodium chloride, cures meat, inhibits the growth of bacteria and changes the flavour of the food.
Smoking food	When fish and meat are hung in wood smoke, the chemicals from the smoke inhibit the growth of microorganisms. They also change the flavour of the food.
Fermentation of food	Cheese, wines, beers and yoghurt contain specific microorganisms; these microorganisms inhibit the growth of other microorganisms that would cause food spoilage.

▶ **Table 17.10** *Continued*

Control technique	Features
Drying of food	Removal of water prevents microorganisms from reproducing. It may alter the texture and flavour of food; examples include dried fruit and biltong.
Altering pH of food	Pickling foods in vinegar lowers their pH, inhibiting the growth of microorganisms as their enzymes and other proteins are denatured. The flavour of the food may be altered, but often this is deemed desirable; examples include pickled onions, gherkins, beetroot, chutneys and kimchi. Sometimes, sodium hydroxide is used to raise the pH and preserve food; for example, century eggs.
Changing the storage temperature for food	With refrigeration at low temperatures (4°C or below), most bacteria (except psychrophiles) reproduce slowly, so this prolongs the storage time of foods for several days. Freezing at very low temperatures, e.g. −19°C, slows microbial reproduction, enabling food to be stored for up to 2 or 3 years.
Pasteurisation of food	In this process, liquids, such as milk, beer and wine, are preserved. Milk is heated to 70°C for 15–30 seconds to kill harmful bacteria, cooled to 10°C to prevent any remaining bacteria from growing and placed in sterilised containers before storage at 4°C.
Canning of food	Food is cooked and sealed in sterilised containers that are then heated to kill any remaining microorganisms. Examples: baked beans, chopped tomatoes.
Excluding oxygen from food	Excluding air deprives any microorganisms of oxygen, preventing them from respiring. Methods include vacuum packing, jellying (sealing the food with a layer of jelly (protein)) and potting (sealing the food under a layer of butter).

Ⅱ PAUSE POINT   Distinguish between antiseptics and disinfectants.

> Hint   Explain differences and similarities and refer to specific examples.
>
> Extend   Discuss the methods for preventing food spoilage by microorganisms.

Measuring microbial growth

There are various ways that microbial growth can be measured. You can design and carry out a practical investigation into some factors that affect the growth of microorganisms; for example, their growth requirements or inhibitors. You will need to use aseptic techniques throughout.

To investigate growth requirements, you could inoculate nutrient broth with a bacterium and adjust the pH, or incubate samples at different temperatures. To investigate oxygen requirements, you could inoculate molten nutrient agar as shown in Figure 17.42; once inoculated, allow the medium to cool and then incubate it.

You may use liquid media to investigate some growth inhibitors. For example, you could expose a bacterial culture to heat or an antibacterial chemical (for a specified length of time), then transfer a loopful of the culture into nutrient broth and incubate it at 30°C for 24 hours, before measuring the turbidity.

You may use *solid media*, such as nutrient agar plates, to investigate growth inhibitors. You can spread a lawn of bacteria (as described in 'Lawn plates', under 'Inoculation and incubation') and then impregnate paper discs with a chemical or antibiotic, before placing them onto the agar, too. After incubation at 30°C for 24–48 hours, you should see the zones of inhibition around each disc. A slightly different version is to use a cork borer and make a well in the agar, after spreading a lawn of bacteria. Then, add a specified volume of an antibacterial into the well. Incubate these well plates the *right way up* and, after 24–48 hours at 30°C, you can measure the zones of inhibition. Apart from disinfectants and antiseptics, you could investigate the antibacterial effects of garlic extract, onion extract, lemon juice, tea tree oil and honey.

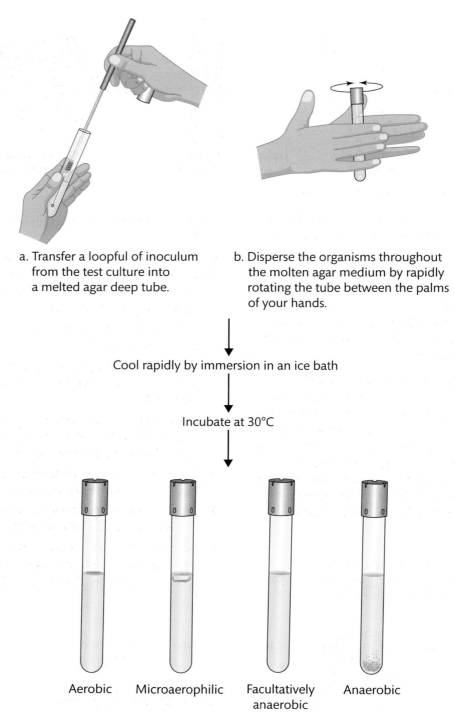

a. Transfer a loopful of inoculum from the test culture into a melted agar deep tube.

b. Disperse the organisms throughout the molten agar medium by rapidly rotating the tube between the palms of your hands.

Cool rapidly by immersion in an ice bath

Incubate at 30°C

Aerobic   Microaerophilic   Facultatively anaerobic   Anaerobic

c. Distribution of growth in nutrient agar shake cultures for the determination of an organism's oxygen requirements.

▶ **Figure 17.42** Inoculating molten media

Colorimetry

The amount of growth in liquid media can be measured using a colorimeter, which measures turbidity (Figure 17.43). A colorimeter shines a beam of light through a sample that is placed in a special plastic container, called a cuvette. A photoelectric cell picks up the light that has passed through the sample and tells you how much light has been absorbed.

To measure turbidity using a colorimeter:

▶ first, you need to calibrate the colorimeter by filling a cuvette with liquid medium that has not been inoculated. Place the cuvette in the special chamber of

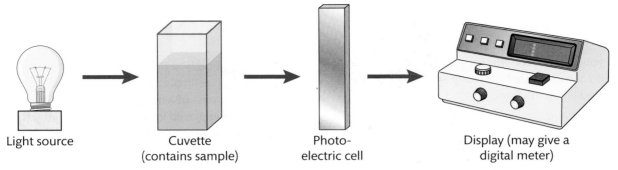

▶ **Figure 17.43** Using a colorimeter to measure turbidity

the colorimeter and use a blue or green filter. As light shines through the sample, set the reading to zero; this sample is clear (no microbial growth) so there are no microbes absorbing the light.

▶ fill another cuvette with a subsample of your cloudy medium that has yeast cells or bacteria growing in it. Place this cuvette in the colorimeter and use the same filter as before. Measure the absorption. The greater the absorption, the greater the growth.

The colorimeter cannot distinguish between particles in the culture medium and cells, which is why you need to calibrate the colorimeter first. It also cannot differentiate between living and dead cells, but if the microorganisms are in their log (exponential) growth phase (as they should be after only 24–48 hours incubation), this should not be a problem. It is a quick and easy method of measurement. You could sample the culture flask at timed intervals; the rate of increase in absorption indicates the growth rate.

Haemocytometer

Yeast cells are large enough to be seen with a light microscope, so they can be counted using a haemocytometer (Figure 17.44). This is a special slide

that has a grid etched into its middle section, below the surface of the slide.

When a special cover slip is placed firmly on the slide, it forms a chamber of depth 0.1 mm, so you can calculate the volume of liquid over each etched square and make a *total cell count* using the steps below.

▶ Shake the tube of liquid medium in which yeast cells have been growing.

▶ Take 1 mL of this liquid and add it to 9 mL of sterile water in another test tube. This dilutes the yeast medium by ×10.

▶ Mix the diluted yeast suspension well and, using a pipette, allow some of it to trickle into the grooves under the haemocytometer cover slip.

▶ Observe the haemocytometer grid under the microscope, using low power first.

▶ Focus on the central grid area where there are 25 squares, each sub-divided into 16 smaller squares.

▶ Count the yeast cells in five of the 25 squares (80 small squares): count the central square and each corner square. The volume of liquid over 80 small squares is 0.02 μL.

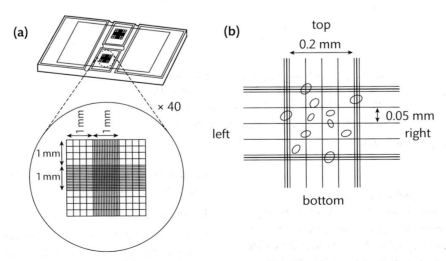

▶ **Figure 17.44** (a) A haemocytometer slide and (b) using a haemocytometer slide to make a total cell count

- If any cells are on the boundaries, only count those on the left and top boundaries and not those on the bottom and right boundaries.

- Now you know how many (*n*) yeast cells are in 0.02 µL and you can calculate how many are in 1 mL of the undiluted liquid medium:

  Number of cells in 1 µL = $\dfrac{n}{0.02}$, which means that in 1 mL of diluted suspension there is $\dfrac{n}{0.02} \times 1000$ and in 1 mL of undiluted suspension there is $\dfrac{n}{0.02} \times 1000 \times 10$.

This method gives the *total cell count* as it counts all cells – alive or dead. It is also very time consuming.

If you want to differentiate between alive and dead cells, a blue dye can be added; living cells actively transport the dye out of them, so if only unstained cells are counted, you can calculate the *total viable cell count*.

If you sample a culture of yeast cells at timed intervals, you could also estimate the growth rate.

Mycelial discs

Many fungi consist of thread-like hyphae. As they grow, the hyphae produce a sort of mat called a mycelium. To measure growth:

- using aseptic techniques, you can cut a disc from a fungal mycelium and add it to some sterile liquid medium (Figure 17.45)

- incubate it for a week at 25°C and at the end of that time you can measure the increase in diameter of the mycelial disc.

Another method is to dry and weigh the mycelium after incubation, to estimate the dry mass. You can compare this with the dry mass of a mycelial disc that is the same size as the one you placed in the nutrient medium, to see how much growth has occurred.

Serial dilution and colony counts

Serial dilution and colony counts give a viable cell count of bacteria; this is an estimation of the number of living cells per mL. If this is done at specified time intervals, this method can also be used to measure the growth rate of a particular species of bacterium.

Transfer a small volume of a bacterial culture, using aseptic techniques, to nutrient agar and spread it out. Where each bacterium in the culture lands on the nutrient agar it will divide by binary fission to give 2, 4, 8, 16 cells, and so on, until it forms a visible colony of bacteria. The colonies can be counted and each one represents a single bacterium.

The number of colonies indicates the number of bacteria in 1 mL of the liquid medium. This number may be very high, so you can carry out a serial dilution of the bacterial culture first. This gives smaller numbers of colonies that can be counted accurately. Figure 17.46 shows how to make a serial dilution.

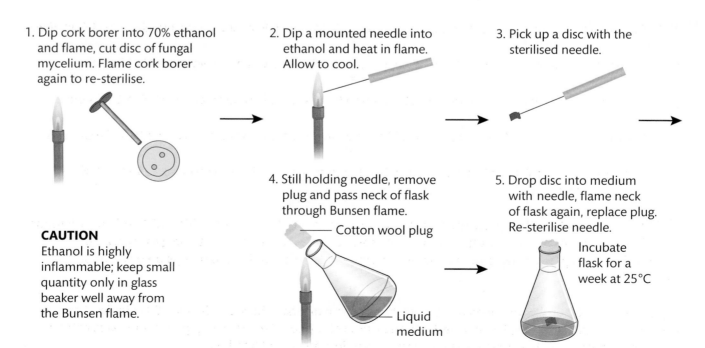

1. Dip cork borer into 70% ethanol and flame, cut disc of fungal mycelium. Flame cork borer again to re-sterilise.

CAUTION
Ethanol is highly inflammable; keep small quantity only in glass beaker well away from the Bunsen flame.

2. Dip a mounted needle into ethanol and heat in flame. Allow to cool.

3. Pick up a disc with the sterilised needle.

4. Still holding needle, remove plug and pass neck of flask through Bunsen flame.
Cotton wool plug
Liquid medium

5. Drop disc into medium with needle, flame neck of flask again, replace plug. Re-sterilise needle.
Incubate flask for a week at 25°C

▶ **Figure 17.45** Inoculating a flask with a mycelial disc

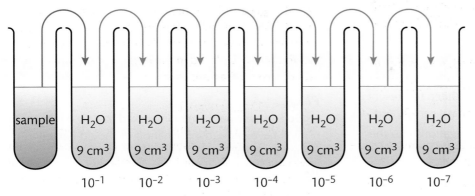

transfer → 1 cm³ pipette 1 cm³ pipette 1 cm³ pipette 1 cm³ pipette 1 cm³ pipette 1 cm³ pipette 1 cm³ pipette
with

sample H₂O 9 cm³ H₂O 9 cm³ H₂O 9 cm³ H₂O 9 cm³ H₂O 9 cm³ H₂O 9 cm³ H₂O 9 cm³

10^{-1} 10^{-2} 10^{-3} 10^{-4} 10^{-5} 10^{-6} 10^{-7}

▸ **Figure 17.46** Carrying out a serial dilution

> **Safety tip**
>
> Use aseptic techniques throughout.

Step-by-step: Serial dilution

10 Steps

1 Inoculate a flask of buffered nutrient broth, pH 7, with a loopful of bacteria, for example, *E. coli*.

▼

2 Mix the contents and, using a pipette, immediately withdraw 1 mL of the solution. Add this to a test tube containing 9 mL sterile water. This gives a ×10 dilution (dilution factor of 10^{-1}).

▼

3 Roll the tube between your palms to mix the contents. Transfer 1 mL from this tube to another tube containing 9 mL sterile water. This gives a ×100 dilution (dilution factor 10^{-2}).

▼

4 Roll the tube to mix the contents. Transfer 1 mL of the ×100 dilution to 9 mL sterile water to give a ×1000 dilution.

▼

5 Roll the tube and transfer 1 mL of the ×1000 dilution to 9 mL sterile water to give a ×10 000 dilution.

▼

6 Roll the tube and transfer 1 mL of the ×10 000 dilution to 9 mL sterile water to give a ×100 000 dilution.

▼

7 Roll the tube and transfer 1 mL of the ×100 000 dilution to 9 mL sterile water to give a ×1 000 000 dilution.

▼

8 Roll the tube and transfer 1 mL of the ×1 000 000 dilution to 9 mL sterile water to give a ×10 000 000 dilution (dilution factor 10^{-7}).

▼

9 Using a different sterile pipette each time, transfer 1 mL from the 10^{-3}, 10^{-4}, 10^{-5}, 10^{-6} and 10^{-7} dilutions onto five separate nutrient agar plates and evenly spread the bacteria using a sterile glass spreader. Anchor the lids in place, make sure the plates are properly labelled and incubate upside down at 30°C for 24 hours.

▼

10 Examine the plates (do *not* open them) and choose those that have between 20 and 200 colonies. Count the colonies. You can divide the plate into sections and use a colony counter and felt tip pen to spot each colony. Then, you can calculate how many bacteria were present in 1 mL of undiluted broth.

PAUSE POINT

Discuss the advantages and disadvantages of colorimetry, haemocytometry and serial dilution for measuring the growth of bacteria.

Hint — Think of the pros and cons of each method.

Extend — How can the growth rate of a multicellular fungus be assessed?

Assessment practice 17.4

D.P6 D.P7 D.M5 CD.D2

For this assessment practice you will need to present a portfolio of practical activities you have carried out.

1 Design, plan and carry out investigations into the growth requirements of microorganisms.

2 Explain how growth inhibitors affect microorganisms.

3 Analyse how the growth of microorganisms is affected by changing environmental factors, such as oxygen levels, temperature, pH, salinity and the presence of antimicrobial substances.

4 Evaluate the aseptic techniques you used to cultivate microorganisms in the exercises above; include reference to aseptic techniques, type of media, methods of inoculation, biocontainment procedures and methods used to estimate growth.

Plan

• Do I know what I am being asked to do?
• Do I have enough information or should I extend my research?

Do

• I know what I want to achieve.
• I can check my work to see where I have gone off task and can make changes to put this right.

Review

• I can instruct someone else on how to complete the task more efficiently.
• I know what I would do differently next time and the approach I would take with the parts that I found difficult this time.

Further reading and resources

Books

Bradshaw, L.J. (1992) *Laboratory Microbiology*, Saunders College Publishing, USA.

Cappuccino, J.G. and Sherman, N. (2016) *Microbiology: A Laboratory Manual.* Pearson.

Stearns, J.C., Surette, M.G. and Kaiser, J.C. (2014) *Microbiology for Dummies.* John Wiley & Sons.

Websites

http://microbiologyonline.org was devised by the Microbiology Society to support the teaching and learning of microbiology in schools and colleges.

https://www.prospects.ac.uk/job-profiles/microbiologist has information on jobs related to microbiology.

https://www.healthcareers.nhs.uk/explore-roles/life-sciences/Microbiology offers ideas about jobs for microbiologists in the health care sector.

THINK ▶▶FUTURE

Sadia Iqbal

Product manager – clinical microbiology

I have a degree in medical microbiology and virology, and a good knowledge of clinical microbiological techniques. I also have some experience in sales. The company I work for specialises in *in vitro* diagnostics, with a focus on infectious diseases. Our products include pre-poured microbiological media, pre-filled vials for blood culture testing, instruments for rapid bacterial detection and automated blood culture microbial detection systems. I spend about half my time in the UK and the rest travelling throughout Europe, as an ambassador for the company and to promote sales. Part of my job involves developing marketing plans and preparing the sales budget. I am also responsible for maintaining quality assurance.

Focusing your skills

I need the following skills to do my job competently:

- outgoing personality
- quick learner
- accurate and ambitious
- sound technical judgement
- analytical skills
- ability to work without supervision and pay attention to detail
- ability to coordinate, manage and complete multiple projects simultaneously
- high emotional intelligence
- strong verbal and written communication skills and ability to communicate at many levels within the organisation
- ability to speak French, German or Spanish and to travel
- strong organisational skills
- able to prioritise, multitask and meet deadlines.

Getting ready for assessment

James is studying for a BTEC National in Applied Science. He was given an assignment as part of his practical portfolio. He was asked to investigate the antibacterial properties of garlic.

How I got started

I gathered all my notes on antimicrobial agents. I used the library to find out more about the antibacterial properties of garlic. I found a reference to an article in the magazine *Scientific American* and luckily my tutor buys this magazine and keeps back copies, so we found the article. I also found some useful websites via the internet.

How I brought it all together

I decided to make an extract of garlic and then use the well method. For this, I made a lawn of bacteria on nutrient agar plates and then, using a cork borer, made a well in the centre. Then, I added a known volume of a known concentration of garlic extract. I needed to tape the lids on in two places and incubated the plates the right way up, for 24 hours at 30°C. My lessons were two days apart, but I arranged with the technician to enter the lab after 24 hours and put my incubated plates in the fridge for a day. I decided to test the garlic extract on one Gram-positive bacterium and one Gram-negative bacterium. I made several replicates for each one, so I could see if the data were reliable and could carry out a statistical test.

What I learned from the experience

Garlic seems to have antibacterial properties.

I should have spent some time researching the most appropriate statistical test to carry out and then practised this test, before using it to analyse my data. I could have extended the investigation by trying other extracts, for example, tea tree oil.

Think about it

1 Have you made a plan with timings so you can complete your assignment by the agreed submission date?

2 Have you researched all the background information about the bacteria and the chemicals you will use?

3 Have you made notes of the sources of your information so you can properly reference all sources? Have you considered all the safety aspects?

4 Are you confident about using aseptic technique and suitable inoculation techniques?

Medical Physics
Applications 21

Getting to know your unit

Modern day surgical techniques are quicker, more effective and less invasive than in the past as a result, in part, of the developments in medical diagnosis imaging techniques. The technology employed in medical diagnosis and treatment changes quickly and the methods now used are becoming safer for the patient but consequently more complex. Study of this unit will give you an insight into the principles used in hospitals for the correct diagnosis of certain illnesses and the techniques used in a variety of treatments. You will gain an understanding of non-ionising techniques such as magnetic resonance imaging, lasers, infrared thermography and ultrasound and also ionising techniques using X-rays, computerised tomography, gamma ray imaging and other methods.

How you will be assessed

This unit will be assessed using a series of internally assessed tasks within assignments set by your tutor. Throughout this unit you will find activities that will help you work towards your assessment. Simply completing these activities will not mean that you have achieved a particular grade, but you will have carried out useful research or preparation that will be relevant when it comes to completing your assignments.

As you complete the tasks in your assignments, it is important to check that you have met all of the Pass grading criteria. You can do this as you work your way through the tasks.

If you are hoping to gain a Merit or Distinction, you should also make sure that you present the information in your assignments in the manner required by the relevant assessment criterion. For example, Merit criteria require you to analyse information and to demonstrate skilful application of procedures, whilst Distinction criteria require you to evaluate your practice.

The assignments set by your tutor will consist of a number of tasks designed to meet the criteria in the table. This is likely to consist of a written report but may also include activities such as:

▶ demonstrating correct and appropriate practical techniques confirmed by observational record and/or witness statement

▶ presenting findings to your peers and reviewing the procedures and applications of your work during class discussion

▶ analysing and reviewing your own performance in a critique which highlights your strengths and weaknesses.

Assessment criteria

This table shows what you must do in order to achieve a **Pass**, **Merit** or **Distinction** grade, and where you can find activities to help you.

Pass	Merit	Distinction

Learning aim **A** Explore the principles, production, uses and benefits of non-ionising instrumentation techniques in medical applications

Pass	Merit	Distinction
A.P1 Explain how the principles and production of non-ionising radiation technologies are used in medical applications **Assessment practice 21.1**	**A.M1** Compare the principles, production and uses of different non-ionising radiation techniques in medical applications **Assessment practice 21.1**	**AB.D1** Justify the choice of non-ionising and ionising radiation techniques in medical applications **Assessment practice 21.1, 21.2**
A.P2 Explain why non-ionising radiation technologies are used for diagnosis and treatment of the human body **Assessment practice 21.1**		

Learning aim **B** Explore the principles, production, uses and benefits of ionising instrumentation techniques in medical applications

Pass	Merit	Distinction
B.P3 Explain how the principles and production of ionising radiation technologies are used in medical applications **Assessment practice 21.2**	**B.M2** Compare the principles, production and uses of different ionising radiation techniques in medical applications **Assessment practice 21.2**	
B.P4 Explain why ionising radiation technologies are used for diagnosis and treatment of the human body **Assessment practice 21.2**		

Learning aim **C** Understand health and safety, associated risks, side effects and limitations of ionising and non-ionising instrumentation techniques in medical applications

Pass	Merit	Distinction
C.P5 Explain the health and safety risks, side effects and limitations of non-ionising and ionising radiation technologies **Assessment practice 21.3**	**C.M3** Compare the health and safety risks, side effects and limitations of non-ionising and ionising radiation technologies in medical applications to maximise the protection of operators and patients **Assessment practice 21.3**	**C.D2** Discuss the consequences of poor health and safety when using non-ionising and ionising radiation technologies and the prevention and safety measures employed **Assessment practice 21.3**
C.P6 Explain how hospitals can employ health and safety measures, when using instrumentation, for the protection of operators and patients **Assessment practice 21.3**		

Getting started

The techniques used in medical physics for diagnosis and treatment are now widely used and accepted by both the medical profession and the public. They are now regarded as essential tools to combat the onset and spread of life-threatening illnesses. Your knowledge of science will enable you to explore and grasp both the principles of the complex technologies used and the precautions needed because of the potentially dangerous high energies used. Conduct a short survey of learners and staff in your school or college to determine the percentage of individuals or their family members who have received diagnosis or treatment from one of the methods outlined in this unit.

Analytical chemistry is an exact branch of chemistry which is performed using an extensive range of laboratory equipment including; glassware, digital and mechanical devices. Using a prepared worksheet and examples of the equipment used and provided by your tutor, attempt to memorise the names of the apparatus. Test your answers with a partner.

A Explore the principles, production, use and benefits of non-ionising instrumentation techniques in medical applications

Not all diagnostic methods use dangerous ionising radiation. In this section you will explore four non-ionising techniques: magnetic resonance imaging, lasers, infrared thermography and ultrasound. For each, you will learn about how they work, their main uses and their benefits and limitations. You will also cover the detail of the instrumentation and the benefits of each method to be able to form your own opinion as to their uses and convey this to others.

Magnetic resonance imaging (MRI)

The human body contains a vast number of hydrogen atoms and different tissues contain different amounts. The nucleus of a hydrogen atom consists of a single **proton** that acts like a small magnet and can be easily affected by a large magnetic field.

> **Key term**
>
> **Proton** – the positively charged particle in the nucleus of atoms.

Instrumentation/production

The MRI scanner (see Figure 21.1) is a large and heavy piece of equipment that uses the most up-to-date technologies and complex computer systems. Most of its weight comes from the large magnet and electromagnets that provide the very strong magnetic fields needed to **polarise** nuclei. These can weigh up to 100 tonnes. The MRI consists of:

▶ **Main magnet** – a large, permanent superconducting electromagnet. (A coil carrying a steady electric current generates a magnetic field running through its centre. Superconducting magnets are not technically 'permanent', though they are long-lasting. They perform the role previously filled by large permanent magnets. If there is an emergency they can be suddenly quenched by blowing off the liquid helium coolant into the outside atmosphere.) They are costly but produce a high field strength of high stability.

▶ **MRI scanner coils** – resistive electromagnets that produces a gradient field where different magnetic strengths over the body help to pinpoint the signals. When scans are being carried out, a knocking sound can be heard when the gradient **field coils** are switched on and off.

▶ **Radiofrequency coils** – produce the input radio waves that excite the nuclei and result in the **nuclear magnetic resonance (NMR)** signal. They need to sustain a power output of up to 1 kW.

▶ **Output signal receiver** – these are essentially tuning coils that pick up the output radio waves from the protons in resonance. There are many types. The type selected is based on the need to produce a specific image, such as large volume areas or more focused tissues. The receiver is linked to a powerful computer that processes the information.

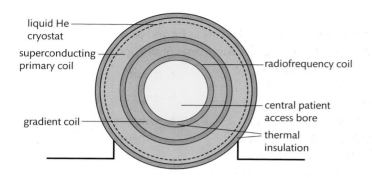

▶ **Figure 21.1** Plan view of MRI scanner

Particles such as protons, **electrons** and neutrons have a property called **spin**, which makes them behave like tiny magnets when they are placed in a magnetic field and rotate with a wobble motion. They show a north and south pole that will line up with the poles of the magnetic field.

Step-by-step: MRI scan

`7 Steps`

1 The patient is placed on a sliding platform into the MRI scanner.

2 A large magnetic field is created around the patient.

3 The alignment of hydrogen nuclei in the body changes, causing them to line up, similar to the effect of a magnet on iron filings.

4 The hydrogen nuclei are then disturbed by a short input pulse of electromagnetic waves of a particular frequency given to certain parts of the body.

5 The hydrogen nuclei flip out of their alignment when the wave gives them extra energy at the right frequency.

6 When the input radiofrequency wave is switched off, one by one the hydrogen nuclei flip back into alignment giving out the same radiiofrequency wave energy that they had absorbed earlier. This 'relaxation' can take several milliseconds.

7 A receiver detects the radiofrequency output and the information is analysed by a well-trained **radiologist**.

MRI principles

Nuclear magnetic resonance (NMR), the process used in MRI, produces a very detailed picture of tissues because the protons can be detected in their exact locations in the body. The process can also provide information about the type of tissue that the protons are in because they re-align at different speeds in different tissues, sending a signal that is particular to the tissue type. The resulting images produced by a powerful computer are very detailed. The photograph shows an MRI scanner in action.

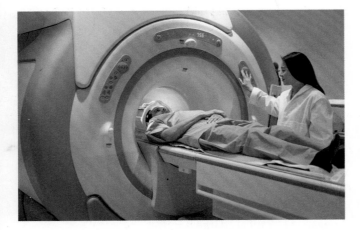

▶ A patient undergoing MRI

How can the radiographer adjust and interpret the image?

▶ Bright and dark areas on the image indicate different chemical surroundings for the protons or different flow properties of the body tissue.

▶ The magnetic field strength, scanner coils and radio wave pulse sequences all affect the clarity of the image. Different detector coils can be chosen to match the size and location of the organ being investigated.

▶ Sometimes a contrast medium (i.e. another chemical that is visible to MRI) can be introduced into the patient. But most MRI scans just look for hydrogen atom protons.

This property allows the particle to absorb energy from incoming electromagnetic wave packets called **photons**, provided that the frequency of the wave is of a particular value.

The energy value of the photon that can do this is directly linked to its frequency. In NMR, this is called the resonance frequency or **Larmor frequency**. This is typically in the range of 12 → 85 MHz, i.e. a radiofrequency.

Key terms

Photon – a particle that is a small 'package' of energy of electromagnetic radiation.

Larmor frequency – the rate of 'precession' or wobble of the proton in the magnetic field.

Energy emitted by the protons is also a radiofrequency. The signal detected can remain for a few milliseconds, or even seconds, and can provide vital information about the type of body tissue.

Diagnostic uses

Generally, the decision to have an MRI scan is dependent on the signs and symptoms experienced by the patient and the overall judgement of the doctor. Because the highest concentrations of protons visible by means of MRI are in soft tissues, and especially in fat and water, an MRI is more suitable for the following areas of the body:

▶ brain

▶ spine

▶ pelvis

▶ joints

▶ abdomen

▶ heart

▶ breasts

▶ other soft tissues and organs

▶ blood vessels.

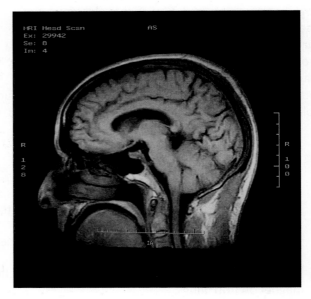

▶ An image of a head using MRI. Note the clarity of the different tissue boundaries.

Benefits of MRI

▶ The process is a non-invasive diagnostic procedure.

▶ There is no contact between the patient and the equipment, apart from being positioned on a suitable sliding platform.

▶ The procedure is painless to the patient.

▶ Images produced show clear differences in soft tissue types with good contrast between them.

▶ The detail in the image is very good, helping the radiologist in diagnosis.

▶ Tissue interfaces (where one tissue meets with another) are clear and well defined.

Lasers

LASER is short for 'Light Amplification by Stimulated Emission of Radiation'. The principle of producing laser light is based on a clear understanding of the electromagnetic spectrum and the way that atoms of a particular element can be put into an **excited state** to release a photon of light energy.

Excited state – the condition of an atom whereby one or more electrons have absorbed energy and move to an energy level above the ground state. This energy level is further from the nucleus of the atom.

To understand how lasers work, we need to look closely at the electromagnetic spectrum of radiation in the range from approximately 1 nm to 1 m. In Unit 1, you learned that the term 'light' generally refers to electromagnetic radiation between these measurements and that light that

is visible to our eyes lies within a band of 400 to 700 nm. Above the 700 nm point, **infrared** radiation begins, and below the 400 nm point is ultraviolet.

Infrared – electromagnetic radiation from beyond the red end of the spectrum emitted by heated objects and having a wavelength of between 800 nm and 1 mm.

Figure 21.2 sets out the basic principles of lasers.

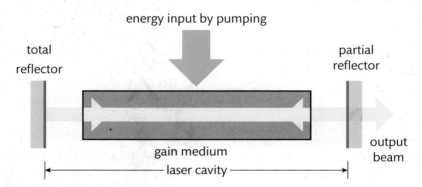

▶ **Figure 21.2** Basic laser principles

Step-by-step: Basic laser principles of operation

5 Steps

1 Energy (known as the pump energy) such as photons of light, electricity or chemicals is introduced into a glass, crystal or gas, also referred to as the gain medium.

2 As the atoms of the gain medium absorb the energy, electrons move from an initial low state of energy to a high state and back again almost immediately.

3 Energy absorbed by the electrons initially is now given off as a photon of light radiation. This is called 'emission'.

4 Since the space inside the laser device is sealed, the photons emitted continually move at high speed inside the tube. If photons collide with another excited atom, then more photons are produced. This is called 'stimulated emission'. Because of this process, light has been 'amplified'.

5 A proportion of the photons are made to bounce back and forth inside the laser tube, while others are allowed to escape, producing a very bright, concentrated beam of light.

Find out information about the types of lasers used for at least three different applications, one of which must be in a medical context. Identify the energy source used and the colour of the light produced and its associated gain medium.

Main types of medical lasers

Laser operating wavelengths used in medicine are as follows:

▶ near infrared: 700–1400 nm

▶ mid infrared: 1400 nm–3 μm

▶ infrared: above 3 μm

▶ visible range: 400–700 nm

▶ ultraviolet: 200–400 nm

If the material of the **gain medium** is different, then it produces a different **frequency** (and hence colour) of light, as shown in Table 21.1.

▶ **Table 21.1** Laser type and visible colour

Laser type	Colour produced
KTP	Green
Ruby	Red
YAG / CO_2	UV or IR

Table 21.2 shows the chemical elements in laser types.

▶ **Table 21.2** Chemical elements and their main characteristics

Gain medium	Applications
Neodymium-YAG (Nd-YAG)	Used in dentistry to treat certain parts of the tooth relating to the softer parts within a cavity. The enamel is not affected because the laser light can be adjusted to reduce power levels.
Yttrium (Y), Aluminium (Al) Garnet (Ga)	This type is often used for the treatment of benign and malignant tumours in bronchial and gastrointestinal areas of the body. It can also be sent through **endoscopes** to difficult parts of the body such as the colon and oesophagus.
Carbon dioxide (CO_2)	A useful tool in the sealing of blood vessels of a range of diameters and in many other applications, including reducing the spread of cells with tumours, sealing nerve endings and facial plastic surgery. The beam is very narrowly directed and there is no damage to surrounding tissue.
Argon (Ar)	This type of laser is used frequently in micro-surgery and ophthalmology. It is also used in the treatment of gastrointestinal ulcers and polyps (small growths in the colon, for example). Other uses include dermatology, where the argon laser is effective at treating skin tumours and removing tattoos.

Key term

Endoscope – a telescope placed inside a long thin tube for insertion into tight, difficult areas.

Therapy uses

Because lasers are able to deliver light energy with high precision, they are very useful for treating eye conditions such as cataracts or myopia, allowing patients to see again without the need for glasses or contact lenses.

Laser treatment in eye surgery can also repair parts of the eye which may need to have blood vessels sealed, or detached retinas 'welded' back in place. It works by fusing the membrane of the interior of the eye against the wall of the eye and has the great advantage of being painless.

Benefits and advantages of lasers in medicine

▶ A narrow light beam produces a perfect incision in the tissue with almost no variation in width or depth.

▶ Plaque in blood vessels can be removed quickly and precisely.

▶ Blood vessels are sealed immediately and tissue is burned (cauterised) during the laser procedure, preventing blood loss.

▶ Precise targeting of the beam helps to reduce damage to healthy surrounding tissue because heat produced by the laser is not conducted well by human tissue. This will reduce scarring.

▶ Deep penetration of some laser light can help to clot the blood quickly or heat up tumours to destroy them.

▶ Pain levels, swelling and scarring are lower than with other methods of treatment.

▶ The laser treatment can be varied as either pulsed or continuous wave. If pulsed, the laser light is administered to tissue in very short bursts, providing higher energy and high temperatures. When continuous, the laser light allows the tissue to absorb heat over a longer period of time. The type of treatment used will depend on the medical condition.

▶ It is monochromatic, meaning that it uses one wavelength.

▶ It has coherence – the photons of light are 'in step' with each other showing a fixed relationship.

▶ It can be focused into a fine beam with a **collimator**.

Link

See *Unit 1: Principles of Applications of Science 1* section C1.

Key term

Collimator – a device, usually made of lead, which makes electromagnetic waves (e.g. gamma rays) parallel.

Infrared thermography (IRT)

The Greek physician Hippocrates (460–370 BC) was probably the first to suggest that diseases or abnormalities in the human body could be identified by skin temperature differences. He tested this idea by using wet mud that he smeared onto patients. If the mud dried quickly on certain parts of the body, it suggested that the skin temperature was higher in that area.

Infrared **thermography** is the method used to detect heat energy (infrared wavelengths) from an object. In medicine, the equipment converts the radiant heat energy produced by the body to a temperature reading and then displays the distribution of temperature in an image that is then analysed.

> **Key term**
>
> **Thermography** – the medical technique for detection of radiant thermal energy (heat) from the body and converting this into a visual image for diagnosis.

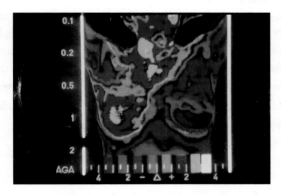

▶ The photograph shows a patient with suspected breast cancer. More heat is emitted from cancer cells because they have a higher **metabolic** activity.

The process of medical thermography allows the surface temperature of the skin from a patient to be measured. Blood flow is regulated by the nervous system and other local factors and the flow of blood in the upper few millimetres of skin provides a useful way to diagnose possible medical problems that may exist in organs that are situated deeper in the body. The process itself cannot measure the exact temperature of organs below the skin, but can often detect temperature changes from tissue damage or tumours that result from **pathological** causes.

> **Key terms**
>
> **Metabolic** – related to a living organism that breaks down food to produce energy and heat. This process is called metabolism.
>
> **Pathological** – condition or complaint caused by a disease.

As the largest and most interconnected organ in the human body, the skin provides information, in terms of temperature, concerning the immediate aspects of a person's bodily condition. Measuring the temperature of the skin will give valuable information to medical staff in 'real' time, rather than after a disease has produced the damage.

Medical thermography has sensitivity to 0.05 °C and, as the temperature of a patient may change significantly in a short space of time, its use is very important. Different diseases or medical conditions, such as bruising, viruses or inflammation, can cause either a drop or rise in temperature in surrounding tissue, and identifying this is vital for doctors to take quick and effective action. For example, diagnosing cardiovascular conditions at an early stage, by looking at inflammation patterns in the body, can allow patients to change certain aspects of their lives to help prevent a possible heart attack or stroke.

❚❚ **PAUSE POINT** Using infrared thermography, which parts of the human body would produce the most intensive bright red colour highlighting strong heat emission?

Hint Consider the flow of blood and relative importance in the activity of the major organs.

Extend Why is it not possible to determine accurately by thermography whether a patient has a serious condition? What are the limits to using thermography to detect serious pathological conditions?

Uses

Infrared thermography can be used in screening programmes, cardiovascular/circulatory disorders and dentistry (see Figure 21.3). It can also be used for:

▶ nerve problem diagnosis, such as arthritis in the vertebrae

▶ vessel disease diagnosis

▶ neurological, muscular and skeletal disease identification

▶ respiratory conditions – bronchitis, pneumonia, flu diagnosis

▶ digestive illness – such as appendicitis and irritable bowel syndrome (IBS).

Screening programmes Breast cancers are identified using the patient's other breast as a 'control'.	**Cardiovascular/circulatory** Detection of reduced blood flow which could result in vascular disease.
Respiratory problems Detection of conditions such as asthma and bovine respiratory disease in calves.	**Dentistry** Temperature distribution in mouth tissue and root canal infections, for example.

▶ **Figure 21.3** Uses of medical thermography

Benefits

Table 21.3 sets out the benefits of using thermography.

▶ **Table 21.3** Benefits of thermography

Benefit	Explanation
Non-contact	Equipment does not have to touch the body to detect the heat emitted and map the body's surface.
Non-invasive	The detector does not enter the body.
No radiation	Heat or infrared waves are non-ionising and do not damage atoms in the body.
Safe on the patient	The equipment detects heat and does not produce it.
Equipment	Generally made up of two main parts: an infrared camera and computer. There are not many controls for operators to worry about.
Very accurate	Can detect to an accuracy of 0.05 °C.

> **Research**
>
> Find out all the uses of thermography in medicine. You should not limit your research to humans, since the technique is also used extensively for animal care by veterinary surgeons.

Ultrasound

You have probably heard the Doppler effect in action. When an ambulance passes by, the sound of its siren changes in pitch when it comes towards you and then moves away from you. The sound waves are bunching up as it approaches and stretching out as it goes past.

This change in wavelength causes a frequency change and hence the pitch of the note is altered. It can help us to work out the velocity of an object or substance and whether it is moving towards us or away from us. In echocardiogram displays, for example, the speed and direction of blood flows in the heart and blood vessels can be indicated on the screen by colours – blue, for blood flowing towards the ultrasound transducer, and red, for blood flowing away from it.

Sound waves that have a frequency of more than 20,000 Hertz (20 kHz) are called ultrasonic or **ultrasound**. These are different from sound waves that we can hear because they have much shorter wavelengths.

> **Key term**
>
> **Ultrasound** – sound waves with a frequency above 20,000 Hertz (20 kHz), inaudible to humans.

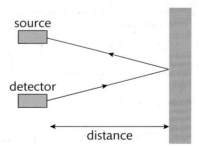

▶ **Figure 21.4** Principle of sound reflection and detection from a surface

An ultrasonic **transducer** converts electrical energy into ultrasound when an alternating voltage is applied. A quartz crystal has a special **piezoelectric** property. This means that it changes shape when stimulated by electricity. If the applied alternating voltage has a frequency equal to the natural frequency of the crystal, it resonates. This produces very large vibrations and ultrasonic waves. Some other crystals and ceramic materials are also piezoelectric, and are often used in modern medical ultrasound.

Transducer – a device that changes a physical quantity (e.g. sound) into an electrical signal or changes an electrical signal into a physical quantity.

Piezoelectric – the property of certain substances, such as a quartz crystal, to change size when it is stimulated by electricity. If an alternating voltage is applied to the crystal it causes a vibration.

TRANSDUCER:
Usually hand held, moved over the surface of the body, contains piezoelectric device which transmits the ultrasound waves through the body.

↓

SIGNAL UNIT:
An electronic processing device which controls the transducer power.

↓

MONITOR:
A computer display screen to visualise the images.

↓

STORAGE UNIT:
A computer data file used to store images.

▶ **Figure 21.5** Instrumentation

In medicine, the process is usually termed sonography or ultrasonography. The technique is performed by a radiologist or a sonographer. High frequency sound waves are directed at tissues in the body to generate an image. The frequencies used are in the **megahertz** range (MHz),

typically from 2 to 15 MHz. Pulses of waves pass through the patient and are partly reflected by boundaries within the body, such as tissue boundaries. These reflected sound waves are picked up by a second transducer that converts the ultrasound to electrical energy. Distances to the tissue boundaries are computed from the time delay between an outgoing ultrasound pulse and the arrival of its echo. Images based on those measured distances are displayed on a monitor. Ultrasound provides very good contrast images for soft tissues. Figure 21.5 shows the instrumentation used.

Key term

Megahertz – measurement of frequencies of 1 million Hertz.

The ultrasound signal from the transducer through the body may be:

▶ absorbed

▶ reflected

▶ refracted

▶ scattered.

When the ultrasound waves are sent through the body, they encounter different boundaries or tissue interfaces and, as a result, can be absorbed, reflected, refracted or scattered to various degrees. The eventual image produced will depend on the density characteristics of the organic matter and the speed of the ultrasound through it.

At a fat/muscle interface, more than 98 per cent of the ultrasound wave is absorbed, so that there is virtually no signal returned to the transducer and no image can be produced. In other examples, such as a muscle/bone interface, up to 60 per cent of the wave can be returned to the transducer, providing a suitable image. When air is present as an interface, almost all the incident ultrasound wave can be reflected back to the transducer. Similarly, hard features such as bone, gallstones or kidney stones all highly reflect ultrasound waves.

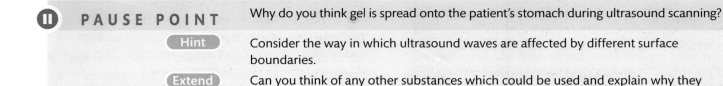

ⅠⅠ PAUSE POINT Why do you think gel is spread onto the patient's stomach during ultrasound scanning?

Hint Consider the way in which ultrasound waves are affected by different surface boundaries.

Extend Can you think of any other substances which could be used and explain why they could be used?

External ultrasound scan

Ultrasound scans performed from outside the body are routine in the monitoring of unborn babies. The transducer is moved over the body surface, sending pulses of ultrasound through different tissue. When the ultrasound pulse encounters a boundary or interface between two tissue mediums, most of the pulse is reflected but a significant amount is refracted. This means that it passes through the boundary into the second medium until encountering another boundary. When this happens, some of the pulse is again reflected and some refracted. This will produce another echo slightly later than the first. A series of echoes will build up, representing different depths of boundaries. When all these various depths are recorded, an image of the foetus can be developed and an image produced on the monitor.

External ultrasound scans are used to determine the condition of many organs including the heart, liver and kidneys. They are also used for the diagnosis of muscular and joint pains.

Internal ultrasound scan

This type of scan involves the insertion of the ultrasound probe into the vagina or rectum to affect a more detailed study of internal organs, including the ovaries and prostate gland. It is also used in monitoring of the foetus where further detail is required.

Endoscopic ultrasound scan

This procedure involves placing an endoscope into the patient's mouth to scan the inside of the oesophagus, stomach and sometimes the pancreas, lungs, gall bladder and liver. Endoscopic ultrasound scans can also be applied to determine the extent of cancer spread to nearby blood vessels or lymphatic glands.

In order for the procedure to cause little discomfort to patients, a local anaesthetic or mild sedative may be administered.

Case study

Ultrasound checklist

Tasmina is an inexperienced ultrasound technician and has received a patient list of incoming expectant mothers who will be prepared for ultrasound during the day. Her role will be to provide departmental support to the specialist sonographer who will be performing the investigation and to ensure that all mothers having the procedure are well informed of the process and of what to expect.

For many of those patients, this will be a 12-week scan, a time when they see their baby for the first time and so Tasmina's reassurance, professionalism and additional knowledge will be greatly appreciated.

The most important points to consider and discuss with each patient will be:
- the process is painless to both mother and baby
- the application of the gel to the mother's belly
- drinking an amount of water before the scan
- the length of time of the scanning process
- heartbeat check
- all-over baby anatomy scan.

Check your understanding

Carry out some additional research to help you with answering these questions:

1 Why is it important to explain to the expectant mother that it is painless?

2 Why is gel applied?

3 What use is drinking water before the scan?

4 Is it important to let the patient know how long the procedure will take? Why?

5 What is the purpose of the heartbeat and anatomy check?

Treatment uses

The ability of ultrasound to cause objects to vibrate is key to understanding how they can be used as a treatment for certain medical complaints.

Kidney stones

Chemical substances contained in urine can sometimes form solid crystals which can develop into larger 'stones' in the kidney. These are called kidney stones and can pass into the ureter (the tube which allows the passage of urine into the bladder ready to be expelled) blocking it up and causing severe pain.

Ultrasound waves are focused onto the stone, causing it to vibrate and shatter into smaller fragments, unblocking the ureter.

Benign and malignant tumours

Ultrasound waves of high intensity have been used in many developed countries to help in the treatment of a variety of tumours and cancers. The ultrasound is of high energy and heats the diseased or damaged tissue enough to destroy it. The process has shown success for cancers in many organs of the body and further developmental research is being carried out.

Diagnosis uses

Echocardiography

In echocardiography, the flow of blood through the heart valves, the motion of the heart as it beats and the amount of blood pumped can be accurately monitored and measured. For example, the aortic valve prevents blood from flowing backwards when the aorta is filled. An echocardiogram can determine whether the valve walls are thinning and if valve replacement is necessary. Further echocardiograms can

monitor the rate at which the heart beats and determine how efficient it is at pumping blood around the body. If this efficiency is low, it can lead to heart failure.

▶ Ultrasound testing of the carotid artery in the neck can determine whether a stroke victim has a decreased blood supply to the brain as a result of a blockage in the arteries.

▶ Ultrasound testing of the walls of the aorta can detect whether they have weakened as a result of the pressure from heart beats over time. This condition is called an aneurysm.

▶ Ultrasound testing of veins can determine if swelling in a person's leg is a result of a blood clot or deep vein thrombosis (DVT).

Assessment practice 21.1   A.P1   A.P2   A.M1   AB.D1

Produce a 'patients' information poster' which outlines the basic principles of MRI, LASER, IRT and ultrasound techniques.

You will need to summarise each section to include:

- basic definition or explanation of the processes involved
- labelled diagrams of the instrumentation for MRI, LASER and ultrasound techniques
- general uses in terms of diagnostic or treatment, where applicable to each technique.

Remember: posters are meant to be eye-catching, clear and informative.

Plan
- What is the task? What am I being asked to do?
- How confident do I feel in my own abilities to complete this task? Are there any areas I think I may struggle with?

Do
- I know what it is I'm doing and what I want to achieve.
- I can identify when I've gone wrong and adjust my thinking/approach to get myself back on course.

Review
- I can explain the results obtained from the task.
- I can apply the activity to other situations.

B   Explore the principles, uses and benefits of ionising instrumentation techniques in medical applications

This section outlines the techniques that are currently used in medicine which apply dangerous **ionising radiations**. You will study the basic principles in the production of ionising radiation, and how these radiations can be applied to provide extremely useful images of the inside of the body for diagnostic purposes. You will also be guided through the uses of X-rays, gamma rays and proton beams as a means of treatment for certain cancers and tumours.

Key term

Ionising radiation - electromagnetic waves or particle beams that have sufficient energy to break chemical bonds, thus producing charged ions in the materials they pass through.

X-rays

X-rays are part of the electromagnetic spectrum and have high frequencies ($3 \times 10^{16} - 3 \times 10^{19}$ Hertz) and short wavelengths (10^{-8} to 10^{-11} m). They have high energies, which makes them very useful in medicine. X-rays are produced when very high-speed electrons stop suddenly when they hit a metal target plate.

Key term

X-rays – high-energy electromagnetic waves that typically have wavelengths of around 10^{-10} m.

They were given the name X-rays because, when they were discovered in 1895 by a physicist named Wilhelm Rontgen, nothing was known about them.

Like all electromagnetic waves that travel through air, X-rays are slightly absorbed or scattered by the presence of atoms and molecules.

X-rays penetrating the body experience scattering to various extents depending on the atomic composition of the organs or tissues and the wavelength of the X-rays. When X-rays enter the body, all tissues and organs absorb a certain amount, but more dense parts of the body absorb the most. In this case the material, e.g. bone, will be brighter on the image, whilst dark areas will indicate less dense tissues which do not absorb X-rays very well.

The procedure for an X-ray is as follows.

▶ The radiographer positions the patient at the X-ray equipment table.

▶ The X-ray recording plate or film holder is placed behind the part of the body to be X-rayed.

▶ The patient is made comfortable and the correct position of the limbs will be checked.

▶ A lead covering may be placed over reproductive organs for protection (lead strongly absorbs X-rays).

▶ The radiographer moves behind a protective screen and activates the X-ray device.

▶ The image is analysed by the radiologist. *Note:* To enhance the image produced by X-rays, a contrast material can be given to the patient orally, rectally or by injecting. These change the way that X-rays (and other electromagnetic waves) interact with the body tissues. This helps to distinguish unhealthy tissues from healthy ones.

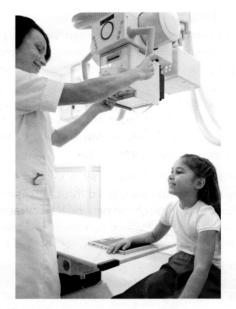

▶ Young person undergoing X-ray diagnosis

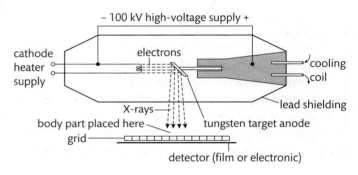

▶ **Figure 21.6** Production of X-rays

The production of X-rays is carried out using an evacuated tube. This is how it works.

Step-by-step: Production of X-rays   6 Steps

1 A filament is heated by the flow of an electric current.

2 Electrons, negative in charge, are given off (thermionic emission) and accelerated towards the anode. This is given a positive charge of up to 100 kV.

3 The electrons' high speed increases their kinetic energy.

4 When the electrons hit the target (usually tungsten on a copper block), a very small percentage of their kinetic energy is transformed into X-rays. Most of their kinetic energy (99.5 per cent) is transformed to heat.

5 The anode is kept cool using circulating oil or water.

6 X-rays pass out through a small window in the lead shielding.

X-ray characteristics

In medicine, care is taken to reduce exposure to patients because of the damage that can be caused by X-rays on healthy tissue. X-rays can be varied in energy to penetrate flesh and not bone, and also to be absorbed by denser diseased tissue so that it shows up clearly on photographic film. Lower energy (longer wavelength) X-rays are more easily absorbed in body tissues than are higher energy X-rays, which have more **penetration**.

To reduce **absorption** of lower energy X-ray photons by healthy tissue, X-rays are filtered through a thin metal plate before going through the patient. If this was not done, the tissues would heat up and become damaged. As with all electromagnetic waves, X-rays spread out as they travel. The overall intensity of the X-rays falls as the distance increases. The exact relationship is an example of an *inverse square* relationship, where the intensity of the X-rays falls to 1/4 of its original value as the distance travelled is doubled.

Key terms

Penetration - the capacity for a radiation to pass through a material while giving up only a small amount of its energy.

Absorption - the capacity of a material to take in external radiation.

PAUSE POINT

Look closely at the diagram that illustrates the workings of the X-ray machine. Quickly memorise the diagram and labels and produce a rough block diagram that explains the operation of the machine in simple stages.

Hint   Highlight each section separately e.g. high voltage, heater, electrons, etc.

Extend   Add a sentence to each stage explaining what is happening.

Research

- Find out all you can about the discovery of X-rays and the scientist most responsible.
- Produce a list with descriptions of the known uses of X-rays in society at present.

X-rays can be used for diagnosis, as shown in Figure 21.7.

Worked Example

The intensity of X-rays follows an inverse square relationship. If the intensity 3 cm from the X-ray source is 100%, calculate the percentage of X-ray intensity which remains at 2× the original distance.

After the original distance, 3 cm, the intensity of X-rays is 100 per cent. If the distance travelled is doubled to 6 cm, the fraction remaining is ¼ of the original value = 25 per cent intensity.

Now find the percentage intensity at 3× and 4× the original distance.

FRACTURES
Checking for broken bones.

DISLOCATIONS
Checking for abnormal bone joint position.

ARTHRITIS/CANCER
Detected through unusual bone or joint conditions.

SURGERY
As an aid to surgeons during surgery.

CHEST
Detection of pneumonia, heart failure, emphysema, tuberculosis, lung cancer, breast cancer screening (mammograms).

FOREIGN OBJECTS
Checking for coins, metal fragments, bullets.

ARTERIES/VEINS
Checking condition of blood transportation using chemical dyes.

▶ **Figure 21.7** Use of X-rays in diagnosis

X-rays for treatment

High energy X-rays are focused at the diseased area of the body in a process referred to as external radiotherapy. The X-rays destroy cancerous cells which cannot repair themselves following the dosage. Normal cells are damaged by the X-rays but can generally repair themselves.

If it is believed by doctors that there is a good chance of destroying the cancer and preserving life, the treatment is called curative. If it is believed that there is no possibility of destroying the cancer, but the treatment may relieve pain, then the treatment is called palliative.

X-rays are used to:

▸ reduce the size of a tumour before surgery so that the length of time and difficulty of the surgical procedure is reduced

▸ kill remaining cancer cells after surgery has taken place and so limit the possibility of additional follow-up surgery

▸ reduce the size of a tumour and the level of pain when there is no possibility of cure. This may help to increase the quality and length of a patient's life as a result of reducing the pain

▸ kill cancer cells without the need for surgery.

Case study

Working with patients in the X-ray department

Gemma and Kelly-Ann are both experienced radiographers and have worked together in the same hospital for the last three years. During breaks and in general conversation with other members of staff in regular departmental update meetings, they have come to realise that there is no suitable guidance document which can be given to patients undergoing either palliative or curative X-ray treatment. In addition to this, they have found that nurses caring for very ill patients who require regular X-ray treatments have very little knowledge of the procedures and effects of X-rays on the human body. They have decided to produce a document which is very informative and comprehensive, but is written in such a way that it can be easily understood. It contains the following information.

1 How X-rays are made
2 The use of X-rays in medicine
3 The procedure followed by the radiographer
4 X-ray characteristics

Check your knowledge

Carry out some additional research to help you with answering these questions.

1 Write a sentence that explains what X-rays are.
2 What happens when X-rays enter the body?
3 How is the damage to healthy tissue reduced during the procedure to kill cancerous cells?
4 What does the 'inverse square law' mean in simple terms?

Computerised tomography (CT) or computerised axial tomography (CAT)

The word tomography is made up of two separate words from the Greek: 'tomos' which means a 'slice' and 'graphe' which means 'drawing'. **Computerised tomography** means that an image is put together by a series of thin sections or cross-sectional images produced by an X-ray scanner. Figure 21.8 shows the principles of a CT scanner.

Key term

Computerised tomography/axial tomography – process by which a three-dimensional image of a body structure is produced from plane cross-section X-ray images along an axis.

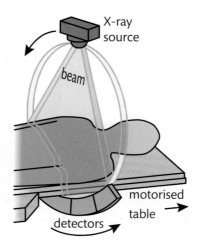

▸ **Figure 21.8** Principles of a CT scanner

Step-by-step: CT scan

8 Steps

1 The patient may be given an injection or meal of a **contrast medium** before the scanning procedure and then moved towards the opening of the CT scanner on a motorised table. How the patient is given the contrast material will be determined by the organs under investigation.

▼

2 The X-ray source is rotated around the circular opening. One rotation takes approximately 1 second.

▼

3 The X-ray beam is spread out to irradiate the part of the patient's body. This beam may be as wide as 10 mm and covers an axial plane. Many planes can be scanned to eventually produce a 3-D image.

▼

4 Many X-ray beams may be undertaken during the time that the patient is in the scanner. The table will be moving through the circular opening as this happens and the patient must remain very still.

▼

5 As the X-rays exit the patient's body, they are detected and recorded immediately, since the detector rotates directly opposite to the X-ray source.

▼

6 The information is processed by a computer. The computer provides a cross-sectional image of the patient's internal organs and tissues. This cross-section represents one complete rotation of the X-ray source and many rotations can help to enhance a 2-D image.

▼

7 If multiple axial planes (at different angles) are scanned, the resulting X-ray scans can be used to create a 3-D image, which helps to produce a valuable additional diagnostic tool.

▼

8 The images are analysed by a radiologist.

Key term

Contrast medium – barium or iodine chemical compounds given to the patient before CT scanning to enhance the appearance of internal organs.

⏸ PAUSE POINT
The CT scanner scans thin slices, one at a time, through a patient's body. If the patient moves slightly during the scans, what effect will this have on the resulting image?

Hint
Consider the effect that movement has when your photograph is taken using a camera or mobile phone.

Extend
The image produced by a CT scan shows different body tissues as areas of variation in brightness due to the amount of X-rays that pass through them. Lighter areas will be tissues that absorb more X-rays. What effect will movement of the patient have on the image between tissue boundaries such as bone and a possible tumour?

Worked Example

A typical single X-ray taken of a suspected fracture may last 1 second and provide an **effective dose** of 0.005 mSv. The average background radiation dosage that a person receives in the UK, depending on the geology of the area, may be 2.7 mSV over the course of 1 year. How long would a patient need to be subjected to X-rays to accumulate the same dosage as in 1 year from natural background radiation?

Natural background radiation dosage in 1 year = 2.7 mSv. The patient receives 0.005 mSv in 1 second.

2.7/0.005 = 540 seconds (9.0 min)

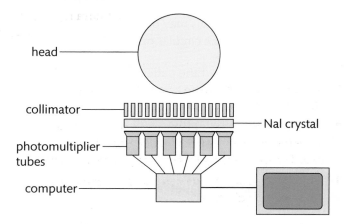

Key term

Effective dose – the absorbed dose of radiation multiplied by a quality factor that depends on the tissue type under investigation.

Diagnosis and monitoring uses

Brain and other tumours

A CT scan can provide information on the size and shape of a tumour to help inform the extent of radiotherapy which may be needed. Abscesses could be drained or a sample of the tissue removed for analysis (**biopsy**).

Key term

Biopsy – an extraction of fluid or tissue from inside the body to determine the type and extent of disease present.

Bone damage

Types of bone damage such as fracture or other problems can be better identified using CT scanning as opposed to standard X-ray. Tissue damage around the bone can also be detected effectively.

The heart

The use of CT scans for identification of heart disease involves injecting the veins of a patient with iodine-based solution that is highlighted by the X-rays to produce a clear image. The CT scan can also identify the level of calcification in the arteries which doctors call a calcium score.

Internal organs

Kidney, liver and spleen damage can occur, due to the nature of the organs. For example, the liver is the largest internal organ and could experience lesions, bleeding and infection, that will need detailed diagnosis rather than standard X-ray. The image produced can also help the surgeon to decide where to place the needle during a biopsy.

Stroke

A stroke is defined as insufficient blood flow in the vessels of the brain that will result in lowered oxygen levels. A CT scan can detect this region in the brain so that immediate treatment can begin.

Gamma ray imaging

The process of gamma ray imaging relies on the production of radionuclides using a special generator to allow the parent radionuclide to decay to a short half-life daughter radionuclide. The addition of chemicals to these radionuclides produces the **radiopharmaceuticals**. Figure 21.9 shows the basic components of a gamma camera.

Key term

Radiopharmaceutical – the name given to pharmaceuticals that contain radionuclides.

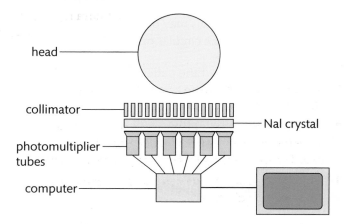

▶ **Figure 21.9** A simplified gamma camera

Table 21.4 lists some common radiopharmaceuticals used in the medical industry:

▶ **Table 21.4** Common radiopharmaceuticals

Radiopharmaceutical	Element and description
^{123}I	Iodine – ready-to-use substance
^{99m}Tc	Technetium – available in multiple forms
^{201}Tl chloride	Thallium – ready-to-use substance
^{67}Ga citrate	Gallium – ready-to-use substance

Principles

- ▶ Radiopharmaceuticals are injected into the patient as a *tracer* that circulates around the body in the patient's blood stream. The radioactive tracers have a short half-life.
- ▶ The radioactive chemicals are absorbed by bone, tissues and other body organs (e.g. lungs, heart, brain, kidneys, etc).
- ▶ Gamma rays are emitted as the substance decays and are converted to electrical signals by the photomultiplier tubes.

▶ The gamma rays from inside the body are detected by the gamma camera, converted to an electrical pulse outside the body and are used to produce a clear image for diagnosis.

▶ The stronger the electrical signal from one particular position, the more localised the position of the radiopharmaceutical.

PAUSE POINT

The half-life of the radioactive tracers injected into the patient are carefully considered so that the radioactivity does not remain for too long a period inside the body, but is active for a sufficient time for the gamma rays to produce a suitable image. If no more gamma rays are detected by the gamma camera after a certain length of time, what does this tell you about the condition of the tracer?

Hint   What happens to the radiopharmaceutical over time, and why does this happen?

Extend   If gamma rays are more localised or concentrated in one area of the body, explain what this means in terms of diagnosis and the physics principles involved.

Generalised diagnosis examples from a bone scan

1 No abnormalities – image shows that the tracer is absorbed uniformly.

2 Bone infarction – little or no tracer absorbed. An infarction is an area of dead tissue caused by loss of blood supply. As a result, no tracer can be absorbed.

3 Infection, fracture, tumour or cancer – bright image indicates high absorption.

Uses of positron emission tomography (PET)

PET uses the same principles as gamma ray imaging in that the radioactive tracer is injected into the patient and the decay process releases gamma rays. However, the type of tracer used with PET produces small particles called positrons which combine with electrons in the body and annihilate each other. The resulting gamma ray photons are picked up by the PET detectors.

In Figure 21.10, the **positron** particle emitted by the radioactive **tracer** inside the body collides with the electrons in the patient's internal organs. The radiopharmaceuticals in the patient emit positrons continually because the protons inside the nuclei decay to produce a neutron and a positron (in beta decay). When an electron and positron collide from opposite directions, they are annihilated and the resulting gamma rays are emitted in opposite directions. The mass of the electron and positron has been converted to an equivalent amount of energy in the form of gamma rays.

Key terms

Positron – a sub-atomic particle of the same mass as an electron but opposite electrical charge (+).

Tracer – in radiotherapy and diagnosis, a substance that can be injected into the patient and easily tracked through the body because of its radioactivity.

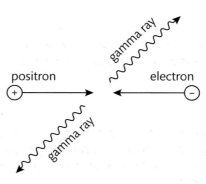

▶ **Figure 21.10** Positron-electron annihilation to produce gamma rays

Radioactive parent isotopes typically used include:

▶ ^{68}Ga (gallium – half-life of approximately 68 minutes)

▶ ^{18}F (fluorine – half-life of approximately 110 minutes)

▶ ^{82}Rb (rubidium – half-life of approximately 1.25 minutes).

Uses of PET include:

▶ **Cancer diagnosis:** The extent of the cancer and detection of recurrent cancer in organs of the body can be determined. This includes cancers of brain, heart, lung and breast.

▶ **Other disorders:** The general function of organs in the body can be monitored. This includes blood flow to the heart, change in shape or size of the organs or infection, memory disorders and seizures.

Benefits

The procedure used in both gamma ray imaging and PET involves the use of high energy and highly penetrating gamma radiation. Since the effect of all ionising radiation on biological cells is damaging, affecting the DNA in the cell and causing mutations, doses must be carefully limited. Gamma rays, being more penetrating, cause less concentrated ionisation than X-rays. The fact that the radionuclides are embedded in pharmaceuticals that are designed to be strongly taken up by particular tissues means that those tissue types can be clearly identified and pinpointed.

Radiotherapy, gamma knife surgery and proton beam therapy

This treatment of mainly malignant diseases is carried out using X-rays, gamma rays or charged particles such as electrons and protons. The procedure requires an accurately calculated dose of ionising radiation over a specified time period that is administered by a radiation oncologist or radiologist. This form of therapy uses high-energy radiations that can provide precision of the radiation beam and so reduce the amount of dosage that the patient will need to endure.

Radiotherapy

This method involves destroying cancer cells by affecting their ability to reproduce and by damaging their DNA. As a result, the cancerous cells are not able to carry out cell division and will die. Cells that are in an active state of cell division don't repair the damage as effectively, so cancer cells that are multiplying quickly are more affected by the treatment than other cell types.

External beam radiotherapy

A **linear accelerator (linac)** is a complex machine that can produce both gamma rays and X-rays by accelerating electrons close to the speed of light.

> **Key term**
>
> **Linear accelerator (linac)** – a large machine that accelerates particles to high speeds in a straight line.

When equipment such as a linear accelerator is used in medical treatment, it is vital that the gamma ray or X-ray beam can be focused accurately enough to be able to destroy the malignant cancer cells whilst causing very little damage to the surrounding cells. Doctors need to work out where to direct the beam, how large it needs to be and what shape is required. To help in directing the beam, some physical devices are used which allow the radiologist to focus the gamma rays into as small an area as possible.

To reduce the effects of radiation in the body and to limit additional damage, the treatment is usually administered daily and over a period of several weeks. This is to ensure that the radiation dose is effectively given to the diseased cells or tumour and also to allow sufficient time for healthy cells to recover.

The radiation oncologist who decides what treatment to give will bear in mind the need to target the diseased cells but also to protect other healthy cells and tissues during the course of the treatment.

Where the treatment given is curative, the approach used will be to:

▶ shrink the tumour before surgery can be carried out

▶ destroy tumours which have not yet spread to other parts of the body

▶ destroy cancer cells which may still be active to eliminate the possibility of recurrence.

Where the treatment given is palliative, the approach used will be to:

▶ relieve the pain caused by reducing the size of the tumour

▶ relieve the symptoms, such as breathing difficulties, caused by the tumour by treatment aimed at shrinking the tumour

▶ give the patient a reassuring sense of care and regular information which will help the patient's overall well-being.

Gamma knife surgery

The word 'knife' is misleading. The process involves the precise targeting of tumours in the brain from a device that is situated on the head like an oversized helmet. It can send multiple, highly accurate beams of gamma radiation to the affected area in the brain, making sure that they converge at the exact point specified by highly skilled radio oncologists.

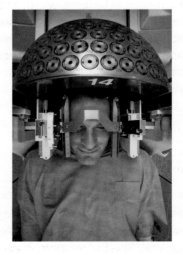

▶ A gamma knife surgery 'helmet' used to direct multiple beams of gamma radiation to an accurate point in the patient's brain

High dose gamma radiation is delivered to the 'target' over a period that may last a few hours. The target is a tumour that will be given a radiation dose using an array of 201 cobalt-60 radiation sources. The beams are focused with great precision and converge at an exact point of the tumour. Planning the dose for treatment is vital if the side effects following the procedure are to be minimised.

The total volume of the tumour to be treated must be carefully determined and decided upon prior to using the **gamma knife**, which has a precision of approximately ±0.5mm. Pre-scanning, using one or more of the methods outlined already in this unit, is used to ensure that the volume of tumour to be targeted is very accurate.

Key term

Gamma knife – the name given to a highly focused beam of gamma radiation used to treat brain tumours.

Reflect

Why are 201 linear accelerators used in the gamma knife helmet? How does this number affect the ability of the gamma knife to target the tumour effectively?

Theory into practice

The gamma knife is regarded as one of the most precise and effective forms of treatment for brain tumours and its use is increasing. The device is normally applied to irradiate volumes of cancerous cells of less than 1 cm³ and typically to very precise tolerances.

As a junior radiographer in the radiology department of a busy urban general hospital, you must continually discuss the method and the equipment used with the patients before the treatment in order to provide a calm and reassuring explanation of the process which they are about to undergo.

Questions normally asked are:

a) How long is the procedure?

b) Is the radiation dangerous?

c) Is the procedure accurate and precise?

d) Will healthy cells be affected?

e) Will I need further treatment?

Use as much information as you can find to help answer these questions about gamma knife treatment. Produce a general report of no more than one side of A4 and include useful diagrams where appropriate.

The procedure is used as a safe and very effective alternative to conventional surgery or when conventional surgery cannot be carried out due to complications in the tumour. The treatment is non-invasive and considerably reduces the health risks to the patient. It is recognised, however, that it is almost impossible, at present, to converge all beams in the exact location intended during the treatment process. This is because when we get closer to the location of the tumoour, the measurements become incredibly small. This small margin of error is acceptable and the medical industry ensures that it does not present a significant problem through clear and highly specified quality control procedures.

Proton beam therapy

These machines must be very large in order to build up the high speeds necessary for proton beams.

Protons are positively charged particles in the nucleus of atoms. They can be accelerated using a particle accelerator called a **synchrotron** to incredibly high speeds and then focused into a very small area. This allows them to be a valuable tool in the treatment of cancerous disease, particularly in the young, underdeveloped brains of children.

Key term

Synchrotron – a circular-shaped device using electrical and magnetic fields to accelerate particles, increasing their kinetic energies.

▶ Large synchrotron in Barcelona, Spain

The advantage of using a beam of protons to kill cancer cells is that once the beam of protons hits its target, it stops. This ensures that healthy tissue damage around the area of the cancerous tumour is significantly less than in other medical physics methods. As a result, proton beam therapy is a suitable method of treatment for cancer that may lie close to important organs or nerves such as the optic nerve attached to the eye.

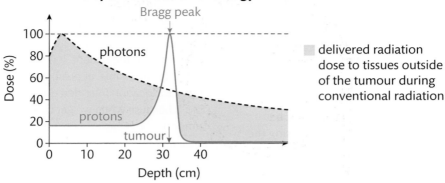

Depth distribution of energy

▶ **Figure 21.11** Depth distribution of energy

The graph in Figure 21.11 shows the comparison of dose treatment between photons (X-rays, gamma rays etc) and proton beam. There is a considerable amount of radiation dose delivered to tissue outside the actual targeted tumour using photon methods.

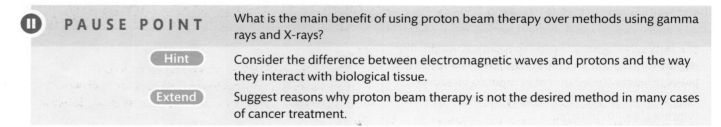

Link

Unit 7: Contemporary Issues in Science

Bragg peak

Treatment methods that use energy from parts of the electromagnetic spectrum use photons to target the tumour. These are essentially a wave of energy and there is usually a greater spread of the energy around the area targeted. The essential difference with proton beams is that the proton is a charged particle that will interact much more strongly with biological tissue and quickly lose its speed.

Scientific tests and calculations have shown that the proton beam will deliver its maximum dose at a specified depth determined by the kinetic energy it is supplied with from the synchrotron. The dose of a proton is at its maximum where it is stopped when it interacts with the target tissue. This is called the **Bragg peak**.

Key term

Bragg peak – the point at which the maximum dose of proton energy is delivered corresponding to the maximum interaction of the proton with tissue.

If the Bragg peak of the proton beam is calculated, scientists can ensure that the tumour will receive its maximum dosage. Healthy surrounding cells do not become affected with any significant amounts of radiation.

PAUSE POINT

What is the main benefit of using proton beam therapy over methods using gamma rays and X-rays?

Hint
Consider the difference between electromagnetic waves and protons and the way they interact with biological tissue.

Extend
Suggest reasons why proton beam therapy is not the desired method in many cases of cancer treatment.

Principles of the synchrotron accelerator

▶ A small amount of hydrogen gas is pumped into the assembly (see Figure 21.12). A large static charge ensures that electrons are removed from the hydrogen atoms, leaving protons.

▶ The electrical potential in the tube allows the positive protons to be strongly attracted to the negative cathode and they accelerate into the first cavity.

▶ The protons enter the circular structure, accelerated each time they pass through the synchrotron and

gaining more and more kinetic energy. This continuous stream of protons can be kept in the accelerator for many weeks.

▶ The magnetic field in the accelerator increases in order to keep the protons in their circular path.

▶ A proton stream is then extracted when required for collision experiments such as in the Large Hadron Collider in CERN, Switzerland, or for use in medical treatment.

Figure 21.13 shows some uses of the synchrotron accelerator.

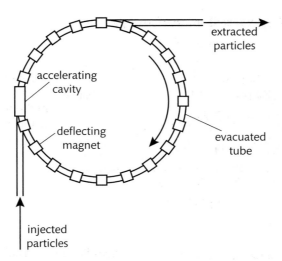

▶ **Figure 21.12** Principle of the synchrotron accelerator. Protons speed up as they go around the large tube and are kept in a central position by a deflecting magnetic field.

Malignant and benign tumours:
Brain, eyes, gastrointestinal, breast, head, oesophagus, lungs and liver, prostate and spine.

Thyroid cancer:
Proton therapy can target the diseased cells with very little effect on the surrounding organs such as the mouth, spinal cord, neck and chest area. It is also suitable for recurring cancers.

Blood and other disorders:
Proton therapy has been shown to be effective in treating blood cancer – leukaemia – and other cancers such as in the lymphatic system.

▶ **Figure 21.13** Uses of the synchrotron accelerator

Assessment practice 21.2

B.P3   B.P4   B.M2   AB.D1

The technology used in medical physics techniques changes rapidly with our understanding of the science involved and improvements in computerised systems.

Medical texts and journals have to be re-written after a few years to keep pace with the advances made. When patients are presented with an explanation of diagnosis or treatment, it can be very confusing and cause concern.

Develop a 'patients' guide' that provides essential information on the methods outlined in this section. It should be in the form of a one page table with the following column headings:
- Ionising method used
- What is it and how is it produced?
- What is the procedure for my scan or treatment?
- How will the process help me?

Plan
- What is the task? What am I being asked to do?
- How confident do I feel in my own abilities to complete this task? Are there any areas I think I may struggle with?

Do
- I know what it is I'm doing and what I want to achieve.
- I can identify when I've gone wrong and adjust my thinking/approach to get myself back on course.

Review
- I can explain the results obtained from the task.
- I can apply the activity to other situations.

C

Understand health and safety, associated risks, side effects and limitations of ionising and non-ionising instrumentation techniques in medical applications

Safety precautions, side effects and risks for operators and patients of non-ionising radiation

Safe operating procedures

The safety of the people operating equipment and patients undergoing the procedure is paramount to the whole procedure and is considered as an absolute priority by all those in the medical profession.

When used appropriately by highly skilled medical staff, ultrasound, for example, is generally considered very safe in its use for diagnosis and imaging, although the energy of ultrasound waves has the potential to damage biological tissue by heating. In rare cases, ultrasound can also cause development of gas pockets in body tissue.

Because of this, for ultrasound scanning during pregnancy, the length of time during which the developing foetus is imaged is kept to a minimum.

Safe operating procedures (SOPs) exist to ensure that all non-ionising procedures are performed with the utmost safeguards against harm to both the patient and the medical staff carrying out the procedure. SOPs differ from one workplace to the next, but the essential aspects will be very similar. In general, medical practitioners who use MRI scanners, medical LASERs, infrared thermographic cameras and ultrasound equipment must comply with the following general requirements before any diagnosis or treatment can take place.

▶ Possess the relevant medical qualifications.

▶ Be thoroughly conversant with the operation of and operational limits of the equipment to be used.

▶ Take full responsibility of the procedure to be carried out within the local medical guidelines and practice.

▶ Identify and accept the scope of use of the procedure to be undertaken and to adhere strictly to the planned procedure.

▶ Follow or develop a formal plan or scheme of work for the procedure to be undertaken, using documented medical guidelines.

▶ Confirm the identity and medical details of the patient prior to the examination.

▶ Understand the biological effects of the technique to be undertaken in terms of the potential hazards.

▶ Minimise the dosage of the technique to ensure that sufficient doses are used to determine appropriate diagnosis, imaging or treatment but are no more than is required.

▶ Have a detailed understanding and knowledge of the equipment to be used and ensure that regular inspection and maintenance of the equipment has taken place.

▶ Ensure that electrical safety tests, repair tests and recording and imaging equipment tests have been carried out in accordance with the required guidelines and within the specified timescales.

▶ Ensure that transducer equipment has been well maintained and appropriately cleaned, especially for inserted probes.

Health and Safety Executive

The Health and Safety Executive (HSE) is the legislative body that is responsible for regulating and enforcing health and safety matters in medical establishments in the UK. The organisation carries out numerous inspections on the use of specific medical instrumentation and equipment as well as the operating conditions of staff, and works with the law to ensure that the findings of its reports are properly enforced where necessary.

In conjunction with this enforcement agency, regulation of medical devices and medicines is the responsibility of the Medicines and Healthcare products Regulatory Agency (MHRA). This is an executive agency of the Department of Health (DoE).

> **Research**
>
> Use the HSE website to find out the current legislative requirements associated with a) ultrasound use, b) LASER use and c) MRI procedures in the medical profession.

Ultrasound

The use of ultrasound scanning has become very common in pregnancy and other diagnosis and treatment regimes and is generally regarded as one of the safest methods currently in practice.

However, health practitioners are very aware of the need to ensure that patients are provided with a duty of care which guards against the possible pain or discomfort which may be attributable to certain methods of ultrasound involving the insertion of the probe into certain orifices of the body.

▶ Scans placed inside the vagina or rectum will be very uncomfortable for the patient. Certain medical conditions may require that a vaginal ultrasound (transvaginal) is needed. These conditions could be abnormal bleeding, muscle tumours of the uterus (fibroids), pain in the pelvic area, thickening of the uterus lining, ovarian cysts or during early pregnancy investigation. Unlike an abdominal scan, the insertion of the probe in the vagina or rectum for between 15 to 30 minutes can cause considerable concern for the patient, but the procedure does not have any known lasting after effects.

▶ Endoscopic ultrasound examination is a very effective method of producing an image of the surrounding internal organs and the oesophagus by combining the ultrasound signal with a microscope in a long thin tube inserted into the patient's oesophagus (or sometimes into the rectum and through the colon). Discomfort or pain felt by the patient can be relieved by the use of sedatives or pain killers and the throat is sprayed with a mild anaesthetic. Sometimes, after the procedure, patients may suffer a sore throat or, occasionally, abdominal pain, mild infection or vomiting. Endoscopic ultrasound is used for:

- pancreatic cysts or pancreatitis
- analysis of the heart
- occurrence of gall stones
- cancer detection.

▶ Certain types of ultrasound scan may require particular aspects related to drinking and eating to be carefully observed by the patient before the procedure is carried out. If the digestive system is being scanned, the patient is instructed not to eat anything for a number of hours prior to the examination. If the scan is to determine the cause of pelvic pain or is part of the early pregnancy ultrasound scanning process, then the patient may be told to drink a quantity of water and retain this in their body until the scanning is complete.

Lasers

Treatment with lasers is used to good effect in micro-surgery, dentistry and ophthalmology, and damages less surrounding healthy tissue than traditional surgery. However, even though it is a form of non-ionising radiation, care is needed when lasers are used by medical practitioners.

Although many of its applications in medical diagnosis and treatment are very effective, there is the possibility that the treatment may not be permanent or the treatment may be incomplete.

Additional problems could also occur. For example, there may be infections following treatment, some scarring may happen and there could be some pain after the initial treatment. Some patients have noticed some significant skin tone changes and also bleeding during and after the procedure.

To safeguard against the risks of using lasers in medical treatment and diagnosis, the exposure limits are regulated. The damage that can occur from the use of lasers during medical treatment is generally related to the eye and to the tissues and skin. Laser light can be inadvertently reflected or misdirected and is called 'stray optical radiation', a particular danger for UV or IR lasers because the radiation cannot be seen and the eye will not blink closed. To prevent the possible unintentional effects of laser light on both the patient and the medical practitioner administering the procedure, the working establishment will have produced its own set of procedures and standards that must be followed by staff. This will include:

▶ the use of protective clothing where normal surgical garments are regarded as insufficient

▶ protective goggles.

Magnetic resonance imaging (MRI)

The use of magnetic resonance imaging is now widespread and is a common means of diagnosis for a number of conditions. The process is regarded as very safe, since the principles do not physically affect the patient. However, as with many new technological devices used in hospitals, there are safety aspects that need to be considered when the procedure is performed.

▶ The patient must be told to remove any metallic items which they may have on them including clothing, coins, belts, etc. The main danger in MRI is the risk of metal objects (especially iron or steel) such as gas cylinders becoming projectiles due to the very strong magnetic fields. This is eliminated by banning all loose metal objects from the entire vicinity.

▶ The patient must be asked a series of questions to ensure that they have not had a cochlear implant, metal clips from brain operations, metallic foreign body lodged in the eye, a cardiac pacemaker inserted, surgery in the last 8 weeks, and that they are not pregnant.

- If the patient is claustrophobic, then the procedure may be adjusted to limit the time in the tunnel. The patient is supplied with a 'panic' button which can be pressed if they feel too uncomfortable.

- The process can be quite noisy especially when the field coils are switching on and off.

ⅠⅠ PAUSE POINT   What would you consider to be the dangers associated with a metal implant or metal fragment that is inside a patient's body during an MRI scan?

      Hint   Consider the strength of the magnetic field produced and the time that the patient is in the tunnel.

      Extend   What could be the effect of an MRI scan on a patient who has undergone recent major surgery?

Safety precautions, side effects and risks for operators and patients of ionising radiation

Safe operating procedures

As with non-ionising radiations, safe operating procedures (SOPs) exist to ensure that all ionising procedures are performed with the utmost safeguards against harm to both the patient and the medical staff carrying out the procedure. In addition to the list provided in the previous sections, medical practitioners who use X-ray equipment, CT scanners, gamma ray and gamma knife surgery equipment and proton beam equipment must comply with the following requirements before any diagnosis, treatment or therapy can take place.

- Have full understanding of the radiation type to be used, the doses applicable and the exposure limits.

- The relevant safeguards which need to be in place to protect the operator.

- The higher dosage factor of using CT scanners.

- The need to wear an appropriate radiation detection badge for the radiation type in use.

- Biological effects of ionising radiation on internal organs, the skin and the eyes.

Medical staff must wear sensitive radiation-detecting equipment (**dosimeters**) when working with ionising radiation. The guidelines are strictly controlled. The types of radiation detectors come in two general forms although there are many variations in use.

- Film badges – these consist of photographic film housed in a light-tight package. Ionising radiation penetrates the packaging and gradually blackens the film. They are 'one use' only and are usually developed and read after a significant period of use – weeks or months depending on the hazard level.

- Ring badges – these contain lithium fluoride crystals that also trap excited electrons within them until the badge is heated to very high temperatures. The energy given off as visible light is proportional to the radiation dose. This is called **thermoluminescence**. They are worn on the hand.

Key terms

Dosimeter – a device for measuring cumulative ionising radiation dose.

Thermoluminescence – the ability of some materials to glow when exposed to certain types of energy over a period of time.

A hospital notice will normally be highly visible in the relevant department as a serious reminder for working with ionising radiation. It may look like the one in Figure 21.14.

Reduce Your Exposure!

Don't take a break in a radioactive materials storage area.

Check your time in an area of radioactive materials.

Look at your dosimeters periodically.

Shield yourself from exposure.

Notify your manager if you are pregnant.

- **Figure 21.14** Hospital warning notice

Health and Safety Executive

The Health and Safety Executive (HSE), as outlined in the previous sections, is also the statutory body which is responsible for regulating and enforcing health and safety matters related to the use and maintenance of ionising methods in diagnosis, treatment and therapy in medical establishments in the UK. The organisation carries out numerous inspections on the use of specific medical instrumentation and equipment, such as highly technical X-ray and CT scanners and gamma ray methods, as well as the operating conditions of staff. It works with the law to ensure that the findings of its reports are properly enforced where necessary. As yet, there are no proton beam synchrotrons available in the UK. These are expected to become operational after 2018 and will be introduced into the HSE guidelines for inspection and maintenance.

Research

Use the HSE website to find out the current legislative requirements associated with:

1 X-rays and CT scanners
2 gamma ray imaging
3 gamma knife surgery in the medical profession.

HSE publication L121 'work with ionising radiation' may help.

Figure 21.15 shows examples of the variation in levels of ionising radiation exposure.

Effect of X-rays, CT, radiotherapy and gamma rays

Ionising radiations used in the diagnosis, treatment and therapy of medical conditions, such as cancer, pose a measurable health risk to patients and to the medical practitioners who perform the task.

The amount of radiation given to patients in diagnosis is dependent on how close vital organs and tissues are to the malignant tumour. There are two terms commonly used by scientists when dealing with radiation doses:

▶ *Absorbed dose* – the amount of energy (in joules) received by a mass of tissue. This is measured in kilograms (kg). It has the unit J/kg and is called the gray (Gy).

▶ *Effective dose* – if the ionising radiation types are compared using the same amounts of energy, alpha particles cause much more biological damage, 20 times more damage than X-rays. In medicine, radiation affects different tissues and organs in different ways and so each tissue or organ has a number that is used as a quality factor. The absorbed dose is multiplied by this number to give the figure for effective dose – also measured in J/kg but called the **sievert (Sv)**.

Key term

Sievert (Sv = SI unit) – the SI unit of effective radiation dose (joules of radiation energy per kilogram of tissue, multiplied by a dimensionless quality factor that depends on the type of radiation).

Mammogram (0.4 mSv)
This is an X-ray scan used for patients for breast cancer screening.

Background (2.7 mSv)
This is radiation received by a typical person from the air or rocks in the UK in one year.

Chest X-ray (0.014 mSv)
Standard X-ray of a patient's chest in one sitting.

CT scan of the spine (10 mSV)
A high dose but regarded as necessary with regards to the benefits involved.

Maximum UK legal annual limit (20 mSv)
A limit for UK workers based on the IRR 1999 guidelines.

▶ **Figure 21.15** Examples of the variation in levels of ionising radiation exposure

Worked Example

A patient requires a course of six chest X-rays over a three-month period to determine the extent and exact nature of the lung cancer that has been caused by prolonged smoking of tobacco. The patient is concerned by the number of X-rays needed and asks if it is safe. Your answer should link the equivalent number of X-rays of the chest to a) the normal dosage of background radiation and b) the dosage for a mammogram.

a) Natural background radiation dosage in 1 year = 2.7 mSv. X-ray dosage is 6 × 0.014 mSv = 0.084 mSv. Compared to the natural annual dosage which is 2.7 mSv, this dosage is 0.084/2.7 × 100% = 3.1%

Find the comparison with a mammogram as a percentage.

Ionising radiation treatment – the downside

Medical health professionals sometimes refer to malignant tissue in the patient as the 'target' and radiation used in this treatment often damages or destroys cells around the target tissue area. DNA within the cells, for example, can be damaged and affect the ability of the cells to grow and divide.

The treatment of malignant cancer is a balancing act. Doctors target the cancerous cells as best they can and try to limit the number of healthy cells that are damaged. Most healthy cells can recover from this treatment if they are allowed sufficient time. This is why doses of radiation therapy are given in short bursts over many weeks. After varying doses of radiation, cells may:

▶ be damaged but repair themselves and operate normally

▶ be damaged but be repaired enough to operate abnormally

▶ die.

The treatment may cause the patient to suffer from side effects such as nausea, skin reactions, hair loss, diarrhoea and tiredness.

Apart from the effects to the patient of the use of ionising radiations in the diagnosis, treatment and therapy of certain medical conditions, there are other possible side effects or experiences that may prove problematic when undergoing the procedures.

▶ Computer tomography (CT) – this process uses a tunnel that houses the X-ray scanners and may cause patients to panic if they suffer from claustrophobia.

▶ Gamma knife surgery – this procedure uses a heavy and sometimes very cumbersome head device which can cause a certain amount of discomfort to the patient. There are fewer side effects to the radiation than radiotherapy, for example.

Assessment practice 21.3

C.P5 C.P6 C.M3 C.D2

The dangers associated with both ionising and non-ionising radiation used in medical diagnosis and treatment are far reaching, producing many side effects and involving particular risks to operators and the patient.

Practitioners in this working environment need to clearly understand the risks and the safety precautions to be followed. The information must be clear and simple for everyone involved.

Develop a large poster or set of leaflets which help to explain, in basic terms and with clarity, the fundamental aspects of the following under the headings: Risks and side-effects for operators, Risks and side-effects for patients, Safety precautions.

- ultrasound
- lasers
- MRI
- X-rays and gamma rays
- proton beam therapy

Plan
- What is the task? What am I being asked to do?
- How confident do I feel in my own abilities to complete this task? Are there any areas I think I may struggle with?

Do
- I know what it is I'm doing and what I want to achieve.
- I can identify when I've gone wrong and adjust my thinking/approach to get myself back on course.

Review
- I can explain the results obtained from the task.
- I can apply the activity to other situations.

Further reading and resources

Websites

www.nhs.uk – useful websites for all the ionising/non-ionising principles and effects of radiation diagnosis, treatment and therapy, proton beam therapy, etc.

www.mhra.gov.uk – information for regulation of radiation.

www.hse.gov.uk – website of the regulation authority for health and safety.

www.npl.co.uk – National Physics Laboratory website for ionising radiation and ultrasound.

www.physicsclassroom.com – tutorials, interactives and more.

THINK ▶FUTURE

Matthew Lewis Radiographer

I work in the radiology section of a very busy general hospital in one of the largest cities in the UK. I have been working as a radiographer for about one year, after studying for my degree and qualifying whilst working in the hospital on an almost full-time basis.

At present I am not working as a diagnostic radiographer. My role involves more of a patient care aspect within the MRI department, although I am also asked to work in other areas of the hospital, including accident and emergency, operating theatre and on the wards, to gain more experience in both the techniques used and in my dealings with patients and other healthcare professionals. Any images produced by the MRI scanner are immediately sent to the highly qualified radiologist who will then be able to determine the patient's condition and the course of treatment available.

My responsibilities are quite extensive and I keep a regular log of my time in other departments and notes that will help me in my continued professional development in the MRI scanning section. When making first contact with a patient, I try to be very reassuring and supportive in order to put them at ease. The whole process can cause anxiety in some patients and this will not help the diagnostic treatment. It's my job to determine how the patient will be prepared for the MRI procedure based on the notes from the initial consultation. This may require a full body scan or simply a head scan depending on the area under study. The supervising radiologist and section manager oversee my work to check on my continued professional progress and understanding of procedures. I'm hoping that my additional study time will allow me the opportunity to specialise in other areas of the hospital and open up my chances of further promotion. This is actively encouraged by the hospital because it helps to ensure that hospital staff remain enthusiastic, involved in the processes and interested in the various medical techniques currently employed.

Focusing your skills

Important procedures

The magnetic resonance imaging section of the hospital can be quite a daunting environment to work in when you are relatively new to the hospital work force. When you become familiar with the staff, procedures and the equipment used, it can become very routine. It is sometimes difficult to remember that you are using equipment at the forefront of our current medical technology and that the method it uses is very difficult to explain or even to understand. Patient care, though, remains our priority and we need to keep a focus on the way that we treat patients and what must be done to ensure that they have a safe and calm experience whilst inside the MRI scanner.

We cannot always assume that all patients have received instructions and so must go through the basic checks firstly to make sure that important points are understood. This includes: checking their personal details on our records, asking them to remove credit cards, jewellery and loose metal objects such as coins, which we place inside a locker for them, then going through a questionnaire with them.

Getting ready for assessment

Rieo is working towards his BTEC National Diploma in Applied Science. The assignment he was given at the start of the first term was based on explaining the principles involved with the production of non-ionising radiation techniques used in medical diagnosis and treatment and then to compare the techniques to assess their suitability for the treatments they provide. The scenario placed Rieo in the position of a radiologist responsible for administering the doses of non-ionising radiation types to patients. The role expects that Rieo will work for different time periods in all relevant departments of the hospital. A full report must be completed.

The report must include:

▸ formal outline of the methods involved in non-ionising techniques

▸ diagrammatic representation of the methods or equipment used

▸ comparison of the methods used and their effectiveness in diagnosis and treatment.

Rieo shares his experience below.

How I got started

First I made sure that my laboratory notebook was available and that I had a clear understanding of what I needed to do. I planned the activity, listing all the non-ionising types of radiation and equipment used. I drew up a table for my list and separated out the key points of each method. I decided to include a variety of sources in my research since some may not provide the correct information or not enough.

Having taken notes from the research sites and listing them as possible reference sources, I began to fill in my table. I roughly outlined the diagrams I would use and listed the essential information and principles in sequence.

How I brought it all together

I began to carefully draw the equipment. For some methods I also used direct downloaded diagrams but carefully annotated them with clear labels and explanations of the principle components used in the equipment and what they are for. I made full note of the references for my report. I then:

▸ finalised the information in the table

▸ added clear notes and completed diagrams for each of the methods outlined

▸ compared the effectiveness of each method, highlighting how each is suitable in the given situation.

After completing most sections in my report and evaluating the work based on my research, I was able to complete an abstract for the report outlining what the activity had told me about the principles used, the reason for using the non-ionising radiation methods and how each method is suitable for the purpose to which it is applied.

What I learned from the experience

I became quite involved with the task and the information that I was beginning to find from research sources. The explanations in physics terms were very complicated, in particular for MRI and lasers, but I was happy with my understanding of how they worked in principle. I also noted that there were many research sources that gave either the same information or very detailed information and learned to ignore some aspects which I thought were too mathematical.

Finding relevant diagrams was not easy. There are many images and photographs of the equipment in use but I needed to produce my own labelled diagrams from a number of different images to get the best result for my report. This was particularly true of the MRI scanner since much of the physics involved in how it works is difficult to visualise. Comparing the methods used was relatively straightforward because each has a certain benefit in terms of how suitable they may be for treating or diagnosing medical conditions.

Think about it

▸ Have you made a clear note of the agreed submission date of the assignment?

▸ Do you have your previous class notes and diagrams to hand in order to use them for reference with your report?

▸ Is your final report written in your own words and referenced clearly where you have used quotations or information from a book, medical journal or website?

CRIME SCENE-D

R

Forensic Evidence, Collection and Analysis

23

Getting to know your unit

Assessment

You will be assessed by a series of assignments set by your tutor.

Forensic science is science relating to the law. Forensic scientists collect evidence in order to work out what happened at a crime scene, and in a form that can be presented in a court case. In this unit you will learn about health and safety issues relating to forensic science, and practise carrying out a forensic analysis.

How you will be assessed

This unit will be assessed by a series of internally assessed tasks set by your tutor. Throughout this unit you will find assessment activity activities that will help you work towards your assessment. Completing these activities will not mean that you have achieved a particular grade, but you will have carried out useful research or preparation that will be relevant when it comes to your final assignment.

In order for you to achieve the tasks in your assignment, it is important to check that you have met all of the Pass grading criteria. You can do this as you work your way through the assignment.

If you are hoping to gain a Merit or Distinction, you should also make sure that you present the information in your assignment in the style that is required by the relevant assessment criterion. For example, Merit criteria require you to analyse and discuss, and Distinction criteria require you to assess and evaluate.

The assignments set by your tutor will consist of a number of tasks designed to meet the criteria in the table. This is likely to consist of a written assignment but may also include activities such as:

- creating a report about how to gather forensic evidence
- processing a simulated crime scene
- a portfolio of laboratory examination forms
- an appropriately structured expert witness statement that includes conclusions, explanations and aspects of probability.

Assessment criteria

This table shows what you must do in order to achieve a **Pass**, **Merit** or **Distinction** grade, and where you can find activities to help you.

Pass	Merit	Distinction

Learning aim **A** Understand how to gather forensic evidence, the integrity to forensic investigation and the importance of health and safety

Pass	Merit	Distinction
A.P1 Describe the procedures used to gather evidence for forensic investigation.	**A.M1** Justify the importance of the procedures used to gather evidence for forensic investigations.	**A.D1** Discuss the importance of documentation methods to forensic investigation and analyse each personnel used to gather evidence in relation to health and safety and to preserve the integrity of evidence. Assessment practice 23.1
A.P2 Outline the roles of crime scene and authorised personnel that attend crime scenes. Assessment practice 23.1	**A.M2** Explain the structure and importance of Scientific Support Units and authorised personnel to forensic investigations. Assessment practice 23.1	

Learning aim **B** Investigate a simulated crime scene using forensic procedures

Pass	Merit	Distinction
B.P3 Carry out a forensic examination of a simulated crime scene, using appropriate forensic procedures to gather biological, physical and chemical evidence.	**B.M3** Justify the forensic procedures used to process a simulated crime scene. Assessment practice 23.2	**B.D2** Evaluate procedures used to process a simulated crime scene. Assessment practice 23.2
B.P4 Describe the forensic methods used to process a simulated crime scene. Assessment practice 23.2		

Learning aim **C** Conduct scientific analysis of physical, biological and chemical evidence

Pass	Merit	Distinction
C.P5 Explain the techniques used in forensic science to analyse physical, chemical and biological evidence. Assessment practice 23.3	**C.M4** Draw valid conclusions from the analysis of physical, chemical and biological evidence. Assessment practice 23.3	**CD.D3** Evaluate the techniques used in forensic science to analyse the physical, chemical and biological evidence gathered and evaluate the findings, including aspects of probability. Assessment practice 23.3
C.P6 Demonstrate analysis of physical, biological and chemical evidence gathered from a simulated crime scene to draw conclusions. Assessment practice 23.3		

Learning aim **D** Be able to justify methods, interpret findings and report on conclusions of forensic techniques and analysis

Pass	Merit	Distinction
D.P7 Draw conclusions from analysis of physical, chemical and biological evidence. Assessment practice 23.4	**D.M5** Explain how physical, chemical and biological forensic analysis justifies the conclusions. Assessment practice 23.4	

Getting started

Forensic science is the application of any branch of science to answer questions of a legal nature and to aid criminal investigations. Forensic science is based on Locard's principle: every contact leaves a trace. By collecting and analysing these evidence traces, we can prove the contact and identify the link between suspect, crime scene and victim. Crime scene investigators (CSIs), also known as scene of crime officers (SOCOs), process the crime scene and recover any potential forensic evidence. In this unit you will learn how to examine and process a simulated crime scene, and how to collect different types of forensic evidence.

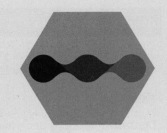

Forensic scientists apply analytical scientific techniques to analyse biological, physical and chemical evidence collected from crime scenes. You will also examine and scientifically test forensic evidence in the laboratory. Forensic experts provide the results and interpretation of their examination in a report for police investigators and the courts, and they may give verbal evidence in court as an expert witness. By the end of the unit, you will be able to document, interpret and present your forensic evidence.

A Understand how to gather forensic evidence, the integrity to forensic investigation and the importance of health and safety

At the crime scene

In this section, you will learn about the initial response to a crime, the personnel that are authorised to enter a crime scene and the importance of early information gathering by all personnel involved. Think about who may not be allowed access to a crime scene and reasons for this.

Scientific support at the crime scene

The role of the **scene of crime officer (SOCO)**, which is also referred to in some police forces as a **crime scene investigator (CSI)**, is to preserve and process a crime scene. They must follow legislation to ensure that all potential forensic evidence is valid and can therefore be presented in court with integrity. They must safely and correctly identify, document, collect, package, label and securely transport evidence, ensuring the **continuity of evidence**.

> **Key terms**
>
> **Scene of crime officer (SOCO)** – an officer who is responsible for evidence collection at a crime scene.
>
> **Crime scene investigator (CSI)** – an officer who is responsible for evidence collection at a crime scene.
>
> **Continuity of evidence** – complete documentation that accounts for the progress of an item of evidence throughout the entire investigation from crime scene to court.

Structure of Scientific Support Units (SSU)

The structure of scientific support units varies between regional forces but in general they consist of the following.

- CSIs attend crime scenes and recover forensic evidence such as fingerprints, DNA and footwear impressions. They also take photographs at crime scenes, some forces have specialist forensic photographers.
- Crime scene coordinators oversee complex crime scenes.
- Crime scene managers are responsible for managing specific crime scenes and investigations.
- Forensic experts are based in laboratories and look at every case and decide which forensic process should be applied, either externally or internally, and carry out analysis in their field of expertise.

Authorised personnel

The first member of personnel to arrive at the scene is called the **first attending officer (FAO)**. This is usually a police officer. Their first task is to assess the scene and establish whether assistance in necessary. The FAO will also search for, and arrest, any suspects if they are still present at the crime scene, and establish if there are any injuries or witnesses. They may even preserve any potential forensic evidence from weathering, for example, by covering a footprint in soil from rain.

First attending officer (FAO) – the first officer at the scene of a crime.

Other authorised personnel

Table 23.1 lists other personnel authorised to enter a crime scene if they require access.

▶ **Table 23.1** Personnel authorised to enter a crime scene

Authorised personnel	Example of reason for access
Crime scene manager	To manage a team of SOCO/CSIs in their evidence collection.
Police officers and Investigators	To collect witness statements from eye witnesses.
Forensic scientists	A forensic pathologist, archaeologist or anthropologist may be called if there is potential evidence relating to these specialised areas.
Paramedic staff	To attend to a medical emergency within the scene, e.g. a victim who is badly injured.
Fire and rescue service officers	To extinguish a fire, and to search and rescue for persons trapped in fire or vehicles.

The investigation

The purpose of the investigation is to establish:

▶ who
▶ when
▶ where
▶ what.

Who

The investigators need to know who the victim is. The victim may be alive and able to give evidence or the victim may need to be identified. The investigators need to also know who the **perpetrator** is. This may be difficult if they are not present at the scene. Therefore, the collection of forensic evidence is important to provide information to narrow the perpetrator down to one individual. It is also important to know 'who' the witnesses are, to provide further information to help 'paint the picture.' They may have seen or heard something in relation to the crime.

Key term

Perpetrator – a person who commits a criminal offence.

When

In order to establish when, investigators need to produce a timeline of events. They should start before the crime took place and look at movements after the crime, again to help 'paint a picture'. The investigators are always looking back at the crime as an event that has already happened, so they must gather as much information as possible to try to understand the events that took place.

Where

Investigators need to understand where the crime took place. For example, where did the perpetrator enter the property, where did they go and where did they leave from? If investigators are dealing with a murder, is it possible that the body was moved to the place where it was found? If so, there may be another crime scene where the murder took place. This may contain very important forensic evidence.

What

Finally, investigators need to understand what the motive was behind the crime, if there was one. They will gather information about past relationship connections between victim and perpetrator to help understand why the crime occurred. They will also record the *modus operandi* (the method of operation). It is a term used by law enforcement authorities to describe the particular manner in which a crime is committed. This can be particularly helpful in connecting different crimes where there are similarities between them, and producing a profile of the perpetrator.

Preservation and recovery of evidence

In this section you will learn the procedures that a SOCO/CSI will perform while processing a crime scene, in order to ensure health and safety, and to collect evidence with integrity. Think about the evidence you may be faced with in your mock crime scene and how you will process your crime scene.

Restriction of the scene and restriction of access

When a SOCO/CSI first arrives at the crime scene, there are a number of security and safety issues that need to be dealt with before the scene and evidence can be processed. Anybody who enters the crime scene could potentially destroy or contaminate evidence by accident if they are not very careful, so access must be controlled. Only authorised personnel are allowed entrance to the crime scene via a suitable **common approach path (CAP)**. To preserve the crime scene and protect the forensic evidence, the boundaries of the crime scene must be identified to isolate the scene. This can be done using barrier tape, vehicles and guards to create a police cordon that encloses the crime scene. A log must also be kept, recording who has had access to the scene, their role and time of entry and exit.

> **Key term**
>
> **Common approach path (CAP)** – the common route used to enter and leave a crime scene. It is used to preserve forensic evidence.

Observation and recording of the scene

A risk assessment will be carried out by the SOCO/CSI to identify health and safety concerns to ensure that they and their team know the **hazards** and how to prevent injury. (Health and safety is discussed later in this unit.) Once the risk assessment has been carried out, authorised people can enter the scene and make a general survey, identifying obvious pieces of evidence and potential points of entry and exit that may have been used by the offender. At this point, the SOCO may protect and preserve any potential forensic evidence, to prevent damage or deterioration. For example, a tent may be used to protect the victim's body from the weather at an outdoor murder scene, or a bin lid may be used to protect a footwear impression in mud.

> **Key term**
>
> **Hazard** – a potential source of harm.

Documenting and recovery of trace materials

The SOCO then has to record the crime scene in its original state, before any evidence is moved or collected. There are a variety of methods for documenting the scene:

1 Note-taking: notes should be specific, detailed, accurate, organised and legible and include the name and signature of SOCOs.

2 Sketches: rough sketches and measurements are made and the location of items of evidence. The sketch should be labelled and include a key and north line showing direction.

3 Photographs: overview photos of the entire scene and surrounding area, including points of entry and exit. Also items of evidence should be photographed to show their original position in the crime scene before they have been moved. Close-up photos of evidence are then taken to record detail. A measuring scale is included in the photo to show the size of each item.

4 Videography: video recordings can be used in addition to still photography.

The information that should be included in the SOCO's crime scene notes includes:

▸ date and times of examination
▸ crime case number/police reference
▸ location/address of crime scene
▸ detailed description of crime scene
▸ location of items of evidence
▸ description of evidence including evidence item/ **exhibit** numbers
▸ time of discovery and sketch of evidence
▸ description of evidence collection and packaging techniques including evidence bag number and time of collection
▸ storage and transportation information.

Prevention of contamination

Contamination is the unwanted transfer of material. This must be avoided at all costs at the crime scene. Individuals can contaminate the scene and evidence at any stage of an investigation, and unwanted transfer of material can also occur between two or more sources of evidence. This is called cross-contamination. Table 23.2 shows a number of prevention measures carried out by CSIs to reduce

▸ **Table 23.2** How contamination can be prevented

How can evidence be contaminated?	How can contamination be prevented?
SOCO/CSI leaving fingerprints, footprints, hairs, fibres or DNA at the scene	Wear full PPE – for a SOCO, this includes suit, gloves, mask and overshoes. There is no eating and drinking at the crime scene and they are not to take in personal items.
Unsealed packaging	Choose correct type of packaging carefully to pack the evidence into and seal carefully to ensure there are no holes where evidence can escape or enter.
Placing two items of evidence into one evidence bag/box (cross-contamination)	Pack items of evidence into separate suitable packaging.
Using old, dirty equipment at a number of different crime scenes	Use new sterile, disposable equipment at every scene e.g. tweezers, swabs, scissors.

contamination. It is so important that contamination prevention measures are carried out, because items of evidence that are deemed as contaminated cannot be used in court, and they may also bring into question the integrity of the other items of evidence in a criminal investigation, which could lead to a non-conviction or a wrong conviction.

Key terms

Exhibit – an object collected for forensic examination.

Contamination – unwanted transfer of material.

The SOCO/CSI must wear full **personal protective equipment (PPE)** to avoid contamination. This must be put on before entry to the crime scene. This disposable protective clothing and equipment worn by SOCOs also reduces the **risk** of injury and protects the SOCO from harm at the crime scene.

Key terms

Personal protective equipment (PPE) – specialised clothing or equipment worn by employees to protect against health and safety hazards.

Risk – a description of the harm that may be caused by a hazard.

▶ SOCO/CSI suit worn to prevent contamination of a crime scene

Theory into practice

1   Identify all the parts of the PPE worn in the photograph.
2   Explain why each is worn and how it avoids contamination.

Methods of collection

Once the crime scene has been recorded, the CSI can start to search the scene intensively for evidence. There are a number of ways to search the scene, following different search patterns, and these will be discussed later in the unit. When items of evidence are discovered, they are photographed **in situ** with an evidence scale (see Figure 23.1) and documented, before the evidence is collected. Details of the evidence, including a description, the location and time of discovery, are recorded. Each item of evidence must be given a unique evidence/exhibit number, usually the initials of the CSI followed by a sequential number. The evidence can then be collected; there are a number of different techniques the SOCO can use, depending on the type of evidence.

Key term

In situ – when an artefact has not been moved from its original place.

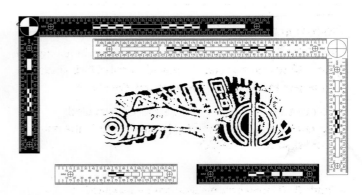

▶ **Figure 23.1** A footprint with a measuring scale

Theory into practice

1   Add to Table 23.3 by researching these method of evidence collection techniques: casting, swabbing, hand picking and vacuuming.
2   Describe each technique.
3   State the types of evidence that each technique may be suitable to collect.

Method of evidence collection	Description of technique	Types of suitable evidence
Shaking	Gently shake item over large piece of paper and collect loose particulate material that falls off.	To recover trace evidence e.g. glass fragments, paint chips, hairs, fibres.
Brushing	Brush surface with clean tooth/paint brush and collect debris on paper/container.	To remove trapped particles from surfaces like shoes, pocket linings, suspect or victim's hair, e.g. gunpowder residue, soil, pollen, hairs, fibres.
Taping	Apply strips of clear sticky tape or gel lifters to surfaces to pick up trace evidence. Sequentially, pull off strips of tape, stick down onto clear plastic acetate sheets and examine with microscope.	To recover fingerprints, fibres and hairs from clothing, car seats, window ledges, edges of broken glass at point of entry, any dry surface.

Packaging and labelling

When the evidence is collected, suitable packaging must be used to protect it, keep the item secure and prevent contamination. There are a range of types of packaging available to the SOCO/CSI, depending on the type and size of evidence, and any health and safety hazards. A common type of packaging used is the tamper evidence bag, a plastic bag imprinted with a unique evidence bag number and an evidence detail label, which is completed by the SOCO at the scene. The bag has a sticky strip to seal it, and if the bag is tampered with in any way, the seal distorts to indicate this tampering. They come in a range of sizes.

These plastic tamper evidence bags are used for small items of clothing, cans, bottles, mobile phones and other small items. Small grip-seal plastic bags are used to package trace evidence such as hairs, fibres and drugs.

There are also small and large brown paper evidence bags for large items of clothing, bed sheets, and curtains, but the SOCO must ensure that these items are not wet as this may distort the paper bag and cause contamination. If any items of clothing were potentially wet and stained, they would be packaged in a plastic tamper evidence bag.

SOCOs/CSIs also have access to hard packaging such as:

▶ weapons tubes to package sharp items to prevent injury, e.g. knives, syringes and scissors.

▶ evidence boxes to protect small and large types of evidence, e.g. guns and mobile phones.

▶ metal/glass air-tight tubes to package small, trace items, e.g. bullets, soil, and hairs. Used for volatile liquids, e.g. petrol or paraffin, to prevent evaporation into air.

▶ Weapon tubes used to securely package knives, for example, found at a crime scene

The packaged evidence must be accurately labelled with the complete details of the item, to maintain the chain of continuity. The following pieces of information should be recorded on the evidence packaging:

▶ evidence number

▶ description and location of evidence

▶ time and date of collection

▶ crime case number

▶ name and signature of SOCO.

Correct packaging and labelling of evidence is essential, as evidence packaged incorrectly may cause harm to others in contact with it, or may cause the evidence to deteriorate or potentially lead to contamination. Packaging that appears to have been tampered with may be dismissed from court, and affect the outcome of the case.

▶ Plastic tamper-proof evidence bags

Storage and transport of a variety of materials

When the evidence has been collected, it must be safely and securely transferred to the evidence store and forensic laboratory. The SOCO will transport the evidence back to the police station/forensic evidence store and book the evidence in. This provides an audit trail and record of each exhibit. From here, the evidence is sent, along with the corresponding paperwork, to the correct forensic laboratory for analysis, using a secure and reliable courier company or police personnel. The storage of evidence must be secure at all times to prevent unauthorised personnel from removing or tampering with the evidence.

While in storage it is important that the evidence is stored under suitable conditions. The evidence must be preserved and protected from deterioration and contamination. For example, blood swabs must be stored between 4 °C and −20 °C, to prevent the blood sample from decaying.

Continuity of evidence

For each piece of evidence submitted to court, there must be thorough and complete documentation of the evidence (notes, sketches and photographs) recording the procedures and methods used to collect it. In addition, a log must be kept showing who was in possession of the evidence at every stage of the criminal investigation. This is either recorded on the evidence bags and boxes themselves, or via a label that is attached to the evidence securely. Together, these maintain the continuity of evidence.

Theory into practice

You are a trainee CSI. You have just finished a day's training and you need to write a plan of the procedures you would carry out if you were faced with the scenario below.

A witness has called the police and reported that they have woken to an abandoned blue car on their drive. The car has mounted their grass and it is very muddy. The witness has reported there is blood on the window screen.

The police report says the car was stolen and was possibly used in an armed robbery.

Write a plan of what you will do to ensure successful processing of the crime scene. Use your imagination about the evidence you may find in a stolen car and how you would collect and package it.

Search patterns

When searching for forensic evidence at a crime scene, the SOCO will decide on the most methodical approach to process the scene and collect the evidence to ensure that no evidence is missed, damaged or contaminated. There are various different search patterns that are commonly used. These vary in name depending on the region of the country where they are used. The use of specific search methods will be determined by the location and size of the crime scene in question.

Quadrant

The quadrant search method (Figure 23.2) is also known as a zone search. Using this method, the crime scene is divided into smaller sections/zones that can be assigned to small teams to search. These sections can be further subdivided into smaller sections to be searched

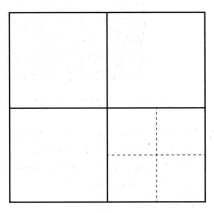

▶ **Figure 23.2** Quadrant search diagram

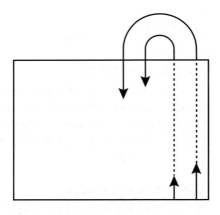
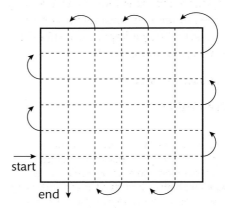

▶ **Figure 23.3** Lane search (above left) and grid search (above right) methods

thoroughly. An example of a crime scene that may be processed in this way is the inside of a vehicle where it can be split into 5 distinct zones, front offside, front nearside, back offside, back near side and boot.

Lane

This is also known as a strip search. In this method, one or more investigators walk in straight lines across the crime scene, collecting the evidence as they go. Investigators walk from one side of the crime scene to the other, returning in the direction that they came from (like swimming lengths in pool lanes). This method is recommended when searching a large area outdoors (Figure 23.3).

Grid

This is similar to the lane search, but the investigators walk across the search they carried out initially, rather than walking just backwards and forwards. They will then change angle so they cover the area like a grid (see Figure 23.3).

Spiral

Spiral searches involve investigators searching for evidence in a widening circle (Figure 23.4). They start from the core of the crime scene and move outwards to the outside.

Wheel

This method requires many investigators to start in the centre of an imaginary circle and move outwards as if they were moving along the spokes of a bicycle wheel (Figure 23.5). This method can cause potential damage to evidence in the centre if care is not taken because the investigators will gather here to start their search.

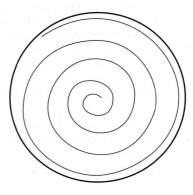

▶ **Figure 23.4** Spiral search

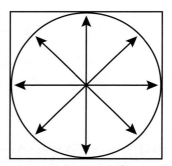

▶ **Figure 23.5** Wheel search

> **Reflect**
>
> Think about a crime scene that involves a stolen car. What might be the best method to search a car, to ensure that you do not miss any important pieces of forensic evidence? (Tip: Think about the layout of a car.)

Health and safety

Crime scenes pose many risks, which could potentially hurt or injure personnel processing the scene. Work at the crime scene is regulated by a number of pieces of legislation that act to protect the employee and other individuals that may be present. CSIs and their employers must follow certain requirements to ensure everybody is kept safe and healthy. Table 23.4 lists some of these requirements.

▶ **Table 23.4** Health and safety legislation

Government legislation	Description
Health and Safety Act 1974	This act covers the majority of the law regarding workplace health and safety. It describes what the employer must do to maintain high standards of health and safety at work and protect anybody at risk from the workplace activities, including their employees, as well as the general public.
Control of Substances Hazardous to Health (COSHH) Regulations 2002	This act requires employers to protect those at risk from hazardous substances used in the workplace. It describes the methods used to reduce occupational illness, including risk assessment, exposure control, incident planning, employee training and health surveillance.
Management of Health and Safety at Work Regulations 1999	This act enforces the employer to carry out a risk assessment of the work place and put in place suitable control measures. It aims to identify health and safety and fire risks, and must be reviewed when necessary.

Case study

Health and safety

Ronnie, a senior CSI, and his team have been called to a road traffic collision to collect forensic evidence to discover the events that led up to the collision. He must ensure that everyone is safe.

Check your knowledge

1 List the hazards that Ronnie and his team may face at a road traffic collision.
2 Explain the risks associated with each hazard.
3 What could the team do to reduce the chance of being injured?

Disposable personal protective equipment

Full personal protective kits must be worn by a CSI whilst they are processing the crime scene. The kit is worn not only to prevent contamination, but also to protect the CSI from any potentially dangerous substances they may come into contact with during their search.

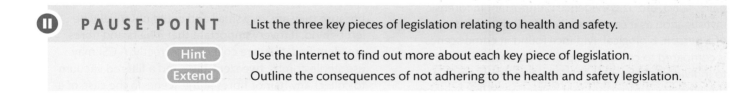

PAUSE POINT List the three key pieces of legislation relating to health and safety.

Hint Use the Internet to find out more about each key piece of legislation.

Extend Outline the consequences of not adhering to the health and safety legislation.

Assessment practice 23.1   A.P1 A.P2 A.M1 A.M2 A.D1

You are a CSI working for your local police force. You have been asked to produce a manual for new CSIs that describes all the procedures used when gathering forensic evidence from a crime scene, and includes an explanation of why they are important. You should also include information about who has authorised access to crime scenes and their involvement with the crime scene. You should:

- describe all the procedures to preserve and recover evidence
- justify why it is important that these procedures are carried out
- describe search patterns used
- justify when these search patterns may be carried out
- describe health and safety considerations
- justify why it is important that health and safety is considered
- outline the roles of authorised personnel who may attend a crime scene and explain the structure of these teams.

You should also provide information in the manual about the importance of documentation throughout the examination of a crime scene.

B # Investigate a simulated crime scene using forensic procedures

In this section, you will learn about the collection of biological, chemical and physical evidence. This will help you to understand how to process your mock crime scene and what to do when you come across different pieces of evidence. Think about the different types of evidence you may come across when dealing with an arson attack compared to a house burglary.

Collection of biological evidence

Biological evidence is any evidence that is from, or was once from, a living organism, for example blood, hairs, saliva, etc. This evidence is important in forensic investigation as it can often allow **individualisation** of a person. Biological evidence found at crime scenes is compared to control samples or databases in order to individualise the people involved in the crime. It is therefore very important that collection of this evidence is done with integrity, and that the evidence is packaged, labelled and stored in the correct way.

> **Key term**
>
> **Individualisation** – to identify a specific individual.

Collection of blood

There are two different types of blood found at a crime scene: liquid (wet) and dried blood. Liquid blood evidence is generally collected using **sterile** swabs. Once the sample is collected, it must be refrigerated or frozen to preserve it, otherwise the sample might not be viable for testing after 48 hours. Dried bloodstains can also be found at the crime scene. If the dried blood is on a moveable item, them the item will be photographed, collected and packaged. If the dried blood is on large immovable objects, then these can either be cut to make them moveable, or distilled water

can be used to dampen the stain before swabbing it. When dried blood is found on clothing, the entire article of clothing should be packaged, labelled and delivered to the laboratory for analysis.

> **Key term**
>
> **Sterile** – free from bacteria and other living microorganisms.

Collection of hairs and fibres

Hairs and fibres at the scene are classed as trace evidence because they are very small and could potentially be left behind. It is very important that if hairs and fibres are identified, they are collected correctly. A CSI may use sterile combs, tweezers, tape and a filtered vacuum to collect any hair or fibres at the scene. In the case of a rape, the CSI or police medical examiner accompanies the victim to the hospital to obtain any hairs or fibres from the victim's body during a medical examination. The CSI seals any hairs or fibre evidence in separate and fully labelled containers for transport.

In some cases, hair may be attached to other exhibits such as clothing or a hat. The larger exhibit will be packaged as it is found. Once the exhibit is received at the laboratory, the scientist processing it will collect hairs and document them as further exhibits from the evidence, and label them accordingly. For example, the exhibit number JHM002A may be given to a hair that is found on a t-shirt with the exhibit number JHM002.

Collection of saliva and semen

Both saliva and semen samples are swabbed using sterile cotton swabs, or the item they are present on (for example, a bedsheet) will be seized for examination. Both evidence types are commonly found during sexual assault cases on the body and on clothes.

Collection of bones

Bones are usually handpicked (using gloves) at the scene to stop any unnecessary damage occurring to the bone, as this could lead to inaccurate observation and interpretation of markings. For example, if large forceps were used to pick a bone up, it could add two indentations on the bone that could lead an expert to interpret they were marks from a struggle. Bones are photographed in situ and then packaged in large plastic evidence bags, which are then fully labelled.

Collection of fingerprints

Fingerprints found at crime scenes are usually **latent** and need to be enhanced in order to be collected. Chemical enhancement of fingerprints is discussed later in this unit. Fingerprints can also be powdered to enhance them and make them visible at the crime scene. An aluminium or black powder is normally used, depending on the background material. Once the powder has adhered to the latent fingerprint, crime scene lifting tape is used to remove the enhanced print. This is fixed to a clear acetate sheet, which is labelled and sent for analysis.

> **Key term**

Latent – present but not visible.

▶ Collecting fingerprints from the scene

Collection of chemical evidence

Chemical evidence is any evidence that contains chemicals that may be found on people, objects or in solutions at the scene. This type of evidence includes drugs, paint, explosives, gunshot residue, fibres, soil samples and other chemicals.

Drugs: classes A, B and C

Drugs of abuse are defined as illegal drugs, or prescription/over-the-counter drugs used for purposes other than those for which they were intended. The Misuse of Drugs Act 1971 is aimed at preventing unauthorised use of drugs of abuse. It categorises drugs into three groups (A, B and C) based on the harm the drug can cause if misused (see Table 23.5). Category A are the most dangerous drugs, and category C are the least dangerous. These categories are used in court to determine the penalty for misuse of a drug. The second piece of legislation is the Misuse of Drugs Regulations 1985, which was amended in 2001. These regulations categorise drugs into five groups and describe the requirements concerning the legitimate prescription, distribution, production, record keeping and storage of drugs. Drugs of abuse can be broadly classified according to their effect on the central nervous system (CNS) and their impact on the activity of the brain. For example, stimulants like cocaine and amphetamines arouse and stimulate the CNS, depressants such as heroin, alcohol and tranquilisers have a depressing effect on the CNS and inhibit brain activity, and hallucinogens e.g. cannabis, ecstasy and magic mushrooms alter perception and mood, without stimulating or depressing the CNS.

▶ **Table 23.5** Legal classification of drugs of abuse

Class of drug	Drugs included	Penalties for possession	Penalties for dealing
A	Crack cocaine, cocaine, ecstasy (MDMA), heroin, LSD, magic mushrooms, methadone, methamphetamine (crystal meth)	Up to 7 years in prison or an unlimited fine or both	Up to life in prison or an unlimited fine or both
B	Amphetamines, barbiturates, cannabis, codeine, ketamine, methylphenidate (Ritalin), synthetic cannabinoids, synthetic cathinones (eg mephedrone, methoxetamine)	Up to 5 years in prison, an unlimited fine or both	Up to 14 years in prison, an unlimited fine or both
C	Anabolic steroids, benzodiazepines (diazepam), gamma hydroxybutyrate (GHB), gamma-butyrolactone (GBL), piperazines (BZP), khat	Up to 2 years in prison or an unlimited fine, or both	Up to 14 years in prison or an unlimited fine, or both

When drugs or poisons are found at the crime scene, they are seized in their entirety. Drugs and poisons can be tested at the scene with the correct presumptive drug test kits or they can be sent to a toxicology laboratory to be tested, to determine their identity and components. Either way, a small sample will be taken from the stock sample and packaged in small evidence collection tubes or sample tubes and sent securely for analysis.

Firearm discharge residue (FDR)

Firearm discharge residue (FDR) is expelled out of the muzzle of the firearm when it is fired. This can be swabbed with a sterile swab from the gun itself, from suspects and from surfaces that may have been close to the position where the firearm was discharged. FDR can be tested directly from the items, or the items can be swabbed and the swab would be packaged, labelled and sent for analysis.

> **Key term**
>
> **Firearm discharge residue (FDR)** – residue deposited on the surfaces near someone who discharges a firearm, e.g. on their clothes and skin.

Collection of physical evidence

Physical evidence is any evidence that has not come from a living or once living organism, and does not contain chemicals. This type of evidence includes:

- footprints
- tool marks
- firearms
- bullets
- documents and mobile phones.

Footprints

Two- and three-dimensional footprints can be left at crime scenes. The print left behind can give vital evidence about a perpetrator and can be used to **profile** suspects, as footprints can indicate the height and sex of the person, for example. Like fingerprints, footprints can also be latent and may need enhancing. Sometimes they can be obvious to the naked eye, due to blood or mud, for example. In either case, they need to be collected so that they can be compared to other shoe prints as required.

> **Key term**
>
> **Profile** – provide data and information on a person.

Two-dimensional footprints

Footprints can be left on many different surfaces such as paper, tiles and cars. Special lighting techniques, including oblique lighting, can be used to uncover hidden impressions. This involves shining a light source diagonally at the ground, so the ridges of any impression create a shadow, which identifies ground that may have been disturbed.

Two-dimensional footprints can also be made in dust and light dirt. These impressions can be lifted using **electrostatic lifting apparatus** (ESLA). Electrostatic dust print lifting devices work by placing a high voltage electrode into contact with a plastic film. Charging the plastic film creates electrostatic adhesions and draws the film onto the surface bearing the print. The dust particles are attracted to the film because of this charge and stick to it. The plastic film retains a charge after the device is turned off, retaining the particles of dust. The contrast between the light coloured dust and the black film can then be photographed.

> **Key term**
>
> **Electrostatic lifting apparatus** – a device used to lift latent dust footprints using a high voltage and plastic film.

Three-dimensional footprints

A footwear impression in soil is an example of a three-dimensional impression. It is not possible to send the entire soil area for analysis, so the CSI must make a cast of the impression at the scene. When a CSI discovers a footwear impression, it must be photographed first and then casted, as follows.

- Water is added to the pre-measured casting powder in a Ziploc-type bag. It is kneaded together until the consistency is thick, like pancake batter.
- The mixture is poured over the impression so that it flows in without causing air bubbles. Once all the impression is covered, it is left to set.
- Once it has set, the cast is carefully lifted out of the mud. Without cleaning the cast or brushing anything off it (as this would destroy any trace evidence), the cast should be placed into a cardboard box or paper bag to be securely transported to the evidence store and then the laboratory.

▶ Footwear cast from a crime scene

Tool marks

Tool mark impressions may also be left at the crime scene – for example, on a window or a door where a perpetrator has gained access or caused criminal damage. Windows and doors are too large to be recovered for forensic analysis in most cases, so a cast must be taken of the tool mark impression left behind at the scene. The CSI can make a silicone-rubber cast. There are two types of tool marks a CSI might find at a crime scene.

▶ Impressed: the tool mark produced is an impression of the tool's shape. This mark is made when a hard object contacts a softer object without moving back and forth (for example, a hammer mark on a door frame).

▶ **Striated:** the tool mark is a series of parallel lines. It is produced when a hard object comes into contact with a softer object and moves back and forth (for example, pry marks on a window frame from a screwdriver).

> **Key term**
>
> **Striated** – having a series of ridges or linear marks.

During tool mark analysis, experts can determine the sort of tool that produced the tool mark impression, and hence whether a tool that has been subsequently seized as a reference sample could have been used at the crime scene and therefore made the mark in question.

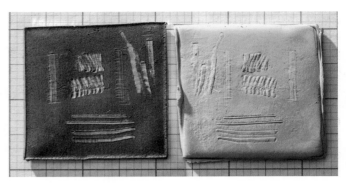

▶ Tool mark cast from a crime scene. The metal plate on the left is test sample of a weapon, and the yellow cast on the right is the cast of the test.

Firearms

A firearm is defined as a 'lethal barrelled weapon of any description from which any shot, bullet or other missile can be discharged' (Section 57 of the Firearms Act 1968). Firearms can be used to shoot, beat and intimidate people, resulting in injury or death. There are three different types of firearm:

▶ handgun

▶ rifle

▶ shotgun.

Firearms must be collected, stored, examined, packaged and labelled by experienced people to make sure that the weapon is safe, not loaded and cannot cause any harm.

Bullets and cartridges

Firearms fire bullets. The bullet and its case and cartridges from shotguns are often found at crime scenes where a firearms offence has been committed. They must also be collected. Hand picking is the general method so that no further marks are made by using an instrument such as tweezers, as this would decreases the validity of the evidence in court.

▶ Bullet and bullet casing collected at a crime scene

Documents and IT

Documents and IT (e.g. mobile phones, computers, tablets and CCTV) that need to be forensically analysed will be hand-picked wearing full PPE to avoid contamination. They will be packaged in various sized plastic tamper-proof evidence bags and sent for analysis.

 PAUSE POINT   List the three types of firearm.

> **Hint**   Use the Internet to find out more about each firearm and how it works.
>
> **Extend**   Think about how CSIs would need to collect and store the different types of firearms if they came across them at a crime scene.

Other important investigative considerations

Prevention of contamination, security, and producing full records at the scene are of upmost importance when collecting evidence for forensic analysis. If these tasks are not carried out correctly, the integrity of the CSI's work will be at risk, not just in the case in question, but in all other cases they have worked on and will work on. The prosecution must have confidence that the evidence collected is the same as the evidence presented in court. They must be sure that contamination did not occur in anyway at the scene, as this makes the evidence invalid and stops the prosecution from building a strong case. If evidence is proved to have been contaminated it is deemed **inadmissible** in court and is therefore not able to be used as evidence during the case. This could lead to the inability to prosecute a suspect. There must also be confidence in the security of the evidence and its chain of continuity. If the evidence has any gaps in security or the chain of continuity, again it would be inadmissible in court.

> **Key term**
>
> **Inadmissible** – not accepted as valid.

Assessment practice 23.2   `B.P3` `B.P4` `B.M3` `A.D2`

You are CSI and you must carry out some training on a simulated crime scene to collect biological, physical and chemical evidence. When you have finished you will be asked to write an account of what you did and why. Your account should:

- document the crime scene
- photograph the crime scene
- preserve any forensic evidence
- decide on an appropriate search method
- prevent contamination by wearing appropriate PPE e.g. CSI suit
- make detailed accurate crime scene notes
- identify, photograph, record, collect, package and label biological, physical and chemical evidence.

You should also evaluate the procedures you carried out, suggesting ways in which you could improve in future.

Plan
- What is the task? What am I being asked to do?
- How confident do I feel in my own abilities to complete this task? Are there any areas I think I may struggle with?

Do
- I know what it is I am doing and what I want to achieve.
- I can identify when I have gone wrong and adjust my thinking/approach to get myself back on course.

Review
- I can explain what the task was and how I approached the task.
- I can explain how I would approach the hard elements differently next time (i.e. what I would do differently).

C   Conduct scientific analysis of biological, chemical and physical evidence

In this section, you will learn about the techniques used to analyse biological, chemical and physical evidence. The techniques used by forensic scientists are based on scientific theory and require you to understand the fundamentals behind each technique in order for you to draw valid conclusions and interpret the results with confidence.

Biological evidence techniques

Blood group analysis

Blood is the most common type of body fluid found at crime scenes. It is usually associated with violent crimes, such as murder or sexual or violent assault. However, blood may also be present at more common crime scenes, such as at burglaries, muggings and vehicle theft scenes.

Forensic serology involves the examination and analysis of blood found at a crime scene or on an item of evidence. There are a number of different laboratory techniques the forensic scientist can use:

▶ **Presumptive** colour tests may be used to get a preliminary indication that the sample is blood.

▶ Blood typing methods can then be used to identify the blood type.

▶ Blood may yield **DNA** evidence which can be analysed to give information on the identity of the blood source.

Blood is a circulating tissue composed of fluid plasma and different cells. It moves around the body in blood vessels and is circulated by the action of the heart pumping. The major function of blood is to transport oxygen necessary to life throughout the body. There are approximately 4.5–6 litres of blood in the average adult human body. The most common type of blood cell is the red blood cell (RBC), or **erythrocyte**, and its role is to deliver oxygen to body tissues via the haemoglobin protein molecules inside red cells. Mammalian RBCs are flat, circular, and depressed in the centre (see Figure 23.6). They lack a nucleus and organelles, and are approximately 6–8 μm in diameter.

▶ **Figure 23.6** Human red blood cells with no nucleus

> **Link**
>
> Go to *Unit 1: Principles and Applications of Science 1* to find more information about the circulatory system.

> **Key terms**
>
> **Presumptive test** – a test that gives a probable indication of results but needs confirming.
>
> **DNA** – deoxyribonucleic acid is a nucleic acid found in cells that carried genetic information.
>
> **Erythrocyte** – red blood cell.

The RBC membrane is composed of lipids (~44%), proteins (~49%) and carbohydrates (~7%).

Red blood cell antigens

Human and animal red blood cells have **antigens**, **extracellular glycoprotein** structures, located within, and sticking out from, the red blood cell (RBC) membrane. These surface antigens instruct a specific blood-type characteristic to the cells, which determines a person's blood group. Variations in the antigens on RBC membranes causes different blood group types. There are only a small number of different blood group types, and the population can be divided into groups of people who share the same blood types. The identification of blood group cannot identify an individual, but it can help eliminate those who possess other blood group types from the inquiry. For example, if a blood sample is recovered from a crime scene and laboratory analysis establishes the type is group AB, it can only be linked to any person who has type AB, which is a large group of people. However, if the suspect is blood type O, they can be eliminated from the investigation as they could not be the source of the AB type blood.

> **Key terms**
>
> **Antigen** – molecule present on surface of red blood cells which can stimulate an immune response, e.g. formation of **antibodies**.
>
> **Antibody** – protein in blood serum produced by the body's immune system that reacts with antigens on red blood cells.
>
> **Extracellular glycoprotein** – a carbohydrate and peptide chain that exists in body fluids outside of cells.

The original blood typing system, the ABO system, was identified by the Austrian biologist Karl Landsteiner in 1901. He categorised human blood into four groups, A, B, AB and O, based on the presence or absence of either or both antigen 'A' and antigen 'B' on the surface of RBCs.

▶ An individual with only A antigens present on the surface of their RBCs is categorised as blood type A.

▶ Those who have only B antigens present are categorised as type B.

▶ If A and B are both present, the blood type is AB.

▶ If neither of the antigens are present, the blood type is group O.

Case study

ABO blood grouping

If you need a blood transfusion, your blood type will be tested to determine your blood group. It is very important that patients are given the correct type of blood, otherwise there could be fatal consequences. If a patient has blood group A, they have A antigens on their red blood cells and anti-B antibodies in their blood serum. If the patient is given blood type B with B antigens present, the body's antibodies will react with the new incoming B antigens and the blood will clot in the body. People with AB blood type have A and B antigens on their RBCs. They have no antibodies in their blood. They can accept any type of blood as A and B antigens in the new blood will not cause a reaction. People with type O have no antigens on their RBCs and their blood can be given to anybody, but they can only receive blood type O.

Check your understanding

1 Produce a table to show the blood groups and the antibodies and antigens present.

2 Add a column to the table to show which blood type each blood group can receive.

3 Add a column to the table to show which blood type each can donate to.

Since the discovery of the ABO blood typing system, 15 different antigen systems based on blood groups have been identified. These are inherited independently of each other, and an individual may have any combination of blood antigens.

Forensic blood typing analysis determines which antigens are present on the RBC surface, and therefore identifies the blood group of a blood sample found at a crime scene. The serological technique consists of testing antibodies and antigens present in the blood. Antibodes are produced in the blood as an immune response to foreign antigens. They bind to specific antigens and render them harmless. A and B antigens and antibodies are used in forensic techniques to analyse the ABO blood group. A small amount of blood is tested with anti-A **antibody serum**. If the A antigen is present, the antibodies bind and the cells clump together, causing a visible, cloudy **precipitate** to form in the blood sample. This is known as agglutination (see Figure 23.7). The blood is then tested with anti-B antibody to determine the presence or absence of the B antigen, and anti-Rhesus antibody can also be used to test for the Rh D antigen.

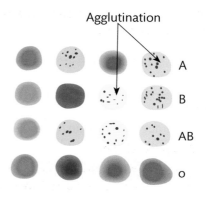

▶ **Figure 23.7** Agglutination when blood grouping test is carried out

The Rhesus antigen system is another blood typing system that was discovered by Karl Landsteiner and Alexander Wiener in the 1930s. This antigen was originally identified in Rhesus monkeys. Subsequent serological testing demonstrated that approximately 85% of the human population have the Rhesus antigen on their RBC surfaces. Individuals with the rhesus antigen on RBC surfaces are 'Rhesus positive' (Rh+), and individuals who lack the antigen are 'Rhesus negative' (Rh-).

Blood group types can be classified from the results in Table 23.6.

▶ A precipitate forming when adding anti-A antibody serum to a blood sample but no precipitate forming when anti-B antibody serum is added would indicate the blood sample being blood group A.

Key terms

Antibody serum – blood serum which contains antibodies that react with corresponding antigen.

Precipitate – a cloudy solid suspension forming in a solution.

▸ **Table 23.6** Shows results from blood grouping test

Blood group	Antigen present on RBC	Reaction with anti-A antibody serum producing a precipitate	Reaction with anti-B antibody serum producing a precipitate
A	A antigen	Yes	No
B	B antigen	No	Yes
AB	A and B antigen	Yes	Yes
O	No antigens	No	No

▸ A precipitate forming when adding anti-B antibody serum to a blood sample but no precipitate forming when adding anti-A antibody serum would indicate the blood sample being blood group B.
▸ A precipitate forming when adding anti-A and anti-B antibody serum to a blood sample would indicate the blood sample being blood group AB.
▸ No precipitate forming when adding anti-A and anti-B antibody serum to a blood sample would indicate the blood sample being blood group O.

Table 23.7 shows possible results when testing for the presence of the Rhesus antigen.

▸ **Table 23.7** Rhesus positive and negative results from blood grouping test

Blood group	Antigen present on RBC	Reaction with anti-D antibody serum producing a precipitate
Rh+	D antigen	Yes
Rh−	No antigen	No

▸ A precipitate forming when adding anti-D antibody serum to a blood sample will indicate the blood sample being blood group Rh+.
▸ No precipitate when adding anti-D antibody serum to a blood sample will indicate the blood sample being blood group Rh−.

Case study

Blood group analysis

A violent fight takes place between two gangs of youths in the street and the police are called. When the police arrive, some offenders flee the scene of the fight. The police barricade the street and Scene of Crime Officers attend to examine the crime scene for forensic evidence. The SOCOs identify a number of blood stains which they recover for forensic analysis. The following day, police arrest four suspects and their blood is taken for reference. In the laboratory, blood group analysis is carried out on the crime scene and suspect's blood samples.

Check your understanding

Examine the results of the analysis below, establish each blood group, and determine whether any of the suspects could have been involved in the fight. Think about who a crime scene sample could have belonged to if it does not match a suspect.

Key: + = positive result, precipitate formed; − = negative result, no precipitate.

Origin of blood	Reaction with anti-A antibody	Reaction with anti-B antibody	Reaction with anti-Rh antibody	Blood group type
Crime scene sample 1	−	−	−	
Crime scene sample 2	+	+	−	
Suspect 1 blood sample	−	+	+	
Suspect 2 blood sample	−	−	+	
Suspect 3 blood sample	+	+	−	
Suspect 4 blood sample	+	−	+	

So if no precipitate formed when adding anti-A and anti-B antibody serum to a blood sample, the blood sample would be blood group O. If a precipitate also formed when the anti-D antibody serum was added the blood group would be O Rhesus positive (O+).

Reflect

1   Blood is tested in equipment containing 'wells'. If you observed precipitate in well A, well B and the Rh well, what would you conclude the blood group of the blood was?

2   If you observed precipitate in well B and the Rh well, what would you conclude the blood group was?

3   If you saw no precipitate in any of the wells, what conclusion would you come to?

Genetics

Another type of biological evidence involves genetics. The genetic information is stored in DNA in a genetic code; genes are made up of a sequence of **nucleotide** bases that determine the **amino acid sequence** of proteins. Different sequences of amino acids code for different proteins in the human body. These protein **coding regions**, or **exons**, are very similar in different individuals. However, not all of the DNA code is used to produce proteins.

Key terms

Nucleotide – one of the structural components, or building blocks, of DNA.

Amino acid sequence – the order in which amino acids are connected by peptide bonds in the chain of polypeptides and proteins.

Coding region – portion of a gene's DNA that codes for protein.

Exon – protein coding region.

Link

Go to *Unit 11: Genetic Engineering* for more on genetics.

Although the human DNA code is 3.3 billion base pairs in size, only a fraction of this, about 2–3%, is used to produce specific proteins. Non-coding **intron** junk DNA makes up 97–98% of the DNA code. These non-coding regions do not possess any genetic function and mutations can occur in these regions that do not affect protein production (although many introns are thought to play a part in gene expression and regulation). These mutations alter the genetic code and are responsible for the small amount of differences between two individuals' DNA sequences.

Key term

Intron – non-coding region in DNA.

These differences between individuals are termed **polymorphisms** and are the key to **DNA profiling**. One of the differences, or polymorphisms, that occurs in the non-coding DNA is due to a mutation causing a short sequence of DNA, called a core repeat sequence or unit, to be repeated over and over again. Approximately 60% of the non-coding DNA consists of short sequences that are repeated many times. This is known as satellite DNA, and within this there is a region known as a minisatellite, which is usually a sequence of 20–50 base pairs repeated anywhere between 50 and several hundred times. These occur at more than 1000 locations throughout the human genome and are known as **Variable Number Tandem Repeats (VNTRs)**. A microsatellite is even smaller, it consists of 2–4 bases repeated 5–15 times, and these are called **Short Tandem Repeats (STRs)**. Professor Sir Alec Jeffreys and his team at Leicester University discovered the patterns in the non-coding DNA in 1984.

Key terms

Polymorphisms – most common type of genetic variation among people.

DNA profiling – using a small set of DNA variations that is very likely to be different in all unrelated individuals.

Variable Number Tandem Repeats (VNTRs) – randomly repeated short sequences of DNA that vary in the number of repeats, and therefore size, between different individuals.

Short Tandem Repeat (STR) – a microsatellite region in DNA of 2–4 bases repeated 5–15 times.

Restriction Fragment Length Polymorphism (RFLP) technique

The RFLP technique was developed in 1984 and was the first type of DNA profiling method used to analyse evidence samples in forensic laboratories. Although the RFLP DNA profiling technique revolutionised the analysis of forensic evidence, it had disadvantages. For example, the technique took a lot of time to carry out and get results, it was not sensitive and did not work if the DNA had deteriorated, and it needed large, good quality DNA evidence samples.

Short Tandem Repeat (STR) DNA profiling technique

A new technique of DNA profiling has since been developed, based on the same VNTR principles as RFLP profiling, except it uses Short Tandem Repeats (STRs). This more recent method involves the use of the **Polymerase Chain Reaction (PCR) technique** which allows the analysis of very small quantities of DNA, for example from a single cell. PCR targets specific regions of DNA containing STRs, and this DNA is copied millions of times. The amplified DNA fragments can then be separated by size and analysed.

> **Key term**
>
> **Polymerase Chain Reaction (PCR) technique** – a technique used to make multiple copies of a segment of DNA.

There are four main stages to produce a DNA profile.

Stage 1: DNA extraction

The DNA must be extracted from the tissue sample. To extract DNA, the cell membranes, structural materials and proteins and enzymes must be destroyed so the DNA can be released and separated from the rest of the cell components. The sample is mixed with lysis solution, which contains detergent and proteinase K. The detergent disrupts the cell membrane and the nuclear envelope, causing the cells to burst open and release the DNA. The DNA is still wrapped tightly around proteins called histones so the proteinase K cuts the histones apart, which in turn frees the DNA. Concentrated salt solution is added. This causes any proteins and cell debris to clump together. An Eppendorf tube (a small plastic tube specifically used in a centrifuge) is then placed into a centrifuge; while inside, the tubes spin at high speed, the heavy clumps of proteins and debris will sink to the bottom of the tube while the DNA will remain evenly distributed throughout the sample. Isopropyl alcohol is added because DNA is not soluble in isopropyl alcohol so it becomes visible to the naked eye, and can be separated from the rest of the mixture.

Stage 2: DNA amplification

Once the target DNA is extracted, PCR is carried out to produce lots of DNA from the tiniest original sample (see Figure 23.8). In addition to the DNA sample DNA primers, dNTPs, polymerase enzyme and PCR reaction mix are added.

1   DNA primers – short sequences of DNA that recognise and bind to either side of the STR of interest. The synthesis of new strands of DNA starts from these primers.

2   dNTPs (deoxynucleotide triphosphates) – the nucleotide building blocks of DNA used to extend the DNA molecule from the primers and produce the new complementary strand (i.e. DNA synthesis). Each nucleotide triphosphate (dATP, dTTP, dCTP, dGTP) consists of a base (A, T, C or G), a sugar and a phosphate group.

3   Polymerase enzyme – this enzyme anneals (sticks) to and moves down the DNA template strands. The 5′ phosphate group of new nucleotides adds to the extending primer at the free 3′ sugar carbon at the end of the growing chain. It uses base-pairing rules to insert the correct complementary nucleotides (i.e. A opposite T, G opposite C). The polymerase enzyme used in PCR is *Taq Gold*®, a heat-stable, recombinant enzyme produced by the bacterium *Thermus aquaticus*, which lives in hot water springs and is able to survive at extremely hot temperatures.

4   PCR reaction mix – including stabilising and preserving buffers and magnesium, a co-factor necessary for the DNA polymerase enzyme.

The PCR technique uses a series of repeated cycles of heating and cooling to produce exponential amplification of the DNA, so the number of DNA molecules doubles with every PCR cycle.

▶   Separating the strands: The DNA–PCR mixture is heated to 95 °C for 60 seconds to denature and break apart the two strands of the double-stranded DNA in the target DNA molecule. The weak hydrogen bonds between **complementary base pairs** of the double stranded DNA break at this temperature and the DNA molecule unzips into single strands.

▶   Annealing of the primers: The temperature is then lowered to 45 °C for 60 seconds so the primers can bind to the single-stranded DNA complementary sequences on both sides of the DNA loci of interest.

▶   Synthesis of DNA: Then the temperature is raised to 72 °C for 60 seconds to enable the Taq polymerase enzyme to copy the target DNA and add new nucleotides onto the extending primers, producing new strands of complementary DNA.

> **Key term**
>
> **Complementary base pairs** – during replication, the nucleotide base sequence of one strand of DNA is known, so the new strand produced is complementary in base sequence to the template strand.

Once the new double-stranded DNA is produced, the temperature is raised again to 95 °C to denature the new

DNA molecules and the DNA synthesis process starts all over again. The PCR cycle is repeated a specific number of times depending on the nature of the biological samples. Usually 28 or 34 cycles are used. As the DNA products of each cycle serve as templates for the next round of amplification, repeated cycles result in the exponential increase in the number of new DNA double-stranded molecules produced.

Stage 3: DNA separation and visualisation

The amplified section of DNA ranges in size, due to the differences between STR repeat region sizes. The DNA fragments are therefore separated according to size using the gel electrophoresis technique, where the DNA fragments move through a gel medium under the influence of an electrical current. As the current is applied, the negatively charged DNA fragments begin moving through the gel towards the positively charged **anode**. The gel essentially acts as a type of molecular sieve, allowing smaller molecules to travel faster than larger fragments.

As the PCR technique produces such a large quantity of DNA, it is possible to stain the agarose gel directly to observe the DNA fragment pattern. The gel containing the separated DNA fragments is soaked in a staining dye, which interacts with the DNA bands and makes them visible. The most common stains used are methylene blue, which turns the DNA bands a dark blue colour, and ethidium bromide, which makes the DNA bands fluoresce under UV light.

Stage 4: DNA analysis

The DNA band pattern produced on a STR gel has only one or two bands at each VNTR loci, which shows the size of the repeat regions on both VNTR **alleles**. If the VNTR locus is heterozygous, two bands of different size are seen, and if the VNTR is homozygous, only one band is seen, as both alleles are the same size. However, the band would be twice the thickness. This would then be compared to another DNA sample of interest – for example, from a suspect. Unknown crime scene DNA samples can be analysed next to suspect reference samples to determine whether there is a DNA band match (see Figure 23.9).

> **Key term**
>
> **Anode** – the positively charged electrode by which the electrons leave an electrical device.

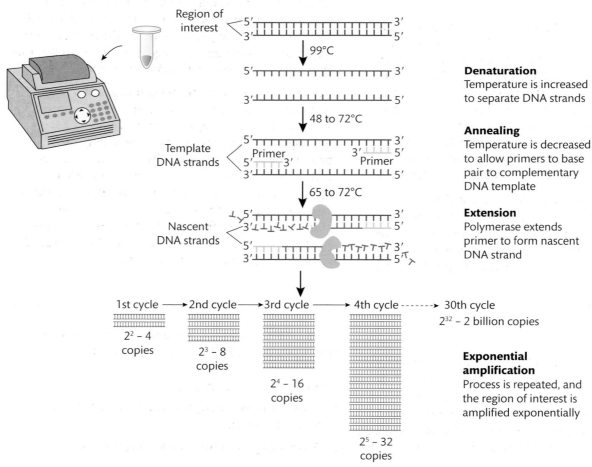

Region of interest — 5' / 3' — Template DNA strands — 99°C

Denaturation
Temperature is increased to separate DNA strands

48 to 72°C

Annealing
Temperature is decreased to allow primers to base pair to complementary DNA template

Primer / Primer — 65 to 72°C

Nascent DNA strands

Extension
Polymerase extends primer to form nascent DNA strand

1st cycle → 2nd cycle → 3rd cycle → 4th cycle ------> 30th cycle

2^2 – 4 copies

2^3 – 8 copies

2^4 – 16 copies

2^5 – 32 copies

2^{32} – 2 billion copies

Exponential amplification
Process is repeated, and the region of interest is amplified exponentially

▶ **Figure 23.8** PCR

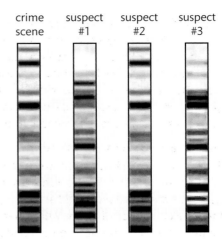

▶ **Figure 23.9** DNA profiles from crime scene blood sample and three suspects

Key term

Allele – one member of a pair of alternative forms of a gene that occupy the same locus on a particular chromosome.

Theory into practice

Figure 23.9 shows a DNA profile from a blood sample swabbed at an armed robbery. There are also three reference DNA profiles: Suspects 1, 2 and 3. Decide which reference sample matches the crime scene sample and produce a conclusion from this.

Ⅱ PAUSE POINT   Describe the stages involved in making a DNA profile.

Hint   Think about how we get the DNA out of cells and how we make this bigger in order to be able to see it.

Extend   Explain why each stage is carried out.

Fingerprints

Fingerprints are the most common type of evidence found at crime scenes and are associated with a wide range of crimes. Fingerprints are biological characteristics that are unique for every individual. The primary goal of a criminal investigation is identification, so fingerprints have high evidential value as they can uniquely associate a person with a crime scene or evidentiary item. Fingerprints are one of the last features to be lost from the skin during decomposition and, in certain circumstances, can be used to identify dead bodies months, or even years, after death.

There are two fundamental principles that allow forensic scientists to use fingerprinting as a means of criminal identification. First is the fact that fingerprints are unique: they are an individual characteristic, different from person to person, and finger to finger. Fingerprints are developed while the foetus is in the womb and are formed by a mixture of genetic and environmental factors. Although we inherit certain fingerprint features, the pressure in the womb also determines how fingerprints form, so that even identical twins have slightly different fingerprints.

The second principle is the fact that fingerprints are permanent. Once formed, they never change during an individual's lifetime, they only get larger as you grow. Although it is impossible to change your fingerprint pattern, many criminals have tried to do so. Superficial damage to fingerprint skin heals and grows back in exactly the same shape and form. If an injury penetrates deeply enough into the skin, a permanent scar will form eradicating the fingerprint pattern. However, the scarring itself becomes a new individual characteristic for identification, as no two scars will be alike.

Case study

Fingerprints

American gangster John Dillinger carried out at least 11 bank robberies in the 1920s and 1930s. In an attempt to evade capture, he tried to destroy his own fingerprints by applying corrosive acid to them. However, when his fingers healed, faint ridge markings were still visible and, when Dillinger's prints were taken at the morgue after he was shot to death, they were still comparable to his fingerprints on record from a previous arrest.

Check your understanding

1 Explain why damaging fingerprints would also lead to identification.

2 Explain why twins have unique fingerprints.

Fingerprinting is a system of identification based on skin ridge patterns on fingertips. Friction skin ridges are a series of elevated lines of skin of different sizes and formations called ridges (or hills), the long deep grooves in-between are called furrows (or valleys). Friction ridge skin is present on the surface of palms, palm side of fingers and thumbs, and soles of feet, and is designed by nature to provide us with a firm grip and to resist slippage.

Basic fingerprint patterns

The flow of the skin ridge surface on a fingertip determines the fingerprint pattern. According to Henry's System of Classification, there are four types of fingerprint pattern:

▶ loop
▶ arch
▶ whorl
▶ composite.

Each pattern type contains sub-groups which possess the same basic characteristics and similar differences among patterns.

Loops are the most common type of pattern. The fingerprint ridges flow in from one side of the finger, curve round and exit on the same side of the finger (Figure 23.10). Loops are categorised according to the slope of the ridges: the loop can slope down to the left or the right.

Arches are the simplest and rarest types of prints, where the ridges flow in from one side of the finger, rise in a wave formation and exit on the other side of the finger (Figure 23.11).

Whorls are the most complex type of prints, where the fingerprint ridges curve round in a complete circuit. Figure 23.12 shows an example.

Composite prints include any combination of two or more fingerprint patterns or unusual patterns that do not fit in any other group. They are sometimes categorised as accidental whorls.

Ridge counting

An easy way to distinguish between two different loop patterns is by counting the number of fingerprint ridges present within the pattern. To do this, you need to identify the three key features of a loop pattern, which are:

▶ delta (the point on a friction ridge at or nearest to the point of divergence of two type lines – see Figure 23.13)
▶ core (the centre area of a fingerprint)
▶ type lines (the two innermost ridges which start parallel, diverge, and surround or tend to surround the pattern area – see Figure 23.14).

loop

▶ **Figure 23.10** Loop fingerprint pattern

arch

whorl

▶ **Figure 23.11** Arch fingerprint pattern

▶ **Figure 23.12** Whorl fingerprint pattern

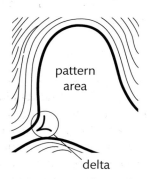

pattern area

delta

▶ **Figure 23.13** Loop patterns showing location of delta at divergence of type lines

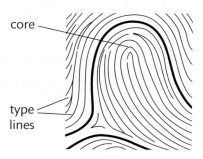

core

type lines

▶ **Figure 23.14** Loop pattern showing highlighted type lines surrounding the pattern area

Once these features have been identified on a loop pattern, it is possible to calculate the ridge count. To determine the ridge count, an imaginary line must be drawn between the delta and the core of the loop pattern. The ridge count can then be established by counting from the delta the number of ridges that cross the line (the core or delta are not included in the count).

A forensic scientist can use all of these fingerprint features to classify an individual's fingerprint pattern. For example:

▶ if only one delta is identified within a pattern, the print can be categorised as a loop
▶ if none are present, the print is an arch
▶ if two are identified, the pattern is a whorl.

As there are only four different types of fingerprint pattern, every individual's fingerprint patterns fall within one of these groups, and large numbers of the population have the same pattern. This is a class characteristic and examining the pattern alone will lead to many comparison matches. Loops are the most common pattern; 60–65% of the population have loops. Therefore many people could potentially match a loop fingerprint mark found at a crime scene. In contrast, arch patterns are much less common and only 5% of the population could potentially be the source of the fingerprint mark. This makes finding an arch pattern of higher evidential value as it is easier to identify the individual (see Table 23.8).

Discussion

Using a magnifying glass, examine your fingerprint ridges and determine the fingerprint pattern on each of your fingers. Look for the presence of deltas and a core to help you. Are your patterns common or rare? Pool your results with your classmates and count in total how many loops, arches and whorls there are in your classroom population. What are the fingerprint pattern proportions in your population? Do they match the statistics seen in the general population?

Fingerprint minutiae

Although fingerprint ridge pattern is a class characteristic, the number, location, flow and formation of individual ridges within the pattern makes a fingerprint unique to an individual. Fingerprint ridges have a number of different types of characteristics, or minutiae that can be used to identify and compare fingerprints. For example:

▶ a ridge ending is where each ridge starts and stops
▶ a bifurcation is where one ridge splits into two.

Some typical ridge characteristics are shown below (see Figure 23.15). When a fingerprint mark is recovered from the scene and a suspect's recorded prints are available, a side-by-side comparison is carried out to determine whether the same number, location and type of minutiae are present in each mark (see Figure 23.16). Until recently, forensic fingerprint experts had to find 16 matching minutiae in two fingerprints to identify them as from the same source. However, this led to many fingerprints being declared as not matching, even when the fingerprint examiner was convinced it was a match. Today, there is no minimum number of minutiae necessary to identify a match, and the decision is left to the discretion of the trained and experienced expert.

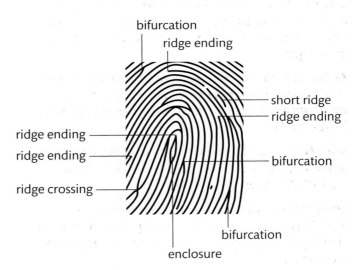

▶ **Figure 23.15** A fingerprint with various minutiae labelled

▶ **Table 23.8** Characteristic features of each fingerprint and their proportions in the population

Fingerprint pattern	Proportion in population (%)	Type lines	Delta	Core	Ridge count
Loop	60–65	Yes	One	Yes	>1
Arch	5	No	None	No	0
Whorl	30–35	Yes	Two	Yes	>1
Composite	Rare	Yes	At least two	Yes	>1

Identifying fingerprint patterns

A car theft is reported to the police and, later that day, a car matching the victim's description of the stolen vehicle is located abandoned in a car park. A scene of crime officer examines the car and recovers a partial fingerprint from the interior of the car. The fingerprint is sent to the forensic laboratory for analysis where it is scanned into the fingerprint database. This database contains millions of records of fingerprints from criminals convicted of a crime. A computer compares the scanned image to the stored fingerprint records and matches a previous offender's prints to the crime scene print.

Check your understanding

1 Confirm the match by identifying the fingerprint patterns, and the location and type of ten minutiae present in both of the fingerprint patterns in Figure 23.16.

▶ **Figure 23.16** A partial fingerprint recovered from the car (left) and the suspect's fingerprint found on the database (right)

Hair and fibre identification and analysis

Approximately 100 hairs are shed each day by an individual onto clothing and items in our environment, and this makes hair a very common form of forensic evidence. In addition, loose animal fibres can easily be shed from the clothes that we wear into our environments. Both of these types of evidence may be transferred during physical contact between a suspect and victim or crime scene, and can be used to reconstruct the events of a crime. Hair and fibre evidence may be associated with a range of crime scenes, such as murder, sexual or violent assault or burglary, and any other crime where contact is made. The recovery and analysis of hairs and fibres transferred during a crime can indicate contact with surfaces or individuals, and so where individuals may have been, and can associate a suspect with a victim or a crime scene.

Animal hairs are natural fibres and these are also analysed in the forensic laboratory for criminal cases such as theft of animals and furs, or illegal breeding, and may be an additional source of linking evidence in any human case where the victim or suspect has a pet.

Hair and fibre evidence can be transferred between suspect and crime scene or victim in one of two ways.

▶ Primary transfer: this is when hairs are transferred directly from the region of the body where they are growing (e.g. the head) onto an item of clothing, seat cover or floor.

▶ Secondary transfer: this is when hairs do not transfer directly from the area of growth, but from a location where hair has already fallen (e.g. from an item of clothing onto the floor or from a seat cover onto an item of clothing).

A suspect could carry on their clothes a hair from an innocent person (primary transfer) and transfer it to the crime scene (secondary transfer), making the innocent person look as if they were present at the scene. When assessing hair evidence, the possibility that hair is present at a scene due to secondary transfer must be considered.

Hairs are defined as slender outgrowths (appendages) of the skin of mammals. Human hair consists of a hair fibre which grows from a root, and projects from the skin's surface (see Figure 23.17). The root at the base of the hair is embedded in the skin, and ends in a soft thickened hair bulb which sits in sac-like pit, called a follicle. Hair growth takes place at the dermal papilla at the bottom of the hair follicle, which contains nerves and a blood supply stimulated by hormones. Blood provides the follicle with essential nutrients, e.g. B vitamins, to nourish the hair root and support the follicle. The follicle cells produce new hair cells which multiply, and move up and away from the papillae. As they do so they become cut off from their supply of nourishment and they start to produce keratin protein which hardens their structure. The cells begin to change; they flatten, line up and squeeze together into tightly packed layers. They lose their nuclei and cell contents and fill with the hard protein keratin. It imparts strength to the hair, as it is resistant to enzyme and acid attack and does not degrade under environmental stress.

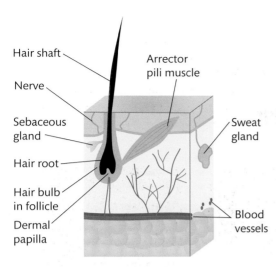

▶ **Figure 23.17** Cross section of hair from skin

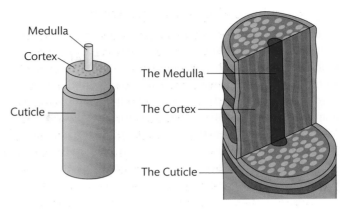

▶ **Figure 23.18** Microscopic structure of hair

Each animal species, including humans, possess hair with distinguishing characteristics, for example the length, colour, shape, root appearance and internal microscopic features. These internal features show the same basic structures and **morphology** shared by every individual, although there is some considerable variability in the arrangement, distribution, and appearance of some of these microscopic hair characteristics between individuals. These variations allow a forensic scientist to differentiate between hairs from different individuals and animal species. Hair evidence is valuable evidence. It can provide strong corroborative evidence for placing a suspect at a scene. However, morphological characteristics cannot be used to individualise the hair, and an individual can only be associated with hair evidence and not identified by it.

> **Key term**
>
> **Morphology** – the structure.

Each hair fibre has three main microscopic structures (see Figure 23.18):
- ▶ cuticle, a hard outer layer that protects the hair shaft
- ▶ cortex surrounding the medulla which is the main bulk of the hair
- ▶ soft medulla at the centre of the hair.

Cuticle

The cuticle is a clear, protective, outside covering which gives hair resistance to chemicals and enables hair to retain its structural features over long periods of time. The cuticle is not continuous, but formed by separate overlapping scales that always point towards the tip end of the hair. At the hair tip there may be no scales left or fragments may have chipped off edges causing hairs to split. This damage is usually seen in dry hair, or hair that has not been trimmed for a while, and is caused by brushing, drying, and the environment. There are three basic types of scale structure or pattern, coronal, imbricate and spinous. These patterns are an important feature for species determination – distinguishing human hair from animal and distinguishing between animal species.

> **Research**
>
> The hair cuticle scales point towards the tip of the hair. You can feel the scales against your fingers if you pull a hair between your fingers from the tip to the root end. This pushes the scales in the opposite direction to which they lie and will feel rough under your fingers. Pulling from the root to the tip pushes the scales flat and feels much smoother. Research the different types of cuticle below and copy and complete Table 23.9 to describe each type.
>
> ▶ **Table 23.9** Types of cuticle
>
> | Type of cuticle | Description | Appearance | Scale index (scale/cm) | Types of animals |
> |---|---|---|---|---|
> | Coronal | | | | |
> | Imbricate | | | | |
> | Spinous | | | | |

Medulla

The medulla is the central, inner core of the hair. It is a canal-like structure of cells surrounded by and running through the centre of the cortex. Not all hairs have medullae, and the 'degree of medullation' varies. The presence and appearance of medulla vary from individual to individual and even between hairs of a single individual. Human head hairs usually have no or fragmented medullae, and rarely show continuous medullation. Most animal hairs have continuous or interrupted medullae. The Medullary Index (MI) measures the width of the medulla relative to the diameter of the hair fibre. Human hair usually has a MI value of less than a third, while animal hair usually has a MI value of a half or greater.

Cortex

The cortex is the bulk zone of hair underneath the cuticle. It is made up of spindle-shaped (long, thin rod-shaped) cortical cells aligned in a regular helical arrangement, parallel to the length of the hair, with air spaces present (cortical fuzi). The cortex is randomly embedded with oval-shaped melanin pigment granules. When new hair is being produced, Melanocyte cells in the hair follicle produce melanin. This gives hair its natural colour. Pigment granules contain two types of melanin which, together, create all natural variations in hair colour. Hair becomes grey and white with age because melanocyte cells cease to function and pigment no longer forms within the cortical cells.

Comparing hair evidence

The forensic analysis of hair evidence usually involves the use of comparison microscopy, where the structure and characteristic features of questioned hairs are compared to known hair samples. A number of the characteristic microscopic features are examined and a comparison is made between the hairs to determine whether the features are similar or different. If the questioned and known hairs have the same microscopic characteristics, this is consistent with them originating from the same source, which means the suspect's hairs were transferred to the crime scene. If the crime scene hairs are microscopically different, they cannot be associated with the source of the known hairs, the suspect. If there are similarities and slight differences observed between the questioned and known hair, then no conclusion can be reached as to whether the questioned hair originated from the same source as the known hairs.

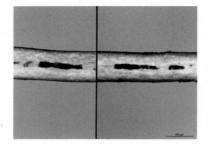

▶ Comparison microscope image of suspect and crime scene hair

A large number of hairs must be taken from the suspect from all areas of the head, to ensure there is a representative control sample. This is because there is variation between hairs from the same individual, called intrapersonal variation, as not every hair from the same person is exactly the same. This is one reason why hair fibre evidence cannot be used for identification. However, when collected with an adequate number of reference control samples, it can provide strong circumstantial evidence. A hair examiner can provide a large amount of information from analysis. Here are some examples.

- ▶ The cuticle pattern.
- ▶ The medullary index.
- ▶ Hair colour, size, distribution and concentration of pigment granules.
- ▶ Is a root present? Did hair fall out or was it pulled?
- ▶ The condition of the hair and tip. How was the hair cut, treated or damaged?
- ▶ Is it human or animal hair?
- ▶ Somatic origin of hair/What part of the body is the hair from?
- ▶ If animal, which species?
- ▶ Does questioned hair recovered from scene compare to reference hair from known individual?

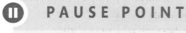 PAUSE POINT   Explain the structure of hair.

 Hint   Close the book and see if you can list the types of evidence and information that a forensic scientists can gain from analysing hair.

 Extend   Research how a comparison microscope is used to compare two hairs.

Bone and skeleton physiology

Forensic anthropology is the study of bones or human remains. Forensic anthropologists assist in searching for, locating, excavating and recovering buried or surface remains at a scene of a crime, and are often called to scenes of mass disasters and war graves to analyse bone fragments and teeth to identify victims. Forensic anthropologists specialise in the analysis and identification of human skeletal remains for legal and humanitarian investigations. In many murder, suicide, accident and mass disaster cases, the bodies are recovered a period of time after the death has occurred. This means that the tissues of the body have decomposed and so traditional means of identification (for example, visual identification of the victim or a forensic post-mortem of the body) are not possible. Bones often survive the process of decay by many years and provide an important form of identification after death.

A forensic anthropologist is a bone specialist who applies standard scientific techniques developed in physical anthropology to identify decomposed, mummified, burned or dismembered human remains. They examine skeletons to identify victims. By establishing the biological profile of the skeleton, they can identify:

▶ gender
▶ height
▶ age at death.

The forensic anthropologist may be able to determine the cause, manner and mode of death, by identifying fracture patterns and trauma to the bones, and may be able to estimate the time of death.

Link

See *Unit 8: Physiology of Human Body Systems* for more on the structure of the skeleton.

The human skeletal system is the rigid frame that supports the body and soft organs and is composed of 206 separate named bones in the adult skeleton. Bones perform several other important functions, including protection of the organs, to act as a mineral storage, the formation of blood cells and to allow movement of the body by acting as levers for the muscles. Bones come in an assortment of shapes and sizes, depending on their specific functions. The adult human skeleton is divided into two parts and bones are classified into various shapes depending upon visual examination (see Table 23.10).

1 The **axial skeleton** forms the long axis of the body. It includes the bones of the skull, vertebral column and rib cage. It is involved in protecting, supporting, or carrying other body parts.

2 The **appendicular skeleton** attaches limbs to the axial skeleton. It includes the bones of the upper and lower limbs and the girdles (shoulder bones, hip bones), which aid movement and manipulation of the environment.

Key terms

Axial skeleton – the long axis of the human body.

Appendicular skeleton – the limbs attached to the axial skeleton.

▶ **Table 23.10** Classification of bone shape

Type of bone	Description	Examples
Long	Considerably longer than wide, elongated shape, hollow shaft plus two closed ends	Humerus, femur, ulna
Short	Roughly cube shaped, approximately equal length and width	Bones of wrist and ankle
Sesamoid bones	Shaped like sesame seed, special type of short bone that form in tendons, vary in size and number in different individuals	Patella
Flat bones	Thin, flattened, usually a bit curved	Sternum (breastbone), scapulae (shoulder blades), ribs, most skull bones
Irregular bones	Have complicated shapes that fit none of preceding classes	Vertebrae and hip bones

When the forensic anthropologist is sent human remains for forensic investigation, they are expected to report conclusions that include if the bones are human or animal, the names of the bones that are present, are the bones from one or more individual, the biological profile of the individual, e.g. sex, age, height and build, and if there is any trauma or anatomical anomalies that can help to identify the victim.

Determination of sex

There are a number of key obvious differences between the male and female skeleton, and most of these differences can be seen in the size of the skeleton and the shape of the skull and pelvic girdle (Figure 23.19). Firstly, the skeletons of males are generally larger in size and more robust than females. Secondly, the skulls of males usually have a wider jaw, squarer chin, more sloping forehead and more pronounced eyebrow bones than females. Finally, differences in shape can be seen in the male and female pelvis, as the female

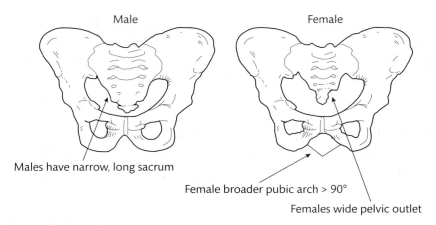

Male

Female

Males have narrow, long sacrum

Female broader pubic arch > 90°

Females wide pelvic outlet

▶ **Figure 23.19** Male and female pelvis with differences labelled

pelvis is designed for childbirth. Females have a wider subpubic angle and sciatic notch and broader pelvic inlets compared to males. The forensic anthropologist is able to make a direct observation of an unknown adult skeleton and establish the sex by visually examining these age-related changes.

Determination of stature

The most common method for estimating living stature (how tall/build) from a skeleton is to measure the lengths of the long limb bones, as long bone length is proportional to height.

Determination of child age

The simplest method for determining the age of a child at death is by measuring the lengths of the long limb bones, the femur, tibia and fibula of the leg and the ulna and radius arm bones. There is a direct relationship between the lengths of the bones and age: as the age of the child increases the length of the bones increase at a specific rate. During the first two years after birth, the rate of growth is very quick, but as children get older, their rate of bone growth slows down.

In addition to bone growth, the teeth of children are changing. The first set of 20 baby milk teeth are called the primary or deciduous teeth. These are only temporary and children shed their primary teeth between the ages of 6 and 12 years (see Figure 23.20). The primary deciduous teeth are replaced by the secondary set of 32 permanent teeth which normally begin to appear between the ages of 6 and 18 years. The rate at which primary teeth erupt through the gum and fall out and the eruption of the secondary teeth follow a typical pattern. They do not develop or fall out all at once but in specific stages at certain ages. It is possible therefore to make an accurate estimate of the age of a young person's skeleton by establishing the number and type of teeth present in the jaw bone.

Theory into practice

1 Create two tables to show the age of eruption of primary teeth and secondary teeth from Figure 23.20.
2 A small skull is recovered from a crime scene and the forensic anthropologist records the presence of the:
   - central incisors
   - lateral incisors
   - cuspids
   - first molars in the upper and lower jaw bones.

Using the data in Figure 23.20, estimate the age of the individual whom the small skull belonged to.

Determination of adult age

The methods used for estimating the age of adult skeletons are based on a number of characteristic **degenerative changes** that occur to fully grown bones over time. Usually, the ends of the ribs that join with the breastbone, the sternum, are examined for a number of age related changes. The surface of the rib end bone changes from smooth cortical bone to become granular and pitted with many holes, and the end of the rib wears down, changing from a rounded protruding surface to a flat and then indented U-shaped surface with an irregular edge contour. In addition, the costal cartilage that joins the rib end to the sternum **ossifies** to bone as age increases and this can be seen as a rim forming around the end of the rib. With time, this rim becomes thinner and more spiky and jagged. This same process occurs at the **pubic symphysis**.

Key terms

Degenerative changes – negative changes that occur over time.

Ossifies – turns into bone tissue.

Pubic symphysis – joint located between the left and right pubic bones.

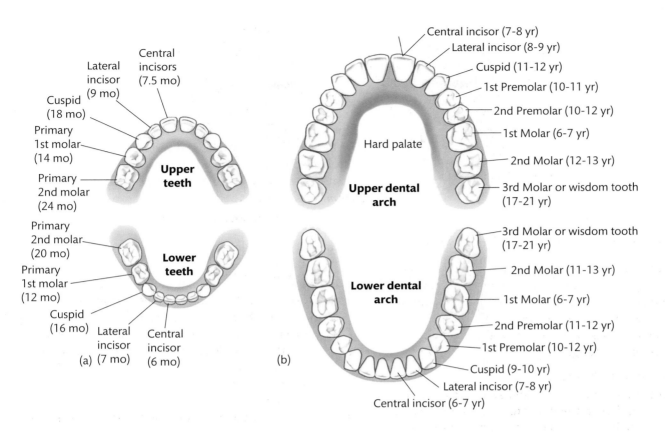

▶ **Figure 23.20** (a) Primary and (b) secondary teeth; age when they erupt

Chemical evidence techniques

Many types of evidence sent for forensic analysis have a chemical nature and forensic chemists use a number of chemical analytical techniques to analyse the evidence they receive to gain results.

Chromatography

Chromatography is a collective term for a number of similar laboratory instrumental techniques in which **mixtures** are separated into individual pure components. Chromatography can be applied to analyse different types of mixtures for a range of criminal investigations and is used in many different types of forensic laboratories, including the following examples.

▶ Forensic toxicology laboratories use chromatography to detect alcohol and drugs in victim or offender blood samples.

▶ Forensic environmental laboratories use chromatography to determine the level of illegal pollutants in a lake or river.

▶ Forensic chemistry laboratories use chromatography to determine the purity of a drug sample and identify the drug and cutting agents.

▶ Forensic questioned document laboratories use chromatography to compare components from pen ink or printer ink samples.

▶ Forensic fire investigation laboratories use chromatography to analyse samples of accelerant used to start fires at arson scenes.

> **Key term**
>
> **Mixture** – two or more substances mixed together, but where no chemical reaction has occurred.

> **Link**
>
> Go to *Unit 2: Practical Scientific Procedures and Techniques* for more on chromatography.

There are a number of reasons why a forensic scientist would want to separate a mixture.

1 They may want to analyse and examine liquid or gas mixtures and their components, and their relationship to one another.

2 They may need to identify a mixture and components or purify and separate components to isolate one of interest for further investigation.

3 They can also quantify (determine the amount) of mixture and components present in a sample.

There is a range of chromatographic techniques which analyse different forensic samples; inks, dyes, pens, markers, highlighters, lipstick, clothing dyes, food colouring (e.g. in sweets), coloured pigments in plants, flowers, drugs, accelerants:

▶ paper chromatography
▶ column chromatography
▶ thin-layer chromatography (TLC)
▶ high-performance liquid chromatography (HPLC)
▶ gas chromatography (GC).

Paper chromatography

Paper chromatography is a technique that separates mixtures into their different components. A small sample of a mixture is placed on a **porous** surface (**stationary phase**), such as filter paper. The paper is placed in contact with a liquid solvent (**mobile phase**) which dissolves the mixture sample (e.g. ink). The molecules present in the mixture distribute themselves between the stationary and mobile phase, depending on how well the components dissolve in the solvent. The solvent moves through and up the paper and components of the

mixture are carried with the solvent a certain distance unique to each component. They move along the paper with the solvent, but different components will travel further than others. Individual components which dissolve well in the solvent are carried along quickly and move far up the filter paper. However, components with a stronger attraction to the paper than the solvent move more slowly and not as far in the same amount of time. The mixture therefore becomes separated, and forensic scientists can identify the chemicals that are present in various mixtures by comparing them to how known samples behave in the same experiment.

> **Key terms**
>
> **Porous** – absorbent, full of tiny holes through which gas or fluids may pass.
>
> **Stationary phase** – part of the apparatus that does not move with the sample, e.g. filter paper.
>
> **Mobile phase** – gas or solvent liquid e.g. water or alcohol that carries the components of the mixture up the stationary phase.

Investigation 23.1

Chromatography to separate the components of different inks

Carry out basic chromatography using felt tip pens and water. You will need a 100ml beaker, pencil, ruler, 3 different coloured felt tips, filter paper and water.

Steps in the investigation	Pay particular attention to...	Think about this...
1. Place a pencil line at one end of the filter paper using the ruler (about 1 cm from the bottom).	Be careful not to touch the filter paper as the oils in your fingers can affect the results.	
2. Take a felt tip and place a dot on the line in the middle. Then take the other two, and either side of the first dot, mark two other dots (space them out along the line).	Be careful not to place the dot too close to the edge, as the sample may run off the paper.	
3. Add a small amount of water to the beaker so that when you hang the filter paper inside the beaker, the water touches the paper but does not go over the pencil line.		When handling glassware, be aware of safety.
4. Hang the filter paper in the beaker using the ruler or a spill to hang it from (fold the top over, a bit like a washing line).	Ensure that the pencil line does not touch the water. The water must run up through the pencil line in order to carry the ink with it.	
5. Watch as the colours separate showing you the different colours that were mixed together to make the felt tip.		
6. When the water line is about 1 cm from the top of the paper, remove the paper from the beaker and mark where the water line finished. This is the solvent front.		Think about how you might accurately measure the solvent front.
7. Measure the distance moved by each spot to collect data to calculate R_f values.		

Chromatogram analysis

Molecules in mixtures travel at different speeds when pulled along the paper by the solvent and so travel different distances in the same given time. The **retention factor** (R_f) can be calculated for samples separated on a chromatogram. This calculates how far a component has moved compared to how far the solvent moved (the **solvent front**) and provides a quantitative method for identifying components in a mixture.

Substances that are more soluble in the solvent will have an R_f value closer to one, as they will move quickly up the paper therefore travelling further. Substances that are less soluble in the solvent will have an R_f value closer to zero as they will move slowly up the paper, and thus not move as

far. Mixtures will produce the same pattern on the paper as long as the paper, solvent, temperature and time are kept constant. Therefore the R_f values will be identical for the same components in different mixtures. Using different solvents or shortening the time of development would change the patterns so it is important when comparing crime scene and reference samples that all variables are kept constant. Known reference samples are usually separated on the same chromatogram to compare to unknown crime scene samples. This allows forensic scientists to compare R_f values and identify unknown samples. Table 23.11 shows how to work out R_f values. R_f value is calculated by dividing the distance of a component spot (mm) by the distance the solvent front moved (mm).

▶ **Table 23.11** Working out R_f values

Dye component	Distance component spot moved (mm)	Distance solvent front moved (mm)	R_f value
Blue spot	24.0	55.0	24.0 / 55.0 = 0.44
Red spot	48.0	55.0	48.0 / 55.0 = 0.87

Theory into practice

Drug samples are collected as evidence from a crime scene and sent to the laboratory for investigation. Two of the samples, evidence A and evidence B, are separated using paper chromatography. The samples are mixed with a little solvent and the dissolved samples are spotted onto the filter paper.

Known pure samples are also added to the chromatogram for reference and comparison. The paper is placed in a beaker containing a small amount of methanol solvent and left to develop for 15 minutes. After this time, the chromatogram is removed from the solvent and the distance that the solvent front and samples have moved is measured. The solvent front moved a distance of 102 mm.

1 Work out the R_f value for the components in Tables 23.12 and 23.13, and identify the components present in drug A.

▶ **Table 23.12** Known pure reference sample results

Component	Type of sample	Distance solvent moved (mm)	R_f value
Lignocaine	Cutting agent	23.0	
Starch	Cutting agent	36.0	
Cocaine	Drug	60.0	
Amphetamine	Drug	94.0	

▶ **Table 23.13** Drug A evidence sample results

Component	Distance solvent moved (mm)	R_f value	Identification of component
1	36.0		
2	94.0		

▶ **Table 23.14** Different types of chromatographic techniques

Chromatography technique	Type of samples separated	Mobile phase	Stationary phase	Common use
Paper	Dried liquid samples	Liquid solvent	Filter paper strip	One of the most common types of chromatography used to analyse pen inks, lipsticks, food and fabric dyes
Thin layer (TLC)	Dried liquid samples	Liquid solvent	TLC sheet – glass/plastic plate covered with a thin layer of silica gel	Used to analyse the dye composition of fibres, inks and paints and to detect pesticide or insecticide residues in food
Liquid, e.g. High Performance Liquid Chromatography (HPLC)	Liquid samples which may incorporate insoluble molecules	Liquid solvent	Column composed of silica or alumina gel powder or suspension of solid beads in a liquid	Used to test water samples for the presence of pollution for example, in lakes. It is also commonly used to analyse metal ions and organic compounds in solutions and to analyse blood found at a crime scene
Gas, e.g. Pyrolysis Gas Chromatography (PGC)	Vaporised samples and gas mixtures	Carrier gas e.g. nitrogen, hydrogen or helium, is used to move gaseous samples	Column composed of a liquid or of absorbent solid beads	Used to detect bombs in airports and to analyse fibres on a person's body. Also used to test for the presence of accelerants in arson cases and residue from explosives (PGC) and to analyse body fluids for the presence and level of alcohol and illegal substances

Thin-layer chromatography

Thin-layer chromatography also separates dried liquid samples using a liquid solvent (mobile phase). However, instead of paper as the stationary phase, it uses an absorbent material, for example a thin layer of silica dried onto a flat glass, plastic or metal plate. The sample mixture is placed at the bottom of the plate and the plate is inserted into the solvent and this works in the same way as paper chromatography.

Spectrometry

Ultraviolet spectroscopy

Ultraviolet (UV) fluorescence spectroscopy involves the spectroscopy of electromagnetic radiation in the UV region. UV light radiation lies just above the violet region of visible light on the **electromagnetic spectrum**. It has shorter **wavelengths** than visible light but longer wavelengths than X-ray radiation. The UV spectrometer uses a beam of UV light which is absorbed by molecules of **organic compounds**. The energy in the UV light excites the electrons, which vibrate and move to a higher energy level.

Link

Go to *Unit 2: Practical Scientific Procedures and Techniques* for more on spectrometry.

Key terms

Electromagnetic spectrum – the range of all types of EM radiation.

Wavelength – the distance from one peak of a wave to the next corresponding peak.

Organic compounds – chemical compounds with one or more atoms of carbon covalently linked to atoms of other elements.

The excited molecules collide with each other, which causes them to lose energy. The electrons drop to their original ground states, releasing the rest of the energy as low energy fluorescent light. Molecules drop to different vibrational energy levels and the emitted fluorescent light will have different frequencies. The wavelength and amount of fluorescent light emitted is characteristic of the molecules in a material. The UV spectrophotometer records the intensity of fluorescence emitted producing an excitation spectrogram of fluorescent light intensity against UV wavelength. Analysis of the different frequencies helps to identify forensic substances such as heroin. However, this technique does not provide a definitive result because there could possibly be other drugs or materials that have a UV absorption spectra similar to the sample being tested.

Infrared spectroscopy

Infrared (IR) light is a type of electromagnetic radiation that lies between the visible light and microwave regions of the electromagnetic spectrum. IR light has wavelengths between about 1 millimetre and 750 nanometres. We cannot see infrared radiation, but we can sometimes feel it as heat. IR spectroscopy measures the wavelengths of light a material absorbs in the IR spectrum. It uses IR radiation to analyse forensic evidence samples, to identify the sample, gain information about structure, and assess the purity of a sample. Covalent bonds in organic substances are not rigid, but continuously vibrate by bending and stretching. The molecules vibrate at a unique frequency which falls within the infrared spectrum. A specific bond in a molecule can only absorb IR radiation at the natural frequency of the bond vibration. The amount of energy it needs to do this will vary from bond to bond, and so each different type of bond in a substance will absorb a different frequency and wavelength (and hence energy) of infrared radiation.

The radiation source of an IR spectrometer emits light at different frequencies in the infrared region of the spectrum. The beam of infrared passes through a sample of the material under investigation. The molecule absorbs some of these frequencies and the emerging beam is analysed and produces an IR spectrum (a graph plotting wave number against percentage of radiation transmittance). The absorption in the infrared region provides a more complex pattern than UV or visible spectroscopy and can provide enough characteristics to specifically identify a substance. Different materials always have distinctively different infrared spectra; therefore each IR spectrum is equivalent to a 'fingerprint' of the substance. IR spectroscopy can be used to analyse a range of types of forensic evidence:

▶ layers of paint in chips from the victim's clothes involved in a hit-and-run accident (the specific make, model, and year of vehicle can be identified from the paint)

▶ synthetic fibre

▶ drugs.

Mass spectrometry

Mass spectrometry is a powerful analytical technique that is used to identify the structure and **chemical properties** of solid, liquid and gas molecules. It is used in the forensic laboratory to quantify and identify the chemical structure of unknown chemical substances. Mass spectrometry can be used to analyse a range of types of forensic evidence in a variety of criminal investigations to:

▶ detect illegal dumping of poisons and toxins in river water

▶ determine whether food or drink samples have been spiked with drugs and poisons

▶ detect and identify the use of steroids in athletes.

> **Key term**
>
> **Chemical properties** – any property of a material that becomes evident during, or after, a chemical reaction.

Colorimetry

Colorimetry can be used to determine the concentration of a sample by establishing the quantity of light absorbed by a substance or solution and the amount of light transmitted through. How much light a substance absorbs depends on the concentration of the sample and is defined by the relationship known as the Beer-Lambert law. This states that absorbance is **directly proportional** to the concentration of a solution: as the concentration increases, the amount of light absorbed increases.

> **Link**
>
> Go to *Unit 2: Practical Scientific Procedures and Techniques* for more on colorimetry.

> **Key term**
>
> **Directly proportional** – increasing or decreasing together.

For example, very weak diluted solutions are light in colour and transparent; they do not absorb much light and transmit most light through the solution. As the concentration increases the number of molecules in the solution increases so more molecules can absorb more energy, and therefore more light.

Colorimetry measures the absorbance of particular wavelengths of light by a specific solution. This is unique for different substances and can be used to identify the material.

Colorimetry works as follows:

▶ The substance/solution is placed into the sample holder of the colorimeter in a glass or plastic cuvette.

▶ A coloured filter complementary to the colour of the solution is passed through the substance.

▶ The atoms of the substance absorb light at particular frequencies and reflect or transmit at other wavelengths.

The spectra graph usually shows one or more typically broad absorption bands, Every type of substance or solution has unique absorption spectra. The results can be converted into energy, which can be used to determine information about the atomic structure of the material.

Chemical presumptive tests

Presumptive tests for body fluids
Presumptive tests are chemical reagent based, colour change, tests capable of detecting the presence of blood, semen and saliva. They are sensitive, quick, simple and cheap techniques that can be used at the crime scene and in the laboratory. Any fluid or stain thought to be a body fluid should be tested with a presumptive test at the crime scene before collection. A sequence of chemicals are applied to suspected stains, and a change in colour would indicate a body fluid is present.

Presumptive tests are not 100% accurate, they only indicate the presence of a body fluid. The results of presumptive tests must be confirmed in the laboratory using conclusive analytical techniques. Presumptive tests always give a positive result in the presence of the specific fluid being tested for. However, presumptive tests can also give false positive results, giving a positive reading or indication when the fluid is not actually present. If a substance contains a chemical oxidant or similar enzyme to the fluid being tested for, it will react and give a positive colour change. The number of substances that give false positive reactions when carrying out a presumptive test determines the test's **specificity**. The fewer substances that will cause false positives increases the test's specificity.

> **Key term**
>
> **Specificity** – the likelihood a presumptive test gives a true positive reaction only.

Presumptive blood tests
Presumptive blood tests react with haemoglobin (Hb) present inside of red blood cells. Haemoglobin is an iron-containing protein in the red blood cells of mammals and other animals. Its role is to transport oxygen from the lungs to the rest of body. The tests are based on the ability of the haemoglobin to catalyse the oxidation of reagents.

Kastle-Meyer presumptive blood test
The Kastle-Meyer test is a presumptive test used to show the presence of blood at a crime scene. Evidence that may appear to be blood is tested to determine if it is actually blood, and not something that just looks like blood.

A spot of the suspect stain is swabbed with a cotton swab to collect a small sample of the substance. The swab is first treated with a drop of ethanol in order to break open any cells that are present. This gives an increased **sensitivity** and specificity. A drop of phenolphthalin reagent is then added to the swab, and after a few seconds, a drop of hydrogen peroxide (H_2O_2) is applied to the swab. If the swab turns pink within 30 seconds, it is said to test positive for blood. Waiting for a period of time over 30 seconds will result in most swabs turning pink naturally as they **oxidise** on their own in the air. This must be taken into consideration when performing the test. The results must be observed within 30 seconds of adding the hydrogen peroxide. The test is non-destructive to the sample, which can then be kept and used in further tests at the lab, such as in DNA analysis.

> **Key terms**
>
> **Sensitivity** – amount by which a body fluid may be diluted and still give a reaction to a presumptive test.
>
> **Oxidise** – undergo or cause to undergo a reaction in which electrons are lost.

Iron in our diet is used to help make haemoglobin in our blood. The haemoglobin transports oxygen around the body. The Kastle-Meyer test relies on the iron in haemoglobin. The iron promotes the oxidation of phenolphthalin to phenolphthalein (oxidation is the loss of electrons). The names of the two chemicals phenolphthalin and phenolphthalein are very similar, but they are structurally different. Phenolphthalin is made by treating phenolphthalein with zinc, which **reduces** it to become phenolphthalin. In other words, phenolphthalin is made by reducing phenolphthalein. Phenolphthalein (pink solution), can therefore be made by oxidising phenolphthalin because oxidising is the opposite to reducing.

> **Key term**
>
> **Reduce** – undergo or cause to undergo a reaction in which electrons are gained.

The chemistry involved in the Kastle-Meyer test is complex. Phenolphthalin is colourless, but when mixed with hydrogen peroxide and blood it oxidises to phenolphthalein and water, and turns pink. The hydrogen peroxide reacts with the haemoglobin in the blood; the haem group of haemoglobin behaves as a peroxidase, reducing the peroxide to water. Phenolphthalin does not directly participate in this process. Instead it acts as an external source of electrons, donating electrons to haemoglobin to convert the phenolphthalin back into the intensely pink-coloured phenolphthalein.

Investigation 23.2

Kastle-Meyer presumptive test for blood

Steps in the investigation	Pay particular attention to...	Think about this...
1. Place the piece of large white paper in front of you with the blood stained exhibit also on the paper, remove the exhibit from the packaging.	Be careful to remove the exhibit over the white paper so that any trace evidence such as hairs or glass will land on the paper and these can be collected as further exhibits.	
2. Fold a piece of round filter paper in four, or take a sterile swab.	Create a point on the filter paper when folding to be able to scrape the blood onto the paper.	
3. Rub the filter paper or swab on the suspected blood stain and place the swab/filter paper into a sterile petri dish.		
4. Apply one drop of ethanol from the dropper bottle to the filter paper/swab over the petri dish.		Think about why ethanol is used first.
5. Apply two drops of phenolphthalin to the swab/filter paper.		
6. Add two drops of hydrogen peroxide to the filter paper/swab.		
7. Observe a colour change before 30 seconds and this is positive for blood.	Record the time as the colour change should happen within 30 seconds to be reliable.	
8. Carry out a control sample. Repeat steps 4–7 on a sterile swab.		

Leucomalachite Green (LMG) presumptive blood test

The LMG test is a non-**toxic**, highly specific and low sensitivity presumptive blood test. It has a number of false positives and identifies 1ml of blood in 10,000ml of water. The LMG reagent, McPhail's reagent, changes from colourless to dark blue-green in the presence of blood and works by detecting the presence of haemoglobin's peroxidase enzyme activity. The LMG test should not be applied directly to a suspected blood stain as the reagent is destructive to DNA.

> **Key term**
>
> **Toxic** – capable of causing injury or death, especially by chemical means.

Luminol

Luminol ($C_8H_7N_3O_2$) is a presumptive test chemical reagent that releases blue chemiluminescent light when mixed with an oxidising agent. When blood stains are suspected of being present but are invisible to the eye, luminol reagent can be sprayed over a large area and examined in the dark.

▶ Enhancement of invisible blood with luminol

Presumptive semen test

Semen evidence may be submitted to the forensic laboratory from rape, sexual assault and abuse cases. It provides important evidence of sexual contact. Semen is commonly recovered from bed sheets, underwear, and clothes, or from vaginal, anal or oral swabs taken during the medical examination of a victim. Semen can survive in the body for a number of days. It can be found in the vaginal tract for up to seven days and in the cervix for up to ten days after sexual intercourse. Therefore samples can be taken from victims up to a week after the crime has taken place. Semen is a whitish, milky fluid, secreted by the gonads (sexual glands) of males and emitted from the urethra (tube in the penis) for fertilisation of the female ova. Semen is composed of sperm cells (spermatozoa) suspended in a liquid medium, the seminal fluid, which contains water, small amounts of enzymes, inorganic salts, vitamins, protein and fructose. The sperm body contains a large concentration of mitochondria that provide energy for sperm and the sperm tail is used for motility and propulsion.

Semen contains high levels of the acid phosphatase (AP) enzyme so suspected semen stains at the crime scene or on items of evidence will be presumptively tested using the AP test.

▶ Positive reactions for the AP test

Presumptive saliva test

Saliva is another commonly encountered biological sample analysed in forensic laboratories. It is often found on the mouth area of drinking glasses, cans and bottles, on food items, bite marks, cigarette butts, stamps, envelopes, chewing gum, balaclavas and on the skin in sexual contact cases. Saliva contains high levels of the salivary amylase enzyme, which digests and breaks down starches in food. The Phadebas presumptive test identifies the presence of amylase activity and is used in forensic laboratories to indicate the presence of saliva. Other body fluids contain low levels of amylase, including blood, semen, sweat and urine, although they usually do not have sufficient quantities to react with this presumptive test. Phadebas is a colourless solution, consisting of insoluble starch microspheres cross-linked to a blue dye. In the presence of salivary amylase, the starch is digested and broken down into simple sugars, releasing the water-soluble dye into the solution and changing the colour of the solution to blue. The depth of colour depends on the concentration of the amylase present, which can be determined by measuring the absorbance of the solution using spectroscopic techniques.

 PAUSE POINT   List the four presumptive tests for body fluids.

> **Hint**   Describe the method for each test and explain the science and result of each test.
> **Extend**   Evaluate the tests in terms of advantages and disadvantages of each.

Presumptive firearms discharge residue (FDR) tests

A number of tests can be carried out to detect the presence of FDR. These colour change tests are simple to carry out, reliable, quick and inexpensive. There are different tests to test for different components of firearms discharge residue.

Modified Griess Test

This detects the presence of nitrate residues expelled from the muzzle of the firearm when it is fired. Nitrate residue is the by-product from the combustion of smokeless gunpowder, the propellant inside the firearm. During the Modified Griess Test, any nitrite compounds that may be present on an item, such as victim clothing, will turn bright orange as a result of nitrate residues being present.

1 Place the evidence face down on the emulsion-coated side of a piece of photographic paper.

2 Soak an unused piece of cheesecloth (cotton cloth) in acetic acid solution. Wring excess acetic acid solution from the cheesecloth and spread the cheesecloth on top of the evidence (with the photographic paper on the bottom, evidence in the middle and acetic acid soaked cheesecloth on the top).

3 Iron the layers with a hot sterile iron. Acetic acid steam penetrates through the layers; any nitrate residue will appear orange.

4 Remove the cheesecloth and separate the evidence from the test paper. The presence of orange on the test paper is the result of a chemical reaction specific for the presence of nitrite residues.

The Modified Griess Test works because:

▶ nitrite residues present in the FDR are exposed to an acetic acid solution and heat to form nitrous acid

▶ the nitrous acid combines with chemicals in the emulsion coated paper to form a bright orange dye.

The dermal nitrate test

This detects the presence of unburned gunpowder and nitrate residues on the hands of suspects. The hands are covered in a hot paraffin wax which is left to harden and then removed. Any FDR residue is transferred from the hands to the paraffin wax. The wax is tested with the chemical diphenylamine, which turns blue in the presence of nitrates.

The sodium rhodizonate test

This test detects if lead firearm residues are present on a surface. The surface is sprayed with sodium rhodizonate solution which will react with any lead present on the surface and will turn to bright pink. Hydrochloric acid solution is added and if this turns blue it confirms the presence of lead.

Research

Produce a table to compare and contrast all the chemical presumptive tests. Include:
- the name of the test
- where it might be used
- the result of the test
- any known false positives.

Chemical enhancement

Fingerprint enhancement

The surface of fingerprint skin is covered in ridges which create our fingerprint patterns. On the top of fingerprint ridges are tiny pores, small openings in the skin from which sweat, produced in sweat glands, is secreted into the ridge surface. Sweat is mainly composed of water, but also contains a number of other substances, including amino acids, proteins, salts and sugars. When a finger touches a surface, the sweat is transferred to it, producing a two-dimensional mirror impression of the fingerprint ridge pattern. These sweat marks are called latent fingerprints and are usually invisible to the naked eye. They must therefore be enhanced to give them colour so that they can be seen, photographed and examined.

There are a number of chemical treatments that can be used at the crime scene and in the forensic laboratory to enhance latent fingerprints. Various chemical reagents react with different components in sweat, causing a colour change. Usually a number of chemical enhancement techniques are used in sequence, to achieve the best, or optimum, enhancement. The technique used depends on the surface that the latent mark is on, as there are different chemical reagents for porous and non-porous surfaces (Table 23.15). Remember that the aim of enhancement is to provide contrast between the fingerprint and the surface it is on, so if the invisible fingerprint was on a white light switch, what colour chemical do you think would be the best to use to make it visible?

Ninhydrin is commonly used for developing latent prints on porous items. It reacts with the amino acid and water-soluble components of sweat, producing a bluish-purple or pink visible fingerprint mark. The blue coloured substance (Ruhemann's Purple) is formed by the reaction of ninhydrin with its reduction products: hydrindantin and ammonia. Ninhydrin solution can be applied to the surface of the item that the latent print is on by spraying, painting or dipping. The reaction may be quite slow, and marks may take between several hours and several weeks to develop fully. However, the process can be sped up by heating the finger mark using an incubator.

Table 23.15 Porous and non-porous materials

Type of surface	Description	Examples
Non-porous	Hard, non-absorbent surface	• Metal (furniture/drink cans) • Gloss paint • Varnished/painted wood • Tiles • Glass (bottles, windows, mirrors) • Leather (furniture/bags) • Plastic • Car steering wheels
Porous	Soft, absorbent surface	• Paper, card and cardboard • Untreated and raw wood • Plasterboard • Matt emulsion painted surfaces • Cloth material (curtains, sheets, clothes, sofas and fabric chairs)

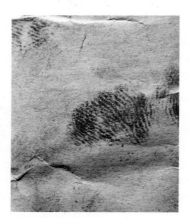

▶ Ninhydrin enhanced fingerprint

Iodine fuming is the oldest technique used to develop latent fingerprints, but is now not commonly used in the forensic laboratory. The technique works best on fresh prints, that are usually no more than a few days old, and can be used on both porous and non-porous surfaces. The technique is non-destructive and can be used in sequence before other treatments, such as ninhydrin treatment of paper.

Iodine is one of the few substances that has the ability to change directly from a solid to a gas when heated, which is the process of sublimation. Latent fingerprints are placed in a closed chamber with iodine crystals. The crystals are heated and give off iodine vapours. Iodine is a halogen and a good oxidising agent. Because of this, iodine itself is easily reduced as it gains electrons. The fingerprint physically absorbs the iodine fumes and the gas oxidises

the fatty acids and oily component of the latent print to produce a yellowish-brown or purple-brown fingerprint. Iodine-developed prints are not permanent and fade over time. However, they can be fixed with α-naphtho- or benzo- flavone solution which darkens the print.

▶ Iodine fuming enhanced fingerprint

Research

There are other fingerprint enhancement techniques used by forensic scientists:
- silver nitrate
- amido black
- superglue fuming (cyanoacrylate fuming).

Use the Internet and books to help research the techniques. Produce a leaflet to help new budding forensic scientists like yourselves. Include in the leaflet information about the fingerprint enhancement techniques listed above. Where possible, include pictures of the equipment used to enhance the print. Identify which types of surface the techniques are best to be used on. Describe how the test is done, and explain why the test works, relating it to the component in sweat that it detects. Get pictures to show what the results would look like after using each technique. Remember to make this leaflet eye catching and informative.

Toxicology: the science of poisons

Toxicology is the study of **poisons** and drugs of abuse, where they have originated from, their properties, effects and antidotes. Forensic toxicologists analyse blood, urine and other tissues in the laboratory to detect the presence of poisons and drugs. A poison is defined as a substance that, if taken into a living organism, will cause injury, illness or death. There are many different types of poisons, which cause different effects on the human body, and they can be classified in a number different ways. They may be grouped depending on their chemical structure, composition and properties or their effects on the human body or even how they are analysed (see Table 23.16).

▶ **Table 23.16** Different poisons

Type of poison	Description	Example
Anions	Some anions are poisonous; they are harmful or deadly to living organisms.	Weed killers, insecticides and bleaching agents. One of the most poisonous anions is cyanide ion.
Corrosive poisons	Corrosive poisons are acids (e.g. hydrochloric acid), alkalis (e.g. potassium and sodium hydroxide) or other substances (e.g. heavy metal salts and strong detergents) that cause physical destruction or 'burning' of the skin and body tissues through direct contact. Corrosive poisons are commonly ingested, causing severe surface and tissue damage to the mouth, oesophagus, stomach and intestines.	Bleach
Gaseous and volatile poisons	A number of gases are poisonous to living organisms and can cause injury and death if inhaled. Some gases are corrosive or irritant. Chlorine and mustard gas cause blisters and attack the eyes and lungs. Nerve gases break down the action of the nervous and respiratory system. Other **asphyxiant** gases, such as carbon monoxide and hydrogen cyanide, affect the bloodstream, restricting the body's ability to absorb oxygen.	Chlorine and mustard gas, carbon monoxide and hydrogen cyanide. Volatile poisons include: cigarette lighter refills, hair-sprays, deodorants, air-fresheners, cleaning products and nail-varnish removers.
Metal and metalloid poisons	A number of metals are poisonous to humans. They can enter the body via ingestion, inhalation or absorption through the skin and mucous membranes, and are stored in the body's soft tissues, where they compete with other ions and bind to proteins, potentially resulting in damage to organs throughout the body and multi-organ system failure, leading to death. Symptoms of metal poisoning include vomiting, diarrhoea, abdominal pain, muscle cramps and paralysis.	Lead (Pb), lithium (Li), mercury (Hg), thallium (Th) and arsenic (As)
Pesticides	'Pesticide' covers any chemical or natural substance used to control, reduce and kill organisms considered to be pests. Ingestion, inhalation or skin contact with pesticides causes a range of symptoms depending on the type of pesticide, but commonly including headache, runny nose and eyes, increased saliva, vomiting, diarrhoea, sweating, general weakness, seizures, shallow breathing, dizziness, abdominal pain or cramps and convulsions.	Insect killers (insecticides), mould and fungi killers (fungicides), weed killers (herbicides) and rat and mouse killers (rodenticides)
Toxins	Toxins are naturally occurring poisonous substances produced by living organisms. Plants, animals, fungus and microorganisms can produce toxins. Some alkaloids have medicinal properties and others, for example cocaine and heroin, are drugs of abuse.	Plant alkaloids, snake and insect venom, carcinogenic aflatoxins produced by the *Aspergillus* fungus, rat poison

Key terms

Asphyxia – a condition caused by the inability to breathe, where there is a large decrease in the amount of oxygen in the body and an increase of carbon dioxide.

Poisons – substances that cause disturbances in organisms, usually by chemical reaction.

Drugs of abuse

Drugs of abuse are defined as illegal drugs, or prescription/over-the-counter drugs used for purposes other than those for which they were intended. Forensic toxicologists investigate a range of drug samples seized

as evidence, to identify and quantify the drug present. UK law charges and sentences suspects depending on the type and amount of drugs seized. The methods of analysis are as follows:

▶ Instrumental analysis: Poisons and drugs of abuse are analysed in the forensic laboratory to determine their identity, composition and quantity of their components. A range of tests can be used and usually a sequence of testing is used to confirm the identity of a substance.

▶ Visual examination: The first stage of analysis involves observing the physical properties of the sample with the naked eye or microscope. Characteristics such as colour, shape, dimensions and markings can provide valuable clues about a drugs identity.

- Presumptive tests: Poisons and drugs of abuse can be tested using chemical presumptive tests to indicate the presence of certain chemical substances. The advantages of these tests include that they are cheap, quick, accurate and reliable, and as they are based on colour changes, the results are easy to determine. During the test, a small amount of the evidence drug sample is tested with chemical reagents. These produce characteristic colours (see Table 23.17) if a chemical reaction occurs with the substance being tested. Presumptive tests for drugs and poisons are not specific as they can also produce false positive results. However, they give the forensic scientist information that allows them to choose the correct type of confirmatory analytical test.

▶ **Table 23.17** Different presumptive tests for drugs of abuse

Reagent used	Drugs detected	Colour change observed
Marquis	Heroin, morphine and most opium derivatives	Purple
	Amphetamines and methamphetamines	Orange-brown
Mandelin	Amphetamine	Blue-green
	Heroin	Brown
	Cocaine	Orange
Dille-Koppanyi	Barbiturates	Violet-blue
Duquenois-Levine	Marijuana	Performed by adding three different solutions to suspect vegetation. Positive result shown by purple colour in chloroform layer
Van Urk	LSD	Blue-purple

Thin layer chromatography (TLC)

Evidence such as blood and urine, which are suspected to contain drugs or poisons, can be separated and analysed using chromatography. Different drugs and poisons appear as different coloured spots on a TLC plate after visualisation, they also travel different distances and therefore have different R_f values. This makes TLC more discriminative than presumptive tests. It can distinguish between a larger number of drugs. However, it is not a confirmatory test and should be backed up with other analytical tests.

Immunoassays

Immunoassays are highly sensitive tests used widely to identify the presence of a drug or poison in a living or dead person. The technique uses the antigen–antibody interaction, in a similar way to the blood typing technique. A specific antibody is used to detect a drug or poison within a body fluid. The substance of interest in the body fluid acts as the antigen. If the substance is present in the sample, the antibody selectively binds with the drug or poison to form an antigen-antibody complex. As the reaction is so specific, this technique is able to confirm identity of a substance.

Specimen collection (ante and post mortem)

Ante mortem specimen collection refers to collection of specimen samples before death and post mortem specimen collection refers to specimen collection after death. Where an individual may have died several hours or days after they have been admitted to hospital, there may be ante mortem specimens available. These are ideal specimens for forensic analysis.

Blood

When analysing ante-mortem blood, it is important to ensure that the blood sample was not taken after a blood transfusion because this may give different readings to that of blood left at the crime scene or make it difficult to use as a comparison sample. If only small volumes of blood are available, then post mortem specimens of blood can be submitted for testing.

Urine

Urine should also be submitted for forensic analysis as it is useful for qualitative analysis and confirmation of blood results. Urine is very useful to give a **quantitative analysis** of alcohol level and drugs. Their metabolites (these are the products once alcohol and drugs have gone through the metabolism) are usually present in urine in much higher concentrations than in blood. The liver is available from almost all post-mortem cases and can be collected easily, it is the primary metabolic organ and is the site where most of the enzyme catalysed reactions take place to break drugs down into their metabolites so drugs are often found in higher concentrations in the liver than in the blood.

Key term

Quantitative analysis – the analysis of a substance to determine the amounts and proportions of its chemical constituents.

Stomach content

Stomach contents should be collected into a plastic container without preservative. This is useful when trying to determine oral overdose, because large amounts of the unabsorbed drug will persist in the stomach.

Bulk and trace samples

Many types of poison and drugs of abuse samples are analysed in the forensic laboratory. These types of evidence are either bulk samples, large enough to be weighed, or trace samples that are so small they cannot be weighed. There is a large range of bulk samples analysed, including:

▶ seized illegal drugs in the form of powders, tablets, plant material, etc.

▶ legal tablet and capsule drugs

▶ poisons

▶ suspicious materials, e.g. alcohol suspected of being spiked with drugs.

Many other types of samples are analysed for the presence of trace levels of poisons or drugs, for example, drug-taking paraphernalia, including syringes, wrapping materials and ashtray contents, and items that have been in contact with drugs, for example, containers, clothing, cutlery, crockery and bank notes. Other trace items that may be analysed include food or drink suspected of containing poison or drugs, and items used in illegal drug laboratories, for example, glassware and solvents.

Physical evidence techniques

Many types of evidence sent for forensic analysis are physical in nature and can establish a link between the scene and its victim or the suspect. Common types of physical evidence include:

▶ firearms

▶ bullets

▶ footwear

▶ tool marks

▶ documents

▶ phones.

Ballistics

Ballistic fingerprinting

Ballistics is the science behind the motion of **projectiles** and firearms, from the inside of the firearm barrel to the outside. Unique marks can be made on bullet cartridges and ballistic investigation can even identify where the shooter may have been standing. When a gun trigger is pulled the hammer mechanism strikes the firing pin inside the gun which initiates the **firing pin** to strike the cartridge primer, this causes the **propellant** to ignite. The ignition of the propellant produces large amounts of gas and pressure, forcing the projectile along the barrel and out of the muzzle of the firearm. Ballistic fingerprinting is based

upon all firearms having unique, individual variations left by machinery during manufacturing and wear and tear through use. Ballistics experts use visual examination and comparison microscopy to identify unique marks and impressions left on the bullet casing and bullets. These marks can be examined, by looking for thickness and angles of striations and other unique characteristics; this can provide crucial links to a specific firearm.

> **Key terms**
>
> **Projectile** – a missile designed to be fired from a firearm.
>
> **Firing pin** – a component inside a firearm that strikes the primer.
>
> **Propellant** – chemical substance used in the production of energy or pressurised gas that is subsequently used to create movement.

Rifling

Rifling is the process of making grooves inside the gun's barrel to make the projectile spin. This increases the accuracy and range of a bullet. Rifling inside the barrel therefore leaves distinct, unique impressions such as grooves, scratches, and indentations known as striations on a fired projectile. When examined, these unique impressions on the fired projectile can be matched to a specific firearm. The areas between the grooves caused by rifling are called the lands.

Firing pin

Marks are also found on the bullet casing. Firing pin impressions are indentations produced as the firing pin of the firearm strikes the primer in the cartridge case. If there are any unique manufacturing marks on the firing pin, these will be transferred to the cartridge case which is expelled at the crime scene of a firearms discharge.

▶ Firing pin marks on firearm cartridge

Breech marks

The breech face rests against the head of the cartridge case and holds the cartridge case in the chamber of the firearm. Breech marks are produced when the explosion forces the head of the casing back onto the firearms breech face. Breech face marks come in all different patterns and they can be matched to test fired cartridges if the suspected gun has been seized or to cartridges found at other crime scenes.

Microstamping

Microscopic markings can be engraved onto the tip of the firing pin or breech face of a firearm with a laser when it is being manufactured. When the gun is fired, the pressure created inside the chamber will mean these markings will be transferred to the cartridge case. If the cases are recovered from a crime scene, forensic ballistic experts can examine the microscopic markings imprinted on the cartridges to link them to the firearm and any other crime scenes the firearm may have been used at.

Ejector marks

The ejector inside the gun is designed to expel the cartridge case from the firearm. This means there is contact between the cartridge and the ejector resulting in another striation that can be used as a means of identification.

 PAUSE POINT List the different impressions that may be used to help link a bullet and cartridge case to a firearm.

> **Hint** Describe how each mark is produced.

> **Extend** Use the Internet and find some images and see if you can identify the different marks present.

Calibre

The term 'calibre' is used to describe the diameter of the barrel and therefore the diameter of the projectile. Calibre is normally measured between opposite lands of a rifled barrel in hundredths of an inch or in millimetres, e.g. 9 mm. Measuring calibre can help determine the make and model of a firearm used (see Figure 23.21).

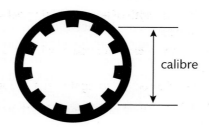

> **Figure 23.21** Calibre measurement

Propellants

A propellant is the material inside the bullet that, when initiated, undergoes a chemical reaction. The reaction produces a large amount of gas quickly and increases the pressure inside the firearm to project the bullet out of the firearm. Gunpowder and smokeless powder (double and single based) are common propellants. A firearm ejects not only a projectile, but also:

▸ partially unburned propellant

▸ combustion products of propellant and primer

▸ minute particles from the barrel, cartridge case and projectile.

This is commonly known as Firearm Discharge Residue (FDR). FDR is extremely important evidence. It settles on nearby surfaces, e.g. clothes, hands, hair and face of the shooter and the target (if they were close enough to the firearm as it was fired). It is tested for with presumptive tests mentioned earlier in the unit.

Case study

Jill Dando

This case illustrates the importance of FDR. Jill Dando was a journalist, newsreader and presenter of the BBC programme *Crimewatch*. On 26 April 1999, Dando was shot dead outside her home in Fulham, London. A local man, Barry George, was convicted and jailed for the murder but was later acquitted on appeal, using FDR evidence. The case remains open.

Check your understanding

1 Research the evidence in this case.

2 Find out why FDR evidence acquitted Barry George.

3 Produce a newspaper article about the case.

Footwear

Footwear marks are a very common evidence type left at crime scenes. Examination of footwear marks evidence is vital to obtain information such as footwear manufacturer, model and shoe size to help profile witness, suspect and victims at a crime scene. This information can be used to help identify the approximate height of the person and the activity that they were performing when the shoeprint was made.

Footwear impressions can be two-dimensional or three-dimensional. They need to be collected in the most suitable way, as discussed earlier in the unit.

Footwear experts will visually compare the tread of the print seized from the scene to a footwear database and to comparison footwear samples if they are available from a suspect's shoes. The experts will:

▶ take measurements of length and width

▶ record any obvious damage to the soles of the shoes

▶ record and measure any distinguishing marks such as logos

▶ look at wear of the shoe's sole as an indicator of how the person generally walks.

Footwear mark evidence can also give information about the activity of the wearer. Impressions left on soft surfaces such as soil can reveal whether a person was walking, running or carrying something heavy when the impression was made.

The aim of the footwear expert is to establish a link between crime scene impression and a specific piece of footwear. A worn piece of footwear will gradually acquire wear and tear on the outsoles. These randomly acquired characteristics will be unique to that specific shoe and may show in crime scene impressions.

Tool marks

Tool marks are again a very common piece of physical evidence left at crime scenes. Examination of tool mark evidence can be used to obtain information about how entry was gained to a property by visually analysing the markings left at the crime scene on windows and doors. This information can then be compared to any weapons seized from suspects. Experts use high powered comparison microscopes to visually examine the minute striations left at the crime scene from casts taken by the CSI and comparison marks made in the laboratory once a weapon has been seized.

Documents

Criminals sometimes attempt to forge important documents such as wills, cheques, passports, loan applications and property deeds. They either alter original documents or create new false documents. Forensic examination of these documents can help identify the culprit and detect deletions, alterations and additions to documents.

Forensic examiners use a number of techniques to investigate documents in question. All analysis is done visually using the naked eye, magnifying glasses and microscopes to identify:

▶ handwriting characteristics

▶ type of machinery to produce the document e.g. photocopier, laser printer or inkjet

▶ changes to the documents

▶ source of the paper used

▶ inks used to write the document.

Handwriting analysis

Handwriting is unique to each individual and has individual features and characteristics that enable us to distinguish between one person and another. Handwriting analysis involves the examination of the construction, proportion and shape of the writing on the document in question. This can then be compared to reference pieces of writing. Although individuals may try to disguise their handwriting, when they are required to write the same piece over and over their natural handwriting will occur throughout. Examples of what handwriting analysts look for include:

▶ capital, cursive or printed text

▶ size – height and width of letters, spacing between letters and words

▶ pen lifts – capital letters connected to lower case letters, retracing strokes

▶ angle – letters slanted left to right or right to left

▶ loop formation – anticlockwise or clockwise, loop join or overhang

▶ diacritics – how 't's are crossed and 'i's dotted.

Signatures are required on documents such as driving licences, passports and bank cards, and they are used as approval or acknowledgement that a document has been made. Each signature is unique to the person who developed it, and although over time it may vary, a signature usually stays very similar throughout life. A signature can be analysed to determine if somebody has tried to forge an important document.

Printed documents

Document examiners are often required to identify the technology used to produce the document in question and furthermore determine the make and model to link a document to a specific machine.

Photocopiers

A photocopier is a very common piece of machinery that is found in most places of work and a document examiner is able to identify if a document has been photocopied and, potentially, which photocopier has been used by looking for two types of marks.

▶ Trash marks — these are spots that appear on the copied document because of dust particles that may be present on the surface of the glass. However, they are only temporary and will change or disappear when used or cleaned.

▶ Drum or mechanism marks — these are regular spaced marks that are produced because of the drum or mechanisms inside the photocopier and these are more permanent.

Laser printers

Laser printers produce high quality printing and are commonly found in offices and work places. Document examiners would be able to determine that the document in question came from a laser printer but, unless there was a fault in the drum that produced a significant mark/repeating pattern, then it would not be possible to determine the exact printer used to produce the document.

Inkjet printers

These are cheaper than laser printers and are commonly found in homes and small businesses. They contain a large number of nozzles inside each print head which spray the ink onto the paper. A document examiner would be able to determine the document in question had come from an inkjet printer, but it is not possible to determine which exact inkjet printer was used to produce the document.

Information technology (IT)

IT is extremely common today. IT that crime scene officers use now usually involves mobile phones, computers, tablets or CCTV. These are all physical pieces of evidence that can help link people to crimes and can provide very useful data.

Mobile phones

Mobile phones and other devices are increasingly being used in all levels of criminal activity. Mobile phones can be seized from a suspect, suspect's property or found at a crime scene. Phone records can be gathered either directly from the phone memory or from the phone provider. Data that can be gathered from a mobile phone includes:

▶ call history and times

▶ messages

▶ contacts

▶ emails

▶ social media usage

▶ Internet browsing history.

By analysing the data, links can be made from it and various lines of enquiries can be followed up.

> **Research**
>
> There are other types of IT that may be used to help a criminal investigation. Research different types of IT, and discuss the types of crime they may feature in and the information/data you may retrieve from them.

Other areas of forensic science

Entomologist

Forensic entomology is the study of insects for legal purpose. There are many ways insects can be used to help gather data about a crime, but the main purpose of forensic entomology is estimating time since death.

When a person dies, their body starts to decompose. Decomposition of a body starts with the action of microorganisms such as fungi and bacteria, followed by the action of a series of insects (arthropods). Bodies decompose at different rates depending on the environmental conditions. For example, if they have been buried, bodies will decompose in a different way to a body that is exposed to the elements such as sun and rain. Dead bodies go through consistent changes, which allow experts to estimate how long that person has been dead. CSIs must collect the insects feeding on a decaying body and send them for analysis, so that the forensic entomologist can estimate the time elapsed since the person died. Forensic entomology is also used because it can also provide data about a person's life before they died. The development of insects depends on specific factors. Drug use can change the lifecycle timing of an insect. For example, cocaine causes the maggots to develop much faster than they normally would.

Insect behaviour can indicate what may have occurred during the crime because flies tend to lay their eggs in moist places in the body, like the eyes and mouth. If eggs or maggots are found on skin that is normally dry, like the arms or legs, it suggests that the skin may have been damaged due to a fall or injury. Finally, the species of insect can provide information about what happened to the body after death. For instance, some insects are found only in some areas, so if a species of insect that only lives in the country is found at a scene in the city, it suggests the body may have been moved sometime after death.

There are limitations to forensic entomology.

▶ Estimation of time of death depends on accurate temperature information; weather changes very quickly, day to day.

▶ Forensic entomology relies on insects, but in the winter there are fewer insects.

▶ It takes time to breed insects and analyse their life cycle so forensic entomology takes time to produce results.

Forensic anthropologist

Forensic anthropologists analyse human remains. They are called on after mass disasters such as earthquakes and tsunamis in order to assist in identifying human remains. They are also used at crime scenes to help recover biological bone evidence and human remains. Forensic anthropologists are experts in osteology, which is the study of bones. Using their knowledge, they analyse human remains to determine how the person died; whether it was suicide, homicide, accidental, or from natural causes. Forensic anthropologists can also determine the age, sex, height, and even the overall health at the time of death. This data assists in criminal investigations.

Forensic profiler

Forensic profiling is the study of trace evidence in order to develop information which can be used by police authorities. It is the process of discovering correlations and links between data gathered and data that exists in databases, so that these links can be used to open up lines of enquiry. A forensic profiler specialises in inductive and deductive reasoning based on evidence from the case to build a profile of the perpetrator. They assist the detectives in gaining a picture of who committed the crime and their motives.

Palynologist

Forensic palynology is the application of the study of pollen and spores to legal matters. It is often used to establish links between objects, people and places based on the identification and analysis of pollen. Pollen is produced and dispersed from plants in order to pollinate and reproduce. The palynologist would ideally visit the crime scene to collect samples. They would conduct a survey of the surrounding plant species, and take photographs. Any evidential samples will be collected from objects or people, and they will also take control samples for analysis. Pollen is analysed under a light microscope to look at pollen grain morphology.

 D

Be able to justify methods, interpret findings and report on conclusions of forensic techniques and analysis

Interpretation of evidence

Draw rational and balanced conclusions from observations

The aim of forensic science is to answer questions of a legal nature by using forensic evidence analysis to provide impartial scientific results in the court of law. The purpose is to establish if the evidence corroborates the statements given by victims, suspects and witnesses. A forensic scientist must state facts in the court room but they are also able to act as an expert witness, to provide their opinion on their interpretation of the results. They are expected to draw a rational balanced conclusion from their observations. They may also be required to give probability of validity which gives an indication of the value of evidence in relation to the crime. Probability is the measure of how likely it is that an event occurred. An expert's opinion is designed to provide the members of the court with a probable or most likely explanation of the evidence to assist the court in reaching their decision.

Forensic analysis involves the comparison of crime scene evidence to reference or comparison samples. We would express results from both using probability estimates. For example, the DNA profile of an unknown crime scene blood sample is compared to the DNA profile of a suspect and the profiles match. The forensic expert would express that it is likely that the suspect is the source of the blood left at the crime scene because they would calculate the chance of the match occurring by chance and give an estimate based on figures at the time – for example, a 1 in 1 billion chance that a random member of the population would match the DNA profile of blood from the crime scene.

Test results and measurements

It is extremely important that when tests are being carried out and results are being recorded they are recorded accurately, including appropriate measurements. It is important that tests are carried out more than once to increase the reliability of the test result and to identify any anomalous results. Control samples should also be carried out to provide evidence that contamination has not occurred and to reduce the possibility of confounding variables. If there are confounding variables it means that the results may be due to more than one factor because all other variables were not controlled.

These results may be presented in working documents, in laboratory record books or printed electronically. They must be clear enough and available for presentation in court.

Presentation of evidence

In addition to producing a case file and writing a correctly structured expert witness statement to present their finding to the courts, CSIs and forensic experts must be available to present their findings verbally in court. They must explain to the jury in layman's terms how they contributed to the criminal investigation, the procedures they followed and the results they obtained. They must remain unbiased and impartial when giving their evidence. They must offer interpretation of their results as well as giving factual information that falls in their area of expertise. They are the only people who are allowed to express their expert opinion in court. In some cases, they will not be called to court to give evidence, so their expert witness statement must be comprehensive, detailed, accurate and organised. It must detail the purpose of the examination carried out, an indication of the results gained and an interpretation of these results relating to the crime.

Provision of forensic science service in England and Wales

Forensic Science Service

The Forensic Science Service (FSS) was a government-owned company in the UK providing forensic services to the police and other agencies in England and Wales. The Government announced that the FSS was closing in December 2010, as they had huge financial losses, and the FSS ceased trading in March 2012. There are FSS archives which include a collection of case files and casework samples, such as microscope slides and DNA samples, that allow old cases to be reviewed in the future. Forensic work is now carried out in police science laboratories or it is contracted out to the private sector. However, in 2015, the National Audit Office warned standards were slipping

and forensic science provision was under threat because police were increasingly relying on unregulated experts to examine samples from suspects and crime scenes. In response the government announced there would be a 'national approach' to forensic science in England and Wales.

Forensic science accreditation

United Kingdom Accreditation Service (UKAS)

UKAS is an accreditation service that supports the criminal justice system through the accreditation of forensic testing laboratories, scene of crime investigation and case work review. UKAS is a highly established accreditation service with accreditation across a broad range of examination and testing activities that take place within forensic science. UKAS accreditation provides assurance and confidence that quality standards are met within forensic science. It also ensures confidence in the competence of forensic laboratories to carry out specific analyses and get reliable, accurate and valid results. UKAS supports the expansion of quality standards within forensic science, and strives to meet the requirements of the Forensic Science Regulator's 'Codes of Practice and Conduct'. This guarantees the evidence being presented to the Criminal Justice System is consistent and dependable.

The Forensic Science Advisory Council (FSAC)

The Forensic Science Advisory Council (FSAC):

- monitors quality standards in the provision of forensic science services
- ensures those supplying forensic science services to the police, including in-house police services, are accredited
- advises on new technologies and applications in the field of forensic science
- ensures compliance with standards relating to national forensic science databases, including the National DNA Database
- provides support to ensure the quality of academic and educational courses in forensic science
- advises on international developments relevant to forensic science quality standards.

Forensic Science Regulator

The Forensic Science Regulator is a government body that has been put together to ensure that the provision of forensic science services throughout the criminal justice system is meeting appropriate scientific quality standards.

Assessment practice 23.3

C.P5   C.P6   C.M4   C.D3   D.P7   D.M5

You work in a specialist forensic laboratory that analyses biological, physical and chemical evidence. You need to carry out analysis on the evidence you collected from your crime scene in learning aim B. You should:

- produce general examination documents to record your observations and results
- describe the techniques carried out and explain the science of the techniques used in your analysis of all evidence
- interpret your results and draw valid conclusions in relation to the purpose of the analysis
- complete an expert witness statement that includes the purpose and results of the examination and an interpretation of results
- include aspects of probability in terms of possible, probable, and likely in relation to the crime scene scenario to assist the court with their decision
- evaluate your working method and results gained.

Your line manager needs a separate report with an evaluation of the techniques used to analyse the evidence, as he is researching new technologies to bring to the company and needs to compare the advantages and disadvantages of the different techniques you have used. Consider:

- ease of use
- subjectivity
- decrease yield
- damage evidence
- cost of equipment
- time
- prior knowledge needed.

Plan

- What is the task? What am I being asked to do?
- How confident do I feel in my own abilities to complete this task? Are there any areas I think I may struggle with?

Do

- I know what it is I am doing and what I want to achieve.
- I can identify when I have gone wrong and adjust my thinking/approach to get myself back on course.

Review

- I can explain what the task was and how I approached the task.
- I can explain how I would approach the hard elements differently next time (i.e. what I would do differently).

THINK ▶FUTURE

Maisy Metcalfe

Senior Crime Scene
Investigator and Trainer

I have been working for Merseyside Police for 15 years, attending crime scenes, collecting forensic evidence and preparing to present evidence in court. I also help to train CSIs in new forensic procedures and train police officers to be forensically aware. This helps to preserve potential forensic evidence when they attend crime scenes before us. People do not realise the level of responsibility we have, not only in ensuring we collect all trace evidence left at crime scenes but that we must avoid contamination and ensure the evidence is secure until it is analysed in the laboratory. The results of forensic analysis on the evidence that we collect help to prosecute or acquit, so it is imperative that the evidence is collected, packaged and labelled correctly.

I can receive a call anytime night or day, depending on when the crime occurs. We are responsible for risk assessing the scene before entry to ensure we are kept safe and no one is injured. We are also responsible for ensuring that the scene is recorded in its entirety and the chain of continuity of evidence is maintained. We do many other administration jobs, such as writing reports, but our main aim is to preserve and collect forensic evidence for analysis.

Focusing your skills

Health and safety

- Identify hazards.
- Understand the risk that each hazard could cause. You should think about specific body parts that may be affected.
- Explain what you could do to reduce the risk from occurring. It is important that you pre-empt potential problems and think in advance what actions you could take in order to limit the risk from occurring.
- Know how you would manage the risk out in the field of work as risks are different in different crime scenes.

Evidence collection

- Identifying potential forensic evidence.
- Knowing how to record each piece accurately.
- Understanding how to package and label.

Transferable skills

- Communication skills are important to ensure that the correct information is conveyed to necessary people, for example, other CSIs and police officers.
- Written communication is extremely important to be able to accurately record observations and give detailed descriptions of the evidence collected.
- You need to be organised; all CSI equipment is sterile and prepared for if you are called out, and you have all the resources you need when you get to the scene.

Getting ready for assessment

Sadie is working towards a BTEC National in Applied Science. She was given an assignment with the title 'Investigate a simulated crime scene and collect biological, physical and chemical evidence' for learning aim B. She had to carry out a forensic examination of a simulated crime scene set up by the assessor and technicians, and use appropriate forensic procedures to gather biological, physical and chemical evidence. She needed to maintain accurate, detailed crime scene notes because she was also required to describe, justify and evaluate her practical procedures. At the crime scene she was also required to:

▶ photograph the crime scene
▶ package evidence collected
▶ label evidence collected
▶ avoid contamination.

Sadie shares her experience below.

How I got started

At the end of each lesson, I filed all my notes in a file, in date order with titles of each lesson/topic. I decided to divide my work into the different learning aims. For each learning aim I had sub-divided my notes into biological, chemical and physical evidence. To help meet the assessment criteria for carrying out the forensic examination, I read over my notes from learning aim A and B the week leading up to my practical assessment. I also wrote a plan of what I thought I would do to help make sure I did not miss any important procedures during my evidence collection.

How I brought it all together

After I had completed my practical examination, I used my detailed crime scene notes to write up the description and justification for my methods and procedures. I did this immediately after my practical examination so that it was fresh in my mind. I decided to lay out my work in line with local University expectations, using size 12 Arial font, 1.5 line spacing and justifying the text. I included a header with my name, date and unit number, and included page numbers for organisation. I also used references throughout my text and included a reference list. To start, I wrote a short introduction about the scenario I had been faced with to introduce the simulated crime scene. For each method carried out and each item of evidence collected, I:

▶ described what I did
▶ explained why I did it
▶ discussed why it was an advantage and thought of how I could improve in future.

What I learned from the experience

I wish I had organised my notes better in class to make it easier to find the information to go in each part. Next time I will highlight important titles and number the pages of my class notes.

I focused on the practical aspect of the assessment by reading the notes before the practical. I could have given a more detailed discussion on the advantages of each method relating it back to the validity and integrity of the evidence. I struggled with the evaluation of my own practice and need to be more critical and creative with what I could do differently next time.

Think about it

▶ Have you made sure you are prepared for the practical examination and know what is expected of you?
▶ Have you clearly provided individual evidence to show you carried out the practical?
▶ Have you broken down the written tasks into small manageable success criteria to ensure you cover all the unit content?
▶ Is your information interpreted and written in your own words? Is it referenced clearly where you have used quotations or information from a book, journal or website?

Glossary

Absolute temperature: or 'thermodynamic temperature' is measured on a scale starting at absolute zero. (Symbol T, SI unit kelvin (K)). Convert temperatures in °C to kelvin by adding 273.15 K, which is the freezing point of water on the kelvin scale.

Absolute zero: the lowest temperature an object can be cooled to, where all thermal energy has been removed from it. There is minimal particle movement and it is in its lowest possible energy state.

Absorption: the capacity of a material to take in external radiation.

Accuracy: the closeness of the readings to the actual value.

Acid: a compound containing hydrogen that dissociates in water to form hydrogen ions.

Acidification (oceans): the lowering of pH levels by the uptake of CO_2.

Activation energy: the minimum amount of energy needed by the reactants for collisions to result in a reaction taking place.

Active transport: the movement of molecules against a concentration gradient, across a membrane.

Addition reaction: a reaction where two or more molecules join together to give a single product.

Adiabatic: means 'no transfer'. It describes a process in a system that is not just well-insulated but is totally thermally isolated, so there is zero heat transfer. Also all work done in the process must be friction free and not create any extra thermal energy.

Afferent arterioles: a group of blood vessels that supply the nephrons in many excretory systems.

Aim: overall general statement of the purpose or intentions of the study.

Akaryotic: having no cell structure, no cytoplasm and no organelles. Consist of nucleic acid and a protein coat.

Alicyclic: a hydrocarbon with carbon atoms joined together in a ring structure.

Aliphatic: a hydrocarbon with carbon atoms joined together in straight or branched chains.

Alkali: a base that dissolves in water to form hydroxide ions.

Allele: one member of a pair of alternative forms of a gene that occupy the same locus on a particular chromosome.

Amino acid sequence: the order in which amino acids are connected by peptide bonds in the chain of polypeptides and proteins.

Amphoteric: substance that can act as both an acid and a base.

Anode: the positively charged electrode by which the electrons leave an electrical device.

Antibody: protein in blood serum produced by the body's immune system that reacts with antigens on red blood cells.

Antibody serum: blood serum which contains antibodies that react with corresponding antigen.

Antigen: molecule present on surface of red blood cells which can stimulate an immune response, e.g. formation of antibodies.

Antimicrobial: a substance that can kill microorganisms or inhibit their growth.

Antioxidant: an enzyme or vitamin added to food to reduce damage due to reactions with oxygen in the air.

Appendicular skeleton: the limbs attached to the axial skeleton.

Appendix: a list of subsidiary material at the end of a report.

Aromatic: a hydrocarbon containing at least one benzene ring.

Asphyxia: a condition caused by the inability to breathe, where there is a large decrease in the amount of oxygen in the body and an increase of carbon dioxide.

Atria: two top chambers of the heart.

Atrioventricular node (AVN): a patch of tissue located in the top of the septum that picks up the wave of excitation from the atria.

Atrioventricular valve: the structure found between the atrial and ventricular chambers of the heart to prevent back flow.

Axial skeleton: the long axis of the human body.

Base: a compound that reacts with an acid to form a salt and water.

Bauxite: aluminium ore.

Bibliography: generally regarded as a list of sources that may or may not have been found useful in providing information for an activity.

Biopsy: an extraction of fluid or tissue from inside the body to determine the type and extent of disease present.

Bragg peak: the point at which the maximum dose of proton energy is delivered corresponding to the maximum interaction of the proton with tissue.

Brittleness: the tendency of a material to fracture under stress. Brittle materials cannot undergo plastic flow and so are not suitable for manufacturing processes such as drawing, rolling, hammering or stamping.

Bundle of His: a collection of heart muscle cells specialised for electrical conduction.

Calcination: heating to high temperature to remove free and chemically bonded water.

Carbon capture: the process of capturing and storing CO_2 from fossil fuel power stations, for example, preventing it from entering the atmosphere.

Cardiac cycle: a complete heartbeat from the generation of the beat to the beginning of the next beat.

Catalyst: a substance that speeds up a reaction, can take part in the reaction but is left unchanged at the end of the reaction.

Chemical properties: any property of a material that becomes evident during, or after, a chemical reaction.

Citation: a quotation or reference from a paper, article, book or specific author.

Coding region: portion of a gene's DNA that codes for protein.

Collimator: a device, usually made of lead, which makes electromagnetic waves (e.g. gamma rays) parallel.

Common approach path (CAP): the common route used to enter and leave a crime scene. It is used to preserve forensic evidence.

Complementary base pairs: during replication, the nucleotide base sequence of one strand of DNA is known, so the new strand produced is complementary in base sequence to the template strand.

Complex ion: a transition metal ion bonded to one or more ligands by dative covalent bonds.

Computerised tomography/axial tomography: process by which a three-dimensional image of a body structure is produced from plane cross-section X-ray images along an axis.

Concentration gradient: the difference in the concentration of a substance between two regions.

Condensate: liquid collected by condensation.

Contamination: unwanted transfer of material.

Contingency plan: a plan or action designed to be introduced in response to circumstances which may or may not actually happen.

Continuity of evidence: complete documentation that accounts for the progress of an item of evidence throughout the entire investigation from crime scene to court.

Contrast medium: barium or iodine chemical compounds given to the patient before CT scanning to enhance the appearance of internal organs.

COSHH: Control of Substances Hazardous to Health Regulations in place for education and industry to limit the exposure to workers of chemical effects.

Cracking: long chain hydrocarbons are broken down into shorter chain hydrocarbons.

Creep: sometimes called 'cold flow', occurs when a material under stress deforms gradually over time. It is more severe in materials that are subjected to heat for long periods.

Crime scene investigator (CSI): an officer who is responsible for evidence collection at a crime scene.

Critical point: the temperature and pressure at which the density of the liquid and vapour phases become identical. Above that critical temperature, only a supercritical gas phase can exist.

Cytoplasm: gel-like substance enclosed by the cell surface membrane; about 80% water; medium in which many metabolic reactions take place; organelles are suspended in the cytoplasm.

Degenerative changes: negative changes that occur over time.

Degrees of freedom: number of variables that are used to make a calculation.

Dehydration: removal of water.

Directly proportional: increasing or decreasing together.

DNA: deoxyribonucleic acid is a nucleic acid found in cells that carried genetic information.

DNA profiling: using a small set of DNA variations that is very likely to be different in all unrelated individuals.

Dosimeter: a device for measuring cumulative ionising radiation dose.

Ductility: the ability of a material to be formed by drawing into new shapes, primarily by means of tensile forces. Ductile materials are generally also malleable.

Effective dose: the absorbed dose of radiation multiplied by a quality factor that depends on the tissue type under investigation.

Elastic hysteresis: occurs in materials like rubber, where internal friction between large molecules dissipates energy producing heat. Loading and unloading of a sample each produces a different stress-strain curve, creating a hysteresis loop, the area of which represents the energy absorbed in the cycle.

Elastic limit: the point on the stress-strain curve, beyond which a material begins to suffer plastic deformation, and so will not completely regain its original dimensions when the stress is removed.

Electrolysis: the decomposition of a compound using electricity.

Electromagnetic spectrum: the range of all types of electromagnetic radiation.

Electron: a sub-atomic particle in all atoms that has a negative electrical charge.

Electron micrograph: photograph of an image seen with an electron microscope.

Electrophiles: species that are electron-pair acceptors and are attracted to areas of high electron density.

Electrophilic addition: a reaction using an electrophile where two or more molecules bond to become one product.

Electrostatic lifting apparatus: a device used to lift latent dust footprints using a high voltage and plastic film.

Endocytosis: movement of bulk material into a cell.

Endoscope: a telescope placed inside a long thin tube for insertion into tight, difficult areas.

Endothermic reaction: a reaction that absorbs energy.

Energy efficiency: is the fraction of the total energy input that is converted into a useful energy output. (The fraction may also be expressed as a percentage: just divide the top of the fraction by the bottom, then multiply by 100.)

Entropy: 'potential for change'. One way to calculate the entropy of a system would be to count how much randomness it contains: i.e. the number of equivalent energy states and positions its particles could be arranged in. The more ordered and compact and high in temperature something is, the lower is its entropy. Things naturally become more randomly mixed up, more spread out and cooler: i.e. their entropy increases. So, at its start, the universe must have had a very low entropy indeed.

Equation of state: an equation connecting the measurable quantities that define the physical state of a system; for example, the pressure, volume, temperature and amounts of material (numbers of molecules) present in the system.

Erythrocyte: red blood cells, containing haemoglobin, that transport oxygen around the body.

Eukaryote: organisms made of cells that contain a nucleus and other specialised structures (membrane-bound organelles).

Eutrophication: a form of water pollution caused when excess fertilisers leach into lakes and rivers. This excess encourages the growth of algae, which then cover the surface, preventing oxygen reaching deeper aquatic plants. The plants die and aerobic bacteria decompose them, using up the oxygen in the water. Animals die or leave the area. Anaerobic bacteria produce methane and hydrogen sulfide.

Excited state: the condition of an atom whereby one or more electrons have absorbed energy and move to an energy level above the ground state. This energy level is further from the nucleus of the atom.

Exhibit: an object collected for forensic examination.

Exocytosis: movement of bulk material out of a cell.

Exon: protein coding region.

Exothermic reaction: a reaction that releases energy.

Extracellular glycoprotein: a carbohydrate and peptide chain that exists in body fluids outside of cells.

Extraction: a means to obtain information from different sources.

Facilitated diffusion: the movement of molecules down their concentration gradient, across a membrane, with the help of carrier proteins. Energy is not required.

Fat soluble: dissolves in fats.

Fatigue: the embrittlement and failure of a material that can occur with relatively low levels of stress if these are repeatedly applied and then relaxed over many cycles.

Fermentation: a metabolic process that occurs in yeasts and bacteria to convert sugar to acids, gases or ethanol.

Field coil: a resistive electromagnet that produces a gradient field over the body helping to locate the radiofrequency signals.

Firearm discharge residue (FDR): residue deposited on the surfaces near someone who discharges a firearm, e.g. on their clothes and skin.

Firing pin: a component inside a firearm that strikes the primer.

First attending officer (FAO): the first officer at the scene of a crime.

Fluid mosaic model: description of the cell membrane structure, a phospholipid bilayer with proteins floating in it.

Free radical: atom with a single unpaired electron.

Frequency: how often a particular value occurs within a set of values.

Functional group: group of atoms responsible for the characteristic reactions of a substance.

Fusion: (or melting) is the change of physical state from solid to liquid. The reverse process is freezing.

Gamma knife: the name given to a highly focused beam of gamma radiation used to treat brain tumours.

Gene: length of DNA that codes for one or more proteins, or that codes for a regulatory length of RNA.

Genetic engineering: a process which allows the transfer of desired genes from one organism to another (not necessarily of the same species) artificially, to enhance the characteristics of the organism.

Genetic modification: the process of altering the DNA in a plant or animal using laboratory techniques.

Genome: all the genes within a cell/organism.

Good clinical practice: a set of quality standards for studies involving human beings.

Good laboratory practice: a system of regulation which ensures that tests carried out in non-clinical laboratories are well planned, reliable and have hazards suitably assessed to reduce risks to the public and environment.

Good manufacturing practice: a system of quality assurance to ensure that medicines and medical products are manufactured to the highest quality standards.

Half-life: the time it takes for the concentration to reduce to half of its original value.

Haloalkane: alkane where at least one hydrogen is replaced by a halogen.

Halogenation: addition of halogen such as bromine or chlorine to an alkane.

Harvard referencing system: a style of referencing system used to mention sources of information which have been looked at to inform your work.

Hazard: a potential source of harm.

Hazcards: a set of documents from CLEAPSS (Consortium of Local Education Authorities for the Provision of Science Services) giving information for storage, disposal and potential risks of chemical and biological substances.

Heat: the quantity of thermal energy transferred during a process.

Heat capacity: the number of heat units needed to raise the temperature of a body by one degree. Also known as thermal capacity.

Heat engines: machines designed to convert thermal energy into useful mechanical work.

Homologous series: family of organic chemicals.

Homolytic fission: formation of a free radical by splitting a bond evenly so each free radical has one of the two available electrons.

Host: organism inhabited by an infecting agent or parasite.

Hydrophilic: associates with water molecules easily.

Hydrophobic: repels water.

Hyphae: cells that grow as long tubular thread-like filaments. A network of hyphae is called a mycelium.

Hypothesis: an explanation, with some evidence, to be further tested by investigation.

Ideal gas: a theoretical model of a gas, where the molecules are assumed to be point particles that take up no volume and that exert no forces on one another in between their elastic collisions. This simplified model makes it easier to calculate their behaviour, based on Newton's laws of motion.

Image: a real image is an image located where light rays from an object are focussed. A real image can be seen on a screen placed at the focal point. A virtual image is one where light rays appear to, but do not actually, come from that image.

In situ: when an artefact has not been moved from its original place.

Inadmissible: not accepted as valid.

Individualisation: to identify a specific individual.

Infrared: electromagnetic radiation from beyond the red end of the spectrum emitted by heated objects and having a wavelength of between 800 nm and 1 mm.

Intron: non-coding region in DNA.

Ionising radiation: electromagnetic waves or particle beams that have sufficient energy to break chemical bonds, thus producing charged ions in the materials they pass through.

Ions: particles that carry a positive or negative charge.

Isomers: two or more compounds that have the same molecular formula but a different arrangement of atoms in the compound and so have different properties.

Isothermal: happens at one fixed temperature. To be reversible, an isothermal process would have to have a negligibly small temperature gradient across the system boundaries, so would be extremely slow.

IUPAC nomenclature: system of the International Union of Pure and Applied Chemistry for naming organic compounds.

Kelvin scale: a temperature scale with absolute zero as zero. The size of one unit (1 K) is the same as one degree Celsius. Water freezes at 273 K.

Larmor frequency: the rate of 'precession' or wobble of the proton in the magnetic field.

Latent: present but not visible.

Latent heat: energy transferred that has the effect of changing the physical state of a substance without changing its temperature.

Leukocytes: white blood cells, of which there are many different types.

Ligand: a molecule or ion that can donate a pair of electrons to the transition metal ion to form a dative bond.

Linear accelerator (linac): a large machine that accelerates particles to high speeds in a straight line.

Lipid: fats or their derivatives, including fatty acids, oils, waxes and steroids; lipids are insoluble in water, but soluble in organic solvents, such as ethanol.

Literature review: a search and evaluation of the available information in your given subject or chosen topic area.

Lobbyist: a person who may or may not be a professional, acting on behalf of businesses or organisations to influence decisions made by government by approaching relevant government officials.

Malleability: the ability to be shaped by means of compressive forces such as occur in rolling, hammering or stamping. Not all malleable materials are also ductile.

Maximum theoretical efficiency (symbol η_{rev}): no real engine can achieve a thermal efficiency higher than that of an ideal reversible engine operating between the same temperatures. (If this were not so it would be possible to create a machine that violated the second law and moved heat from a colder to a hotter body without doing work.) In practice, engine efficiency is always lower than this because all real engines have irreversible processes that produce less net work output and instead output more heat at the low temperature.

Mean: the average of all the numbers within a set of results. It is obtained by totalling the results and then dividing the total by the number of results.

Mean bond enthalpy: the average amount of energy for a mole of a given bond to undergo homolytic fission in the gaseous state.

Megahertz: measurement of frequencies of 1 million Hertz.

Metabolic: related to a living organism that breaks down food to produce energy and heat. This process is called metabolism.

Microorganisms: microscopic organisms that include viruses, bacteria, fungi and some protists.

Mixture: two or more substances mixed together, but where no chemical reaction has occurred.

Mobile phase: gas or solvent liquid e.g. water or alcohol that carries the components of the mixture up the stationary phase.

Mode: the data value that occurs the most often in a set of values.

Morphology: the structure.

Myocardium: the middle and thickest layer of the heart wall, composed of cardiac muscle.

Nanoparticles: particles approximately 10^{-9} m in size which have a large surface area for their volume.

Natural process: any real process that is not driven by an input of work or heat from a source outside the system being studied, but just occurs naturally.

Normal distribution: a set of real value data that, when plotted on a histogram, produces a bell shaped curve. This is called a normal curve and its peak is the mean.

Nuclear magnetic resonance (NMR): the absorption of specific electromagnetic frequencies by atomic nuclei which resonate and, after a short delay, re-emit a radiofrequency.

Nucleotide: one of the structural components, or building blocks, of DNA.

Objectives: stages to be completed to successfully achieve the aim.

Organelle: an organised and specialised structure within a cell; some, e.g. mitochondria, are membrane-bound and are found only in eukaryotic cells. Ribosomes are organelles that are not bound by a membrane and occur in prokaryotic and eukaryotic cells.

Organic compounds: chemical compounds with one or more atoms of carbon covalently linked to atoms of other elements.

Organic compound: compound based on a carbon structure.

Osmoregulation: the control of water and salt levels in the body which prevents problems with osmosis.

Osmosis: the net movement of water molecules from an area of high water potential to an area of low water potential, down a water potential gradient and across a partially permeable membrane.

Ossifies: turns into bone tissue.

Oxidise: undergo or cause to undergo a reaction in which electrons are lost.

Pasteurisation: the use of heat to kill bacteria in food or drink.

Pathological: condition or complaint caused by a disease.

Penetration: the capacity for a radiation to pass through a material while giving up only a small amount of its energy.

Pericardium: a fibrous membrane that surrounds and protects the heart.

Perpetrator: a person who commits a criminal offence.

Personal Protective Equipment (PPE): equipment designed to protect the wearer by limiting the risk of injury or infection. Specialised clothing or equipment worn by employees to protect against health and safety hazards.

Petrochemical industry: industry that produces materials by refining petroleum.

Phase: a separate part of system which is not uniform throughout. For example, a mixture of ice and water is a two-phase system, but a solution of salt in water is a one-phase system.

Photomicrograph: photograph taken through a microscope, to show a magnified image of an object.

Photon: a particle that is a small 'package' of energy of electromagnetic radiation.

Piezoelectric: the property of certain substances, such as a quartz crystal, to change size when it is stimulated by electricity. If an alternating voltage is applied to the crystal it causes a vibration.

Plastic deformation: occurs under stress levels that are sufficient to make the solid material begin to flow, rather like a liquid. When the stress is removed, a change in an object's shape and size remains. This is called a permanent set.

Pluripotent: referring to a stem cell which has the ability to become almost any cell type in the body.

Poisons: substances that cause disturbances in organisms, usually by chemical reaction.

Polarise: producing charged poles in an object (e.g. molecule) that limit its vibrations to a single direction.

Polymerase Chain Reaction (PCR) technique: a technique used to make multiple copies of a segment of DNA.

Polymorphisms: most common type of genetic variation among people.

Population: all items, people or other individual components which have the characteristic which is to be investigated.

Porous: absorbent, full of tiny holes through which gas or fluids may pass.

Positron: a sub-atomic particle of the same mass as an electron but opposite electrical charge (+).

Precipitate: a cloudy solid suspension forming in a solution.

Precision: the degree of uncertainty of a measurement linked to the size of the measured unit.

Presumptive test: a test that gives a probable indication of results but needs confirming.

Primary data: information that you have collected directly during your investigation.

Profile: provide data and information on a person.

Projectile: a missile designed to be fired from a firearm.

Prokaryote: organisms made of cells that have cell surface membranes, cytoplasm and a cell wall, but do not have a proper nucleus containing DNA. Their DNA floats free in the cell's cytoplasm.

Propellant: chemical substance used in the production of energy or pressurised gas that is subsequently used to create movement.

Proton: positively charged particle in the nucleus of all atoms.

Pubic symphysis: joint located between the left and right pubic bones.

Pulmonary circulation: parts of the circulatory system concerned with the transport of oxygen to, and carbon dioxide from, the heart to lungs.

Purkinje fibres: specialised conducting fibres found in the heart.

Qualitative evidence: information gathered about the qualities of a given investigation with no direct measurements.

Quantitative analysis: the analysis of a substance to determine the amounts and proportions of its chemical constituents.

Quantitative evidence: information gathered using measurement of quantities and producing numerical data.

Radiologist: a medically qualified specialist responsible for diagnosing and treating conditions requiring ionising and non-ionising radiological techniques.

Radiopharmaceutical: the name given to pharmaceuticals that contain radionuclides.

Reaction mechanism: the step by step sequence of reactions that lead to an overall chemical change occurring. It shows the movement of electrons in the process.

Reduce: undergo or cause to undergo a reaction in which electrons are gained.

References: a list of sources at the end of a report used to help provide information for an activity.

Refractive index: light changes speed as it passes form one medium, such as air, to another, such as glass. This change in speed can lead to a change in direction of the wave, which is called refraction. Refractive index is the ratio of speed of light in vacuum/speed of light in medium.

Refractory material: material that is physically and chemically stable at very high temperatures, for example, over 3000°C.

Rehydration: addition of water.

Reliability: a measure of how trustworthy the evidence or data may be in an investigation or research; how well a set of experiments, tests or measurements is able to be repeated with similar results.

Repeatable: the consistency of a set of results.

Respiratory minute ventilation: volume of air breathed in or out per minute.

Retention factor: quantitative measure of a component's properties in a mixture.

Reversible process: one which could be fully reversed in time, following the same path and exactly reversing the quantities of heat transferred and of work done. This is an ideal concept that can never be fully achieved in practice.

Risk: a description of the harm that may be caused by a hazard.

Salt: a compound formed by an acid–base reaction where the hydrogen in the acid has been replaced by a metal (or other positive) ion.

Sample: a smaller selection taken from the population which has the characteristics under investigation.

Saturated compound: contains single bonds only.

Saturated vapour: is vapour that is in equilibrium with a liquid phase of the same substance. Any further compression causes more liquid to condense. The vapour cannot exert a higher pressure at that temperature than its saturated vapour pressure (SVP).

Scene of crime officer (SOCO): an officer who is responsible for evidence collection at a crime scene.

Secondary data: information collected by someone else that you have used in your investigation.

Sensitivity: amount by which a body fluid may be diluted and still give a reaction to a presumptive test.

Septum: the dividing wall between the right and left sides of the heart.

Short Tandem Repeat (STR): a microsatellite region in DNA of 2–4 bases repeated 5–15 times.

Sievert (Sv = SI unit): the SI unit of effective radiation dose (joules of radiation energy per kilogram of tissue, and multiplied by a dimensionless quality factor that depends on the type of radiation).

Sinoatrial node (SAN): a patch of tissue found in the right atrium that generates the electrical activity and initiates a wave of excitation at regular intervals.

Slurry: semi-liquid mixture containing fine solid particles.

Smear slide: a slide prepared when a sample (for example, blood, bacteria or sediment) is spread thinly (smeared) on the slide in preparation for examination.

Solvent front: the leading edge of the moving solvent as it progresses along the stationary phase. The distance the solvent moves is measured from the sample baseline to the point where the solvent front stops.

Specific heat capacity: the energy required to raise the temperature of 1 g of a substance by 1 K.

Specificity: the likelihood a presumptive test gives a true positive reaction only.

Spin: a property of nature that causes particles to behave like small magnets in a uniform magnetic field.

Stationary phase: part of the apparatus that does not move with the sample, e.g. filter paper.

Stem cell: undifferentiated cells which can develop into many different cell types.

Sterile: free from bacteria and other living microorganisms.

Strength: also called ultimate tensile stress, is the maximum stress that the material can bear. This occurs just before the material fails and fractures.

Striated: having a series of ridges or linear marks.

Surface area: a measure of the total space occupied by the surface.

Surroundings: everything that is not part of the system e.g. water bath or beaker or aqueous solution that reactants are dissolved in. The rest of the universe, outside the system boundary.

Symbiotic: an interaction between two organisms where both organisms benefit.

Synchrotron: a circular-shaped device using electrical and magnetic fields to accelerate particles, increasing their kinetic energies.

System: the part of the universe whose properties you are investigating. It is enclosed by a boundary defined by you, the experimenter.

Systemic circulation: parts of the circulatory system concerned with the transport of oxygen to, and carbon dioxide from, the heart to body cells.

Temperature: the physical quantity that determines the rate at which heat will flow: from a 'hot' body (system) to a 'colder' (i.e. lower temperature) one. It is directly proportional to the average kinetic energy of the molecules.

Thermal efficiency: another name for energy efficiency when the energy input is in the form of heat. This name emphasises that the temperatures involved are of critical importance to the efficiency.

Thermal equilibrium: exists when two systems are in thermal contact, but there is no net transfer of heat because they are at the same temperature.

Thermodynamic temperature: (symbol T) the modern way of defining absolute temperature, which does not rely on the existence of an ideal gas, but instead uses the amounts of heat transferred in idealised engine cycles.

Thermography: the medical technique for detection of radiant thermal energy (heat) from the body and converting this into a visual image for diagnosis.

Thermoluminescence: the ability of some materials to glow when exposed to certain types of energy over a period of time.

Thermometer: a device with a readily measurable property that varies directly with temperature.

Third person: a means of writing an account of an activity without referring to anyone who performed the activity.

Thrombocytes (platelets): component of blood involved in blood clotting.

Toxic: capable of causing injury or death, especially by chemical means.

Tracer: in radiotherapy and diagnosis, a substance that can be injected into the patient and easily tracked through the body because of its radioactivity.

Transducer: a device that changes a physical quantity (e.g. sound) into an electrical signal or changes an electrical signal into a physical quantity.

Triple point: the temperature and pressure at which all three phases: solid, liquid and vapour, can exist in equilibrium.

Ultrasound: sound waves with a frequency above 20,000 Hertz (20 kHz), inaudible to humans.

Valid: the degree to which the method and results obtained reflect the real results.

Validity: how well a particular investigation measures what it claims to measure.

Vaporisation: is the change of physical state from liquid to gas. The reverse process is condensation.

Vapour: is the name given to a gas that is at a sufficiently low temperature that it will change state to a liquid or solid if its pressure is increased sufficiently. In other words, a gas below its critical temperature (see below).

Variable Number Tandem Repeats (VNTRs): randomly repeated short sequences of DNA that vary in the number of repeats, and therefore size, between different individuals.

Vena cava: a large vein carrying deoxygenated blood into the heart.

Ventilation: the exchange of air between the lungs and the surroundings.

Ventricle: two bottom chambers of the heart.

Virus: submicroscopic infective agent consisting of a protein coat and nucleic acid: either DNA or RNA, but not both.

Water potential: a measure of the ability of water molecules to move freely in solution.

Wavelength: the distance from one peak of a wave to the next corresponding peak.

Work: the work done in a process is the amount of mechanical energy transferred. (Symbol: W, SI unit: Joule (J)). (Be careful not to confuse W, for the energy transfer quantity 'work', with W meaning the power unit 'Watt'.)

X-rays: high-energy electromagnetic waves that typically have wavelengths of around 10^{-10} m.

Yield point: the point where the start of plastic flow causes a change of slope on the stress-strain curve. Iron and steel and a few other metals show a clearly defined yield with a drop of stress, while in many other materials the exact position of the yield point is hard to spot.

Index